Co-Constructing Economic Transformation Through Enterprise, Narrative, and Systemic Design

Christos Christodoulou Volos
https://orcid.org/0000-0002-3188-844X
Neapolis University Pafos, Cyprus

Andreas Masouras
https://orcid.org/0000-0001-8989-3930
Neapolis University Pafos, Cyprus

IGI Global
Scientific Publishing
Publishing Tomorrow's Research Today

Vice President of Editorial	Melissa Wagner
Director of Acquisitions	Mikaela Felty
Director of Book Development	Jocelynn Hessler
Production Manager	Mike Brehm
Cover Design	Jose Rosado

Published in the United States of America by
IGI Global Scientific Publishing
701 East Chocolate Avenue
Hershey, PA, 17033, USA
Tel: 717-533-8845 | Fax: 717-533-7115
Website: https://www.igi-global.com E-mail: cust@igi-global.com

Library of Congress Cataloging-in-Publication Data

LCCN: 2025034255 (CIP Data Pending)
ISBN13: 979-8-3373-4566-6
Isbn13Softcover: 979-8-3373-4567-3
EISBN13: 979-8-3373-4568-0

British Cataloguing in Publication Data
A Cataloguing in Publication record for this book is available from the British Library.

Table of Contents

Detailed Table of Contents

This chapter remakes macroeconomic indicators such as GDP, inflation, and employment levels into narrative tools actively building the public's knowledge, behavior, and policy actions. Contrasting with the perception of economic statistics as objective facts, it utilizes narrative and behavioral economics to describe how these indicators are built, blown up, and read in a communicative context that includes media, politicians, and the public. In historical case studies like the Great Recession, Eurozone crisis, and COVID-19, the chapter illustrates how narrative dynamics drive economic cycles by generating feedback loops from sentiment to behavior and back to data. It argues that the macroeconomy is a reflexive system where narratives generate behaviors that reshape the very metrics that necessitated them. Implications for policymakers, business leaders, and theorists are explored, including the need for ethical stewardship of story, complexity-sensitive frameworks, and critical narrative literacy in order to facilitate economic resilience and transformation.

Reframing Value Narrative Strategy as a Catalyst for Economic Transformation in Entrepreneurial Ecosystems for the upcoming book Co-Constructing Economic Transformation Through Enterprise Narrative and Systemic Design examines how strategic storytelling redefines and operationalizes value within entrepreneurial contexts. Drawing on narrative theory, institutional logics, and systems thinking, the authors develop a conceptual model showing how narratives frame opportunities,

from skill development to industrial strategy — aimed at fostering inclusive growth and providing suggestions for policymakers in India and other emerging nations.

This study examines the influence of communication infrastructures and digital platform dynamics on entrepreneurial success in Greece and Cyprus. Employing a phenomenological methodology, it utilises interviews and observations from sixty entrepreneurs across various sectors. The findings underscore that high-quality infrastructure is both vital and strategically beneficial, as it facilitates global collaboration and market access. Cyprus benefits from government-led infrastructure development, whereas Greek entrepreneurs demonstrate innovation in overcoming limitations. Key to success are multi-platform orchestration, algorithm optimization, and cultural communication strategies that integrate global digital standards with local practices. The research introduces the concept of "ecosystem bridging capabilities," highlighting entrepreneurs as active orchestrators who leverage network effects despite limited domestic markets. These insights contribute to the development of entrepreneurial education and policy aimed at promoting sustainable growth.

This chapter interrogates the constitutive role of education in shaping economic futures in the Global South, arguing that educational systems function both as narrative infrastructures, i.e., constructing societal grammars of value, success, and belonging, and as systemic design spaces, structuring the material conditions under which these narratives are enacted. Integrating insights from narrative economics, systemic design theory, and decolonial pedagogy, it examines how colonial legacies, neoliberal reforms, and alternative pedagogies influence economic subject formation. Drawing on case studies from Asia, Africa, and Latin America, including Bhutan's GNH schools, Brazil's Landless Workers' Movement, and innovation hubs across Africa, it demonstrates that plural, justice-oriented models are operational realities. The chapter advances a framework for scaling such innovations through policy reform, participatory governance, and post-growth valuation systems, positioning education as a generative commons for building equitable, sustainable, and pluriversal economies.

This systematic literature review (1990-2024) of 120 peer-reviewed articles addresses a gap in understanding how media discourse shapes market sentiment through narrative mechanisms. Employing PRISMA guidelines, the study synthesizes the influence of linguistic framing, behavioral pathways, and systemic feedback loops on financial behaviors and macroeconomic outcomes. Findings indicate media both reflects and molds sentiment, disproportionately impacting non-expert investors and influencing algorithmic trading. The review proposes a "narrative resilience" framework to mitigate destabilizing discourse, bridging behavioral economics, communication, and computational social science.

This paper traces the changes in the co-invention patterns of the largest German companies from 2019 to 2023 using patent statistics. The assumption about the withdrawal of R&D activities from emerging economies is tested. Based on descriptive statistics and binary logistic regression, we observe three trends. The first trend concerns the increase of sole patenting activities in the subsidiaries in emerging economies, especially in China. We also observe the decrease in the number of patent filings generated by the headquarters and subsidiaries in advanced economies. There is an increase in co-invention with inventors from external organizations, including by subsidiaries in emerging economies. There is a plethora of possible explanations. One of them is the benefits of exercising a recombinant advantage implying embeddedness in local innovation systems and collaborations with external partners. Another explanation is the nature of corporate decision-making, including loss-aversion that allows for high levels of risk toleration.

Preface

Economic transformation is often discussed as if it were a mechanical process—driven by markets, technologies, or policies acting upon an otherwise passive social world. This volume begins from a different premise: that economies are actively *co-constructed* through narratives, enterprises, institutions, and systemic designs that shape how value is imagined, enacted, and sustained over time. *Co-Constructing Economic Transformation Through Enterprise, Narrative, and Systemic Design* brings together interdisciplinary perspectives to explore how stories, infrastructures, moral imaginaries, and organizational practices interact to produce economic realities.

Across disciplines such as economics, management, communication, accounting, education, and systems theory, the contributors challenge the notion of economic indicators, markets, and technologies as neutral or purely technical artifacts. Instead, they reveal them as narrative-laden systems—embedded in cultural meanings, ethical choices, and power relations—that influence behavior, legitimacy, and long-term development trajectories. From macroeconomic signals and entrepreneurial storytelling to accounting standards, AI-generated narratives, educational curricula, and global R&D networks, the chapters collectively argue that transformation emerges through feedback loops between meaning and materiality.

A central theme of the volume is reflexivity. Economic actors do not merely respond to conditions; they interpret them through narratives that shape expectations, emotions, and strategic action. These interpretations, in turn, alter the very systems they describe—reshaping markets, institutions, and development paths. Whether through media discourse influencing market sentiment, sustainability narratives redefining enterprise value, or algorithmic storytelling reframing managerial imagination, the chapters demonstrate how narratives function as infrastructures of coordination and change.

Equally important is the ethical dimension of economic design. Several contributions interrogate the moral assumptions embedded in dominant growth paradigms and highlight the risks of superficial reform or narrative co-optation. By foregrounding concepts such as dignity, sustainability, intergenerational responsibility, and pluriversal imagination, the volume calls for forms of economic transformation that are not only innovative and competitive, but also just, inclusive, and resilient.

Taken together, this book offers scholars, policymakers, educators, and practitioners a framework for understanding economic transformation as a participatory, narrative, and systemic process. It invites readers to move beyond linear models of change and to engage more consciously with the stories, designs, and infrastructures through which economic futures are continuously made.

CHAPTER OVERVIEW

Chapter 1 – Narratives of Stability: How Macroeconomic Signals Shape Collective Economic Behavior

This chapter reframes macroeconomic indicators—such as GDP, inflation, and employment—not as objective facts, but as narrative constructs that actively shape public sentiment, behavior, and policy. Through historical crises, it demonstrates how economic narratives generate reflexive feedback loops that influence the very data they claim to describe.

Chapter 2 – Reframing Value Through Narrative Strategy for Economic Transformation in Entrepreneurial Ecosystems

Focusing on entrepreneurial contexts, this chapter shows how strategic storytelling mobilizes resources, aligns stakeholders, and disrupts entrenched growth paradigms. It introduces conceptual and empirical tools for assessing narrative impact within innovation ecosystems.

Chapter 3 – Fostering Growth and Prosperity through Sustainable Resource Management and Environmental Conservation

This chapter explores sustainability as a co-created economic narrative, demonstrating how enterprises can integrate environmental stewardship, community engagement, and systematic design to generate resilient and innovative growth models.

Chapter 4 – Creative Destruction and Moral Imagination: Rethinking Market Evolution in the Age of Impact

By positioning economists and entrepreneurs as moral architects, this chapter examines how markets embed values and meanings. It critiques profit-centric models and advances intergenerational responsibility and ethical imagination as foundations for long-term economic design.

Chapter 5 – Accounting as Narrative Infrastructure: Reframing Value Through Systemic Design and Economic Imagination

This chapter reconceptualizes accounting as a narrative and symbolic system that shapes legitimacy and organizational identity. Through ESG and integrated reporting, it shows how accounting infrastructures can support regenerative and inclusive economic models.

Chapter 6 – Algorithmic Narrativity in Entrepreneurship: Reframing Management through Storytelling

Examining generative AI as a co-author of entrepreneurial narratives, this chapter analyzes how algorithmically mediated storytelling reshapes managerial imagination, legitimacy, and strategic orientation in contemporary enterprises.

Chapter 7 – Cultural Narratives and Market Dynamics: Identity, Storytelling, and Consumer Behaviour in the 21st Century

This chapter investigates markets as cultural ecosystems, demonstrating how identity-driven consumption and symbolic meaning influence brand strategy, consumer loyalty, and competitive advantage in a globalized digital economy.

Chapter 8 – Analyzing the Economic Development and Changing Occupational Structure in India

Using historical data and structural transformation theory, this chapter traces India's evolving occupational structure and economic development, offering forecasting frameworks and policy insights relevant to emerging economies.

Chapter 9 – Communication Infrastructures and Platform Dynamics in Shaping Entrepreneurial Success

Based on qualitative research in Greece and Cyprus, this chapter highlights the role of digital infrastructures, platform orchestration, and cultural communication strategies in enabling entrepreneurial success in small and constrained markets.

Chapter 10 – Educational Infrastructures and Pluriversal Imagination for Equitable Economic Futures

This chapter positions education as both narrative infrastructure and systemic design space, examining how decolonial and post-growth pedagogies can cultivate plural, justice-oriented economic futures in the Global South.

Chapter 11 – Exploring Narrative Constructions of Market Sentiment

Through a systematic literature review, this chapter synthesizes how media narratives shape financial behavior and macroeconomic outcomes, proposing a framework of narrative resilience to mitigate destabilizing discourse.

Chapter 12 – Global R&D Strategies of German Multinational Enterprises

The final chapter analyzes shifting co-invention patterns in German MNEs, highlighting the role of global collaboration, local embeddedness, and risk tolerance in contemporary innovation strategies.

Importantly, this book does not offer a single blueprint for transformation. Rather, it proposes a way of thinking and practicing economic change—one that treats narratives as design tools, enterprises as moral and cultural actors, and systems as participatory spaces open to revision. From AI-generated entrepreneurial imaginaries to pluriversal educational models, the volume illustrates that alternative economic futures are not abstract ideals but operational realities already taking shape across contexts.

In closing, *Co-Constructing Economic Transformation Through Enterprise, Narrative, and Systemic Design* invites scholars, practitioners, and policymakers to move beyond reactive adaptation and toward conscious co-creation. By cultivating narrative literacy, ethical imagination, and systemic awareness, economic actors can better navigate uncertainty while actively shaping futures that are not only innovative and competitive, but also just, resilient, and meaningful.

Christos Christodoulou-Volos
Neapolis University Pafos, Cyprus

Andreas Masouras
Neapolis University Pafos, Cyprus

Chapter 1
Narratives of Stability:
How Macroeconomic Signals Shape Collective Economic Behavior

Christos Christodoulou-Volos
https://orcid.org/0000-0002-3188-844X
Neapolis University Pafos, Cyprus

ABSTRACT

This chapter remakes macroeconomic indicators such as GDP, inflation, and employment levels into narrative tools actively building the public's knowledge, behavior, and policy actions. Contrasting with the perception of economic statistics as objective facts, it utilizes narrative and behavioral economics to describe how these indicators are built, blown up, and read in a communicative context that includes media, politicians, and the public. In historical case studies like the Great Recession, Eurozone crisis, and COVID-19, the chapter illustrates how narrative dynamics drive economic cycles by generating feedback loops from sentiment to behavior and back to data. It argues that the macroeconomy is a reflexive system where narratives generate behaviors that reshape the very metrics that necessitated them. Implications for policymakers, business leaders, and theorists are explored, including the need for ethical stewardship of story, complexity-sensitive frameworks, and critical narrative literacy in order to facilitate economic resilience and transformation.

DOI: 10.4018/979-8-3373-4566-6.ch001

1. INTRODUCTION: THE POWER OF ECONOMIC SIGNALS

Numbers in the modern economy do not simply record reality; they construct it. Macroeconomic statistics, such as GDP growth rates, unemployment levels, inflation rates, and consumer sentiment indexes, are not impartial reflections of underlying fundamentals; they are active forces that shape behavior, influence perceptions, and construct futures. These indications are embedded within rich story contexts, perpetually re-interpreted, re-framed, and amplified by policymakers, entrepreneurs, the media, and the public. The work of macroeconomic indicators is therefore well beyond the technicality of spreadsheets and reports; they are essential tools in transformational architecture.

It is in the midst of this process that there lies a key understanding is to be discovered: economic outcomes are deeply entrenched with expectations. John Maynard Keynes recognized as much almost a century ago when he wrote about the role of "animal spirits" in driving investment and consumption patterns (Keynes, 1936). More recently, Robert Shiller (2019) extended this point, arguing that "economic events are largely driven by the stories we tell about them." Economic data here function less as cold fact and more as raw material to be employed to craft interesting, self-reinforcing narratives.

These narratives have real impacts. When the U.S. Federal Reserve telegraphs a rate hike, financial markets react within minutes, even though the policy shift is weeks or months away. When news websites herald an upcoming recession based on a marginal drop in GDP, consumer confidence plummets, inducing lower spending that triggers the recession. The signal strength of economic messages cannot be separated from their statistical payload. They are performative: what is anticipated produces what occurs.

This chapter explores the subtle interdependence between macroeconomic signals, story construction, and collective economic behavior. It argues that macroeconomic communication is policy per se, one with profound effects on entrepreneurial decision, market stability, and system transformation. It is essential that economists, entrepreneurs, business leaders, policymakers, and citizens, as well, grasp how signals are constructed, delivered, and interpreted while they attempt to navigate a more uncertain and complex global economy.

Traditional macroeconomics has generally treated indicators as an impartial reflection of reality. GDP is the measure of output; inflation is a measure of price stability; unemployment levels are the measure of labor market health. But this positivist approach forgets one basic truth: data need interpretation before they can be made sense of. Numbers do not speak. Deciding what to measure, how to measure it, and how to present it necessarily involves value judgments and framing decisions (Espeland & Stevens, 2008). As an example, consider the measurement of inflation. Whether or not to use core inflation (excluding volatile food and energy prices) or headline inflation makes a big difference to how economic welfare is perceived by the general population. Similarly, tension relating to whether GDP is an appropriate indicator of progress for society, highlighted by initiatives such as the OECD's "Beyond GDP" project, arises from raising awareness that what we measure influences what we value (Stiglitz, Sen, & Fitoussi, 2009).

Macro-economic indicators here are not objective truths but socially constructed symbols inserted into broader ideological, political, and communicative contexts. This has long been known in critiques of international economic governance (Stiglitz, 2002). Their meaning is dependent on narratives created by institutional actors, mass media institutions, and cultural norms.

An understanding of the constructed character of economic signals has a second significant implication: communication is an intrinsic aspect of economic policymaking. Central banks, finance ministries, and international institutions have understood for some time that managing expectations is at least as vital as setting interest rates up or down or fine-tuning fiscal budgets.

The practice of forward guidance by central banks is a case in point. Rather than trying to get along with traditional interest rate movements, institutions like the U.S. Federal Reserve today use carefully crafted language to influence market expectations of the future direction of monetary policy. Former Federal Reserve Chairman Ben Bernanke famously characterized monetary policy as "98 percent talk and only 2 percent action" (Bernanke, 2010). By announcing intentions ahead of time, policymakers attempt to stabilize markets, ease transitions, and prevent destabilizing speculation.

But communication strategy does not have to be limited to monetary policy. Governments around the globe applied narrative management to maintain economic confidence through the COVID-19 pandemic as they navigated shutdowns of unprecedented scale. Phrases like "flattening the curve," "temporary disruption," and "build back better" framed the economic crisis as one that was in bounds, temporary, and repairable, thus exhorting firms and households to maintain consumption and investment levels rather than panic.

The economic communication structure, hence, works as a soft infrastructure behind markets. Its strength relies not only on the character of the communicator but also on the cultural alignment of the stories utilized. Communication in line with prevailing social values, like justice, resilience, or innovation, will be more likely to succeed in shaping expectations and behavior.

While policymakers may create the initial economic narratives, it is the established and emerging media that are the primary amplifiers and interpreters of the signals. News articles, opinion pieces, social media, and financial blogs take technical information and make it comprehensible, emotionally relevant stories.

It is not an objective process. Media outlets, with their audience engagement metrics and ideological inclinations, generally select, frame, and privilege economic news in ways that validate some stories and suppress others (Baker, Bloom, & Davis, 2016). A modest dip in a high-profile indicator can be framed as evidence of systemic collapse or a fleeting disturbance in an otherwise solid rebound.

Under the conditions of new media, the speed and scope of media amplification have increased exponentially. Viral narratives can sweep across markets worldwide in hours and contribute to heightened volatility. This dynamic requires policymakers and entrepreneurs alike not only to respond to the content of economic signals but to respond to their narrative ecology, the broader system of information production and distribution that contributes to public understanding and collective action.

The most important feature of macroeconomic stories is likely to be their self-fulfilling nature. Collective belief in an economic story, Shiller (2019) argues, has the power to force the actual conduct that confirms it. If people think there will be a recession sufficiently strongly, they may shut their purses, thereby reducing aggregate demand and triggering the recession they feared.

This cycle of feedback is best observed in entrepreneurial networks. During periods of narrative optimism, say the dot-com boom of the late 1990s or the green tech boom of the early 2020s, venture capital investments flood in, startup formation soars, and innovation bursts. When narratives turn bad, investment dries up, hiring freezes commence, and risk aversion kills the new firms.

The 2008 financial crisis case proves it in spectacular fashion. While the systemic meltdown narrative rode the media wires, consumer confidence (measured by the University of Michigan Index) plummeted by 35%, and OECD business confidence plummeted by 28% over the course of a few months. The ensuing collapse in confidence fueled the slump, extending and intensifying the recession far beyond initial expectations (OECD, 2009). Similarly, during the COVID-19 crisis, tales of resilience and rebound—spurred by government stimulus packages and vaccine technology breakthroughs—appeared to ground consumer sentiment and spur a surprisingly rapid economic rebound in much of the economy by the end of 2020 (OECD, 2021). Thus, to understand macroeconomic behavior, one must be more than a quantitative analyst; one must be narrative literate—reading between the lines of how stories about the economy generate real-world effects.

This review adopts a systematic, interdisciplinary approach to the integration of conceptual analysis, empirical synthesis, and methodological evaluation. Literature was identified via systematic keyword searches ("narrative economics," "macroeconomic narratives," "economic storytelling," "narrative contagion") on websites such as Web of Science, EconStor, NBER, and arXiv and supplemented with scans of milestone books and top economics journals in an effort to incorporate theoretical and empirical work. Inclusion criteria were articles that inaugurated narrative economics (e.g., Shiller, 2017), studies formulating definitions or typologies of narratives (e.g., Roos & Reccius, 2021), empirical studies employing NLP, sentiment tracking, or computational modeling (e.g., Flynn & Sastry, 2024), and formal models that featured narrative contagion or feedback effects. Each paper was charted by narrative type (i.e., central bank narratives, crisis narratives, social media discussions), mechanism and transmission channel (i.e., media amplification, policy discussions, viral contagion), behavioral effect (i.e., confidence shifts, asset prices, consumption), and methodological approach (i.e., qualitative analysis, text mining, agent-based modeling, macro models with narrative indicators) to enable a harmonized classification that aligns narrative types with channels, effects, and leading studies.

Synthesis and critical analysis were focused on the identification of convergence areas and gaps in the literature, with particular interest in causality, precision of measurement, and methodological coverage. Of specific importance was how narrative models are synthesized with traditional macroeconomic models such as rational expectations and signal extraction. Quality control measures favored peer-reviewed research, leading working papers (e.g., NBER), and chapters by reputable scholars. Where definitions differed, a preference was given to those that offered both conceptual richness and empirical tractability, in the interest of clarity and consistency in this review. The format of the outline guarantees the review is not only comprehensive in scope but also methodologically detailed, forming a strong foundation for Sections 2 to 8.

The chapter is structured to lead the reader through a progressively more sophisticated understanding of how macroeconomic narrative is structured and structures economic activity. It begins with the "Introduction," which establishes the significance of macroeconomic indicators as narrative devices and defines the scope of the chapter. The second section, "Understanding Macroeconomic Signals as Narratives," investigates the way economic indicators are being constructed as stories and what psychological and institutional pathways influence how they are received and amplified. The third section, "Communicative Architecture of Macroeconomic Policy," analyzes how central banks, governments, and media producers construct and disseminate economic narratives. This follows on from "Case Studies: Macroeconomic Narratives in Action," where the close reading of turning points in history—such as the 2008 financial crisis, the COVID-19 pandemic, and inflation shocks—is used to show the actual impact that narratives have on economic performance. The fifth section, "The Feedback Loop: Behavior Shapes Data, Data Reinforces Behavior," is conjectural about the recursive dynamic between narratives, behavior, and macroeconomic data. Section five, "The Feedback Loop: Behavior Shapes Data, Data Reinforces Behavior," hypothesizes on the recursive dynamic between narratives, behavior, and macroeconomic data. Next, "Implications for

Policymakers, Entrepreneurs, and Markets" reflects upon how economic actors can effectively negotiate, utilize, or redefine macro-narratives to their advantage or that of the greater good. "Rethinking Macroeconomic Narratives for a Transformative Future," prior to the conclusion, provides blueprints for more sustainable, ethically framed, and equitable narrative-construction and economic storytelling models. Finally, the "Conclusion" synthesizes the findings of the chapter and reiterates its main argument: macroeconomic systems are reflexive in nature, sustained not only by information but by the stories that we build around that information.

2. UNDERSTANDING MACROECONOMIC SIGNALS AS NARRATIVES

Embedding economic indicators into firm, transmissible narratives isn't an auxiliary byproduct; it's the core of how economies work. Where conventional macroeconomic theory perceives data as inputs to rational choice models, more and more interdisciplinary research, from behavioral economics to communication science and sociology, shows that economic signals only become meaningful through narratives surrounding them (Shiller, 2019; Akerlof & Shiller, 2009).

This chapter addresses the process of translating economic information into a story, the psychological and social dynamics of narrative reception and transmission, and the long-term effects of economic storytelling on individual behavior, collective action, and system dynamics.

2.1 From Raw Data to Storylines: How Narratives Form

In the initial stages, macroeconomic data exists in an abstract, typically unobtainable form. The broad public sees no numbers for GDP, unemployment rates, or bond yields in raw statistical terms. Rather, these are mediated, selected, interpreted, and embedded in culturally resonant narratives by communicative actors such as policymakers, journalists, market specialists, and opinion leaders.

The process of converting raw data into narrative formats can be elucidated in several steps:

1. **Selection**: The available data is noticed only in part. For instance, a minor rise in claims of unemployment might be highlighted if the overall scenario is fear of a looming recession.
2. **Framing**: Certain facts are set within certain contexts to be interpreted. Is the rising unemployment being framed as a temporary fix or as an indication of a

fundamental failure in the system? The choice has a profound impact on audience reactions (Entman, 1993).

3. **Story Building:** Bounded facts are extended into longer story lines, which are likely to be emotionally engaging. Headlines like "Economy on the Brink" or "Recovery Gathers Steam" not only inform about the news but construct stories that suggest causality, momentum, and moral tone.
4. **Amplification**: By repetition, news coverage, and online social media virality, some stories get amplified, replacing other potential readings.

This (Data → Selection → Framing → Story → Amplification) is a reinforcement, nonlinear cycle. Early media frames can bias subsequent reporting and interpretation, producing cascading narrative effects (Tversky & Kahneman, 1974; Shiller, 2019).

Incidentally, such dynamics are not only seen during periods of crises, but also during periods of stability or optimism. Incremental productivity improvements in measures during the 1990s tech boom, for example, were inflated into sweeping narratives of a "New Economy" that fueled investment bubbles and redirected policy debates.

Therefore, macroeconomic narratives are not exogenous or random. They are stories, institutionally crafted objects that are used to construct and are, in turn, constructed by institutional agents and mass psychology.

2.2 The Psychology of Economic Narratives

The strength of macroeconomic narratives is that they can engage fundamental psychological processes. Humankind is a story animal (Gottschall, 2012). We do not struggle with complexity by adding up numbers, but by building facts into narrative, emotional stories that give us direction and sense.

Several cognitive heuristics and biases make economic narratives more engaging:

- Availability More memorable events, because they are covered thoroughly in the news or because they are hyped, are regarded as more likely (Tversky & Kahneman, 1974). Excessive media coverage of fear of recession can thus increase the public perception that a downturn is near, even contrary to fundamentals.
- Affect Heuristic: Affect Heuristic: Affective responses to narratives—fear, hope, anger—are more powerful in influencing risk and reward judgments than are statistical chances (Slovic et al., 2004). Fear tales in the 2008 financial crisis, for instance, led the vast majority of consumers and investors to behave much more cautiously than raw data might have.

- Confirmation Bias: Individuals look for and construe messages in a manner that confirms their current views. Thus, once a tale (e.g., "the housing market is unsafe") gains traction, it feeds itself.
- Group Polarization: Within social networks, being surrounded by the same view makes collective views more robust and tales stronger and less susceptible to factually rebuffing them (Sunstein, 2002).

In economic contexts, such psychological inclinations create narratives a remarkable momentum. Even if new information emerges contradicting the dominant stories, dominant stories remain, or get reframed so they no longer contradict themselves.

Such insight into cognitive foundations is helpful to economists, businesspeople, and policymakers who aim to make sense of or influence macroeconomic behavior.

2.3 The Role of Institutions and Cultural Context

Macroeconomic narratives are formed and communicated within institutional and cultural contexts, and cognitive biases at the personal level operate. Different societies, media structures, and political systems shape not only which narratives arise but also how they are read and acted upon.

These institutions, such as central banks, ministries, governments, universities, and media companies, have ontological functions in legitimizing economic stories. A central bank warning of inflation is more credible than a trending TikTok video hypothesizing on hyperinflation, yet both can amass massive followings. Institutional expertise, experience, and legitimacy determine narrative acceptance. For Stiglitz (2002), under globalization, institutional trust is important for whether stories about the economy are credible or not.

Furthermore, cultural values dictate which economic stories find traction. For example:

- In individualistic cultures like the United States, narratives of entrepreneurial perseverance and self-reliance amid downturns are most likely to flourish.
- In collectivist cultures like Japan, stories emphasizing social unity and shared sacrifice during economic downturns could be more effective.
- Those high uncertainty avoidance cultures (Germany, for instance) are shaped by narratives that offer clear-cut blueprints for stabilizing the economy (Hofstede, 2001).

    It is the interaction among cognitive bias, institutional power, and cultural fit that causes macroeconomic narratives to go viral, stick, and transform behavior.

It is the interaction among cognitive bias, institutional power, and cultural fit that causes macroeconomic narratives to go viral, stick, and transform behavior.

2.4 Case Studies: Narratives in Action

To make visible the power of economic narratives to transform, we investigate a number of important historical moments in which narratives heavily impacted macroeconomic results.

2.4.1 The Great Recession of 2008–2009

The September 2008 failure of Lehman Brothers unleashed a world financial panic. Years of underlying fragilities had been built up, yet the narrative of systemic meltdown, powered by symbolic media imagery and apocalyptic expert predictions, ramified behavioral shifts. Consumer confidence fell more than 30% over several months (University of Michigan Surveys of Consumers, 2009), business investment became frozen, and stock markets around the world plummeted.

In this context, the U.S. administration and Federal Reserve initiated massive stimulus measures. Importantly, however, these measures were accompanied by a rhetoric of stabilization: that bold action would restore things to normal. President Obama's framing of the Recovery Act as "the beginning of the end of the crisis" was central in gradually reversing panic (Obama, 2009).

2.4.2 The COVID-19 Pandemic and Economic Resilience

The onset of COVID-19 initiated the sharpest global economic downturn since World War II. Still, by late 2020 and 2021, the majority of economies—especially those with successful governmental communication—began to revive faster than initially projected.

Narratives emphasizing "flattening the curve," "disruption for a temporary period only," and "building back better" maintained consumer and business confidence. Despite record-level unemployment, U.S. consumer spending registered strong gains in the latter half of 2020, fueled by stimulus checks and optimism about vaccines (OECD, 2021).

Here, narrative framing aptly lowered the threat of prolonged depression, highlighting the policy-relevant value of narrative.

2.4.3 Inflation Fears and the Energy Crisis (2022–2024)

Recent increases in energy prices and inflation renewed popular concerns about economic uncertainty. News stories vacillated between "transitory inflation" and "stagflation fears," and they shaped popular expectations and financial market volatility.

For example, the European Central Bank's unambiguous articulation of a "measured and targeted response" prevented the loss of consumer confidence in the Eurozone as precipitously as in previous crises (ECB, 2023). In contrast, where government narratives in such countries were confusing or panicking, popular trust and economic resilience were undermined.

2.5 Implications for Economic Actors

The referencing of macroeconomic indicators as narratives has profound implications:

- **For Economists**: Modeling needs to consider narrative dynamics, formation of expectations, and feedback. Ignoring narrative effects risks systematic forecast errors.
- **For Entrepreneurs and Business Leaders**: Macro-narratives—and the formation of them—are a key form of strategic action. Business organizations that align themselves with the kind of appealing economic stories (i.e., green revolutions, digital revolutions) gain capital and customer backing even in times of crisis.
- **For Policymakers**: Economic policy needs to be preceded by steady, credible communication strategies. Contradictions, mixed signals, or loss of narrative are as likely to inflate crises as soothe them.
- **For the Public**: Getting "narrative literate," i.e., the ability to think critically about economic stories, is becoming a significant civic skill in dealing with volatile macroeconomic times.

3. COMMUNICATIVE ARCHITECTURE OF MACROECONOMIC POLICY

Policy success now depends not just on technical precision but also on the ability to manage stories. The historical assumption of macroeconomic policy action through open, mechanistic channels—interest rate changes, fiscal multipliers, or regulatory moves—is no longer adequate. Instead, the mode by which policies are administered, interpreted, and story-told is what makes them so truly effective (Blinder et al., 2008; Shiller, 2019).

This chapter looks at the communicative dimension of macroeconomic policy and the manner in which central banks, governments, and international organizations design, deliver, and interpret stories to influence expectations and behavior. It will separate the communication strategies edifice, analyze their advantages and disadvantages, and look at innovation in a social media and real-time information age.

3.1 Communication as a Policy Tool

Ever since the 1990s, economists have recognized that the management of expectations is one of the essential pillars of effective macroeconomic policy. Central banks like the Federal Reserve, the ECB, and the Bank of England have increasingly focused on forward guidance—public and open statements about future directions for policy—in order to affect market behavior in ways that supplement or pre-empt mechanical policy actions (Woodford, 2005).

Blinder et al. (2008) have used the term "open mouth operations" to explain the manner in which central banks steer markets through mere talking, instead of taking interest rate or liquidity measures. By carefully designed communications, policymakers try to anchor inflation expectations, stabilize financial markets, and shape consumer and investor attitudes.

The communicative function is thus not a secondary accessory to policy but an essential, strategic tool.

Key elements include:

- **Clarity**: Remove uncertainty of policy intents.
- **Consistency**: No inconsistent signal over time or between policymakers.
- **Credibility**: Past communications that seem consistent with enacted policy actions create trust.
- **Contingency Planning**: Explain what contingency has the right to alter policy paths.

These principles provide the basis for the modern communicative architecture. Bringing them to life, however, demands much more than technical competence.

3.2 The Components of Communicative Architecture

Communicative architecture is the structured policy-issue messaging system design. There are a number of layers, each with a different function:

3.2.1 Core Narrative Design

At the foundational level is the core narrative: the master narrative that defines the institution's worldview of the economy and its policy approach.

For example, during the COVID-19 pandemic, the Federal Reserve's narrative was all about the stories of "extraordinary support," "temporary disruption," and "long-term resilience." These stories guided all subsequent communications, such as rate decisions, speeches, and media interviews.

Central narratives provide coherence and emotional resonance, framing otherwise technical decisions in a format that is understandable by the general public (Shiller, 2019).

Pull Quote: *"We are not just guardians of monetary stability; we are stewards of confidence." — Christine Lagarde, ECB President, 2020.*

3.2.2 Tactical Messaging

Beneath the underlying story are tactical communications: isolated declarations, press releases, meeting notes, projections, and policy briefs. Tactical messages must be aligned with the underlying story but also respond in real time to evolving conditions.

Examples include:

- Interest rate announcements.
- Update of inflation projections.
- Results of bank stress tests.
- Fiscal spending emergency packages.

Effective tactical messaging requires accuracy, on-time delivery, timeliness, and segmenting the audience, recognizing that each constituency (markets, businesses, households) receives information differently.

3.2.3 Strategic Channels

Selecting the proper channels of communication is important. These may be:

- Print Media: Newspapers, TV, and financial newspapers.
- Institutional Platforms: Government portals, central bank websites.
- Testimonies and Speeches: Key individuals like the Federal Reserve Chairperson or the
- The Secretary of the Treasury makes speeches at important events.

- Social Media: Increasingly crucial for reaching younger or less financially literate audiences.

For instance, the Bank of England broke down monetary interventions in simple English in 2020 through brief, cartoon-like videos shared on Twitter and Instagram—far different from its traditionally subdued publications previously.

Central stories provide significance and emotion and frame otherwise technical decisions within a story that makes sense to a broad audience (Shiller, 2019).

Visual Aid:

The flowchart illustrates the cyclical process whereby macroeconomic data feeds back into outcomes in the real world. It begins with the collection of macroeconomic indicators—e.g., inflation, employment, and GDP—and channels them through policy interpretation by institutions like central banks and government agencies. These interpretations are distilled into broad narratives that attempt to explain economic conditions in a coherent and easily comprehensible form. Tactical messaging follows, where communications professionals write these stories to influence individual publics. The messages are delivered via media channels, including news outlets and social media platforms, to influence public opinion and behavior. As the citizenry and businesses adjust their behavior—such as consumption, saving, or investing—in response to these messages, the aggregate behavior creates new macroeconomic outcomes. These outcomes generate new information, restarting the loop. This feedback mechanism highlights how data-driven stories and media framing can, in turn, indirectly influence the economy itself.

Figure 1. The cyclical process whereby macroeconomic data feeds back into outcomes in the real world.

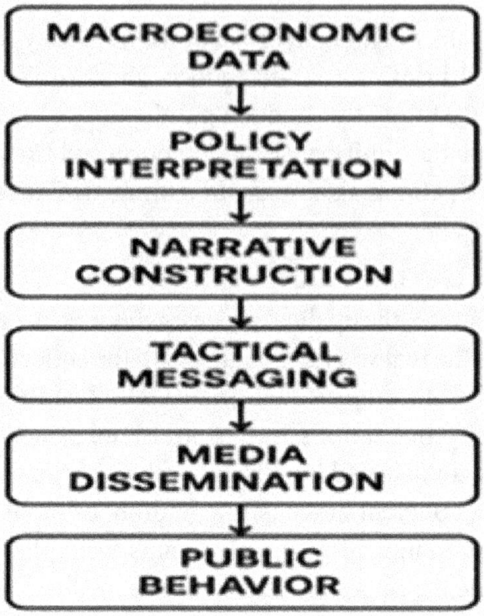

3.3 Managing Expectations: Anchoring and Shifting

One of the main goals of communicative architecture is to anchor inflation, growth, employment, and financial stability expectations. Unbalanced expectations can create suboptimal or even destabilizing behavior, such as premature price rises, wage demands, or withdrawals of investment.

Anchoring Strategies Include:

- Explicit Targets: E.g., the 2% inflation target adopted by most central banks.
- Timelines: Remarks on when policy changes will occur.
- Conditionality: Defining the economic circumstances under which action is to be initiated.

However, actuality typically demands changes in expectations, especially at the time of the incidence of shocks. Policymakers need to then tidily re-label messages without sacrificing credibility.

Case Example: Early in 2021, the Federal Reserve was firmly adamant that inflation increases were "transitory." As inflation continued, the Fed incrementally shifted its rhetoric to emphasize “entrenched inflation risks” and “policy normalization” and stress a string of rate increases in 2022. Although the adjustment was appropriate, the initial miscommunication created reputational issues (Federal Reserve Board, 2022).

3.4 Risks and Challenges in Communicative Strategy

While effective, communication is also hazardous:

3.4.1 Overcommitment

Those strong policy forward guidance commitments, followed by later revocations, run the risk of annihilating credibility once and for all. Trust once lost is not easily regained.

3.4.2 Mixed Messaging

Mixed signals from different officials confuse. This “cacophony risk” is particularly acute where there are conflicting stories from multiple agencies (e.g., monetary and fiscal institutions).

3.4.3 Information Overload

In a 24/7 media and social media age, simply “talking more” is not sufficient. Policymakers must reconcile openness with restraint in narrative, lest they become victims of excessive technical detail that shuts out non-experts.

3.4.4 Narrative Capture

Even policymakers themselves risk becoming captured by prevailing market narratives and acting according to mood rather than fact. These risks promote short-termism and less-than-optimum policy action.

3.5 Innovations and Best Practices

With such challenges, leading macroeconomic institutions are adopting new communication strategies:

- Scenario-Based Forecasting: Illustrating a number of possible futures rather than one forecast enables managing uncertainty without excessive commitment.
- Storytelling techniques: Using metaphors, analogies from history, and relatable examples to describe abstruse concepts (e.g., comparing inflation management to "steering a big ship using a little rudder").
- Engagement Platforms: Engagement Platforms: Holding town halls, interactive webinars, and public consultations to create a two-way dialogue.
- Narrative Stress Testing: Stress testing co-workers internally regarding how diverse stories can play out under several shocks and forecasting communications reactions in advance.

Example: The ECB tested a revolutionary "Narrative Barometer" in 2022, monitoring real-time social media sentiment to gauge the public's attitudes towards inflation and money policy. Tone and messaging were informed by this data, creating a loop for narrative monitoring and narrative release (ECB, 2023).

3.6 Conclusion: Toward Narrative-Aware Policymaking

The communicative aspect of macroeconomic policy is no longer on the outskirts—it's at the center of what policies accomplish. When the world is networked and full of information, narratives are the connecting bridges from policy objectives to economic performance..

Effective communicative strategies require:

- Thoughtful underlying narrative design.
- Strategic agility and simplicity.
- Credible expectation management.
- Preemptive risk management.
- Continuous innovation and adaptation.

In mastering narrative economics and communicative architecture, policymakers are better able to communicate policy, build public trust, and steady volatile macroeconomic circumstances even under conditions of extreme uncertainty.

4. CASE STUDIES: MACROECONOMIC NARRATIVES IN ACTION

While stories are the oxygen of modern macroeconomic management, case studies give us the best view of how their dynamics play out. In this chapter, we look at several milestone examples where macroeconomic narrative really got its way, stabilizing the system in some instances but destabilizing crises in others.

By working through cases, we reveal how narrative framing, communications approach, and public acceptance interact with each other to support or deconstruct policy performance.

4.1 The Great Recession and the "Green Shoots" Narrative (2009)

Context: The 2008 Global Financial Crisis was a catastrophic breakdown in financial and economic stability. In the spring of 2009, as economies continued to contract and unemployment deepened, policymakers and commentators in the media began disseminating bullish caution, using the metaphor of "green shoots" to signal possible signs of recovery.

The term originally gained notice when then-Federal Reserve Chairman Ben Bernanke spoke of "green shoots" of economic recovery in an interview on 60 Minutes in March 2009. The moniker was soon picked up by the media and became a market mood touchstone.

Pull Quote: *"Green shoots are starting to appear, but it's still too early to tell how strong they'll be." In March 2009, Ben Bernanke.*

Narrative Dynamics:

- Core Story: Fragile but emerging recovery.
- Tactical Messaging: Fed meeting minutes, news releases, and speeches all express sage optimism.
- Media Amplification: The metaphor was evocative, highly relatable, and easily translated.

Outcome:

- Equity Markets: The S&P 500 bottomed in March 2009 and began a protracted rally.
- Confidence Indexes: Consumer and business confidence indexes bottomed and rebounded in mid-2009.
- Policy Space: Allowed governments to maintain fiscal support without being buried by panic.

Analysis: The "green shoots" story reduced shared fear without proclaiming false victory. It was applicable; it was fragile but still very optimistic. The equilibrium had to restore appetite for risk and re-anchor expectations.

Table 1. Shifts in Consumer Confidence (2009)

Month	U.S. Consumer Confidence Index	Notable Event
Feb-09	25.3	Pre-" green shoots"
Mar-09	26.9	Bernanke's interview
Jun-09	54.8	Broad media adoption

(Source: The Conference Board)

4.2 The European Sovereign Debt Crisis and the "Whatever It Takes" Moment (2012)

Context: By the summer of 2012, there was existential uncertainty in the Eurozone. Italian and Spanish bond yields climbed to panic levels, and speculation on potential euro exits spread. Confidence in the common currency surviving was eroded rapidly.

It was at this heightened moment of tension that European Central Bank President Mario Draghi made his now-notorious statement:

Pull Quote: *"Within our mandate, the ECB is ready to do whatever it takes to preserve the euro. And believe me, it will be enough." —Mario Draghi, July 2012.*

Unlike the constricting backdrop, European Central Bank President Mario Draghi spoke the now historic words:

- Core Story: The Eurozone is irreversibly.
- Tactical Messaging: Regular releases of the Outright Monetary Transactions (OMT) scheme.
- Media Amplification: Draghi's phrasing was strong, memorable, and immediately went viral.

Outcome: Bond Markets: Spreads between peripheral and German bonds fell sharply after Draghi's statement.

4.3 COVID-19 and the "Whatever Is Necessary" Fiscal-Monetary Narratives (2020)

Context: The onset of the COVID-19 pandemic triggered simultaneous supply and demand shocks. Initial market reactions were catastrophic: the fastest bear market in U.S. history, unprecedented unemployment claims, and collapsing global trade.

In response, governments and central banks worldwide launched extraordinary interventions, but crucially also communicated unlimited commitment to support economies.

Key Narrative Themes:

- U.S. Federal Reserve: "We will not run out of ammunition."
- U.S. Treasury: "We are working at wartime speed."
- European Union: "No one will be left behind."

Result:

- Equity Markets: Sharp V-shaped recovery by end-2020.
- Consumer Sentiment: Unexpected resurgence in mid-2020 amid ongoing health crisis.
- Policy Impact: Stronger large fiscal imbalances and unconventional monetary policy support.

Analysis: The size and assertiveness of the narrative short-circuited more profound spirals of confidence. As virus threats remained, the belief that powerful players were propping up the economy became self-fulfilling. The case showed that macro-narratives could "buy time" for the economy to adjust (Powell, 2020).

4.4 Comparative Narratives in Economies with Varied Institutional Quality

To further embed the power of narrative analysis, this chapter extends beyond the US and Eurozone by discussing macro-narrative forces in developing or non-Western nations. It is from these examples that one can see how institutional trustworthiness, media environments, and cultural norms influence the construction and impact of economic storytelling.

4.4.1. India: Inflation Narratives and Democratic Media Pluralism

Context: India has been facing sustained inflation pressures in the past two decades. In 2013, when consumer inflation had broken through 10%, the Reserve

Bank of India (RBI) implemented a new inflation-targeting approach and followed a public communication strategy.

Narrative Dynamics:

- RBI Governor communication was credible and transparent because of the underlying story, shaping expectations through technical briefings and metaphorical orations (e.g., "inflation is a silent tax").
- Pluralism of media (e.g., English business dailies and regional vernacular television) produced fragmented narrative diffusion, with different sociolinguistic regional variants of narrative uptake.

Outcome: Long-term expectations were anchored by the RBI narrative, but diffusion varied by sociolinguistic region, highlighting the significance of media structure and public trust.

4.4.2. Brazil: Fiscal Narratives in a Context of Volatile Credibility

Context: The government's macroeconomic credibility was dented by political turmoil when Brazil faced the 2015–2016 fiscal crisis. Corruption scandals, such as Petrobras, had caused public confidence in official data to be eroded.

Narrative Breakdown:

- The government announcements of financial discipline were generally not believed, and social media stories blaming manipulation of inflation and job figures gained appeal.
- Institutional mistrust created narrative gaps, and these were filled by opposition discourses, external institutions, and autonomous economists.

Outcome: Supply optimism fell in the presence of relatively modest changes to data. Illustrates the extent to which low institutional trust unravels the performativity of official stories.

4.4.3. Nigeria: Currency Devaluation and State-Controlled Media

Context: The 2016 foreign exchange crisis in Nigeria was severe. Under the pretext of "protection of national sovereignty," the Central Bank first maintained its demand on a fixed exchange rate."

Narrative Strategy:

- The government presented devaluation as "economic betrayal," employing post-colonial rhetoric.
- Black market exchange rates and social media chatter went against the official stance, with Twitter and WhatsApp escalating concerns about depreciation.

Outcome: As a result, the devaluation caught markets off guard, and the breakdown in the alignment between official propaganda and empirical data contributed to the delay. Illustrates how top-down narratives can break down in highly informal media spaces.

4.4.4. China: Coordinated Narrative Management through State-Controlled Media

Context: During the COVID 2020 economic depression, the "stability above all" framing was used by China's government through state media, WeChat in China, and central bank statements.

Narrative Strategy:

- Offsetting positive economic news reports duplicated (e.g., "dual circulation strategy," "return to normal") were synchronized across media.
- Panic or disavowal narratives were suppressed.

Outcome: Consumer confidence bounced back quicker in early 2020 in comparable economies compared to consumer confidence that rebounded quicker, reflecting high congruity of state-media discourse and public behavior, but raising concerns about openness and diversity.

Table 2. Key Insights from Comparative Cases

Case	Institutional Quality	Media Ecosystem	Narrative Impact
India	Moderate-high	Fragmented, multilingual	Moderate, regionally uneven
Brazil	Volatile	Partly free, mistrustful	Weak narrative anchoring
Nigeria	Low	Informal, digitally driven	Narrative dissonance, late correction
China	High (state-directed)	Controlled, coordinated	Strong alignment, low pluralism

4.4.5. Conclusion: Institutional Context Matters

These examples show that the performative potential of macroeconomic narratives is not constant:

- Stories in pluralist, high-trust systems must mediate multiple publics.
- Narrative credibility is weak and will be counterproductive in low-trust or opaque systems.
- Where there is state control, coherence is high, while transparency is sacrificed.

Future policy development and research must take into account how the power of stories depends on institutional capacity, communicative openness, and cultural resonance.

Figure 2.

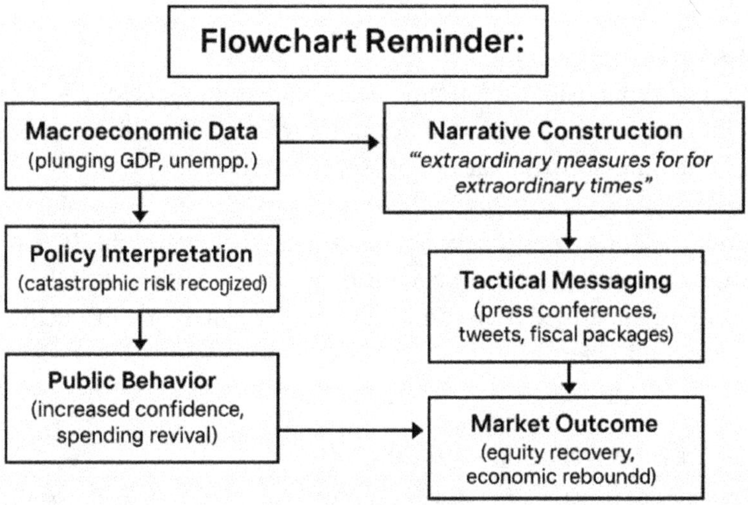

This flowchart graphically illustrates the communications feedback loop that drives economic recovery in times of crisis. It begins with deteriorating macroeconomic indicators, such as the sudden and sharp reduction in GDP and rising unemployment, that portend gross instability. This is followed by policy interpretation, where policymakers realize the catastrophic risks and seriousness of the situation. From which a narrative is constructed, generally framed as "extraordinary measures for extraordinary times," to justify out-of-the-ordinary intervention. This narrative is translated into strategic messaging in the guise of press conferences, tweets, and announcements of fiscal or monetary stimulus. The signals are picked up by the public, and they have the ability to bring about change in public behavior, boosting consumer confidence and triggering spending. Therefore, the market outcome could reflect stabilization or rebound with positive equities and other markers of economic revival. This transformation shows the extent to which well-designed messaging and prompt communication can smooth the spillovers of an economic shock.

4.5 Failure to Control Narrative: Inflation Expectations Post-COVID (2021-2022)

Context: Throughout the early part of 2021, inflation in advanced economies had picked up pace. Initially, central banks—most notably the Federal Reserve—labeled rising prices "transitory," caused by temporary supply bottlenecks.

Pull Quote: "*We expect inflation to be transitory. Bottlenecks will ease." — Jerome Powell, May 2021.*

But inflation was much more resilient. Central banks had to take a radical shift toward tight money during the early part of 2022 when consumer price rises hit multi-decade levels.

Narrative Dynamics:

- **Core Narrative Misalignment:** The fundamental story mismatch in narrative forces is the adamant refusal to believe transitory inflation in the face of mounting evidence to the contrary.
- **Tactical Disorientation:** Disjointed communication from various Fed officials.
- **Media Blowback:** Senior media outlets chastised policymakers for lagging behind the curve.

Outcome:

- **Public Confidence:** Inflation expectations grew less anchored.
- **Bond Markets:** Abrupt sell-offs; 10-year yields surged.
- **Policy Reversal**: Abrupt rate increases, policy tightening.

Analysis: Failure to align the narrative to evolving evidence in time severely undermined credibility. It is imperative that policy narratives change without becoming reactive or frightened (Clarida, 2022).

4.6 Case Data Summary

Table 3. Narrative Effectiveness Metrics

Case	Narrative Type	Public Response	Market Reaction	Outcome Assessment
2009 "Green Shoots"	Hopeful, cautious	Improved sentiment	Equity rally	Positive
2012 "Whatever It Takes"	Bold, defensive	Stabilized confidence	Bond spread narrowing	Highly Positive
2020 COVID Response	Unlimited support	Revived trust	Rapid market recovery	Positive
2021-22 Inflation	Dismissive early, aggressive later	Distrust, confusion	Market volatility	Negative

4.7 Lessons from Case Studies

These case studies offer various crucial lessons, including:

- Narrative Timing: When used early, plausible optimism can stabilize tumultuous circumstances.
- Credibility Is Currency: Unrealistic optimism unrelated to facts annihilates trust in the end.
- Simple, Resonant Messages Win: Metaphor, bold pronouncements, and emotionally resonant language win out over technical jargon.
- Strategic Ambiguity Can Be Useful: Vague commitments ("whatever it takes") offer leeway without fanning panic.
- Narrative Evolution Is Paramount: Successful policy communication must change with shifting realities without compromising on previous messaging coherence.

Lastly, stories are living, and their management is no more than an art or a science. Communication planners, policymakers, and economists will need to work together in constructing stories that are technically sound, emotionally intelligent, and socially sustainable.

5. THE FEEDBACK LOOP: BEHAVIOR SHAPES DATA, DATA REINFORCES BEHAVIOR

One of the most powerful things about macroeconomic stories is that they are not merely passive accounts of reality—they actually produce economic results. The outcomes then influence the subsequent round of stories, creating a feedback loop that reinforces itself.

Traditional economic models, which view agents as passive maximizers responding to exogenous variables, are challenged by this dynamic, where expectations, behavior, and information constantly interact. Instead, it underlines the reflexive character of macroeconomies (Soros, 1987).

In the next section, we investigate the mechanism of this feedback loop, illustrate it by empirical examples, and analyze its implications for policymaking and communications strategy.

5.1 Conceptualizing the Feedback Loop

Traditional macroeconomic models (e.g., IS-LM, AS-AD) draw on economic facts, such as rates of growth of GDP, inflation, and levels of employment, as relatively objective facts conditioned by structural forces and policy interventions. But once narratives condition agent behavior, and behavior feeds back into the economic aggregates, a reflexivity ensues.

We can represent the macro-narrative feedback loop schematically:

Figure 3.

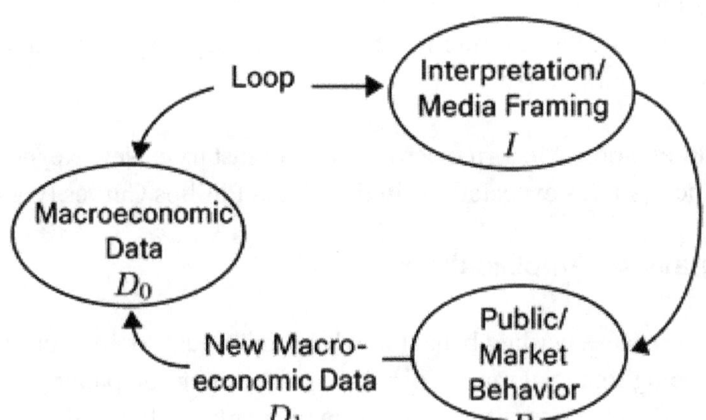

The flowchart illustrates a self-reinforcing macro-narrative feedback loop that begins with some preliminary macroeconomic data (D_0), e.g., GDP or unemployment figures. The data is interpreted and framed by the media (I), which affects how the markets and public perceive the information. Public and market behavior (B) then alters—consumers spend more or less, investors shift assets, etc.—which in turn generates new macroeconomic data (D_1). This additional information feeds back into the cycle, starting the process anew. The model emphasizes how perception and behavior mediate the transformation of economic information into economic outcomes.

Each cycle has the potential to either amplify or dampen shocks depending on the prevailing narrative framework and public confidence.

5.2 Mechanisms of Feedback

There are three key mechanisms that enable this feedback process:

5.2.1 Expectation-Driven Behavior

When agents believe specific outcomes to be likely, what they do actually makes these outcomes more likely—a procedure sometimes called self-fulfilling prophecy (Merton, 1948).

Examples:

- If companies believe a recession to occur (due to negative narratives), they may reduce investment and hiring in advance, thereby producing the recession.
- If consumers believe future inflation to rise, they may accelerate purchasing, driving inflation today.

Model Integration: Modern macro theory, for instance, New Keynesian models, explicitly incorporates expectations in the IS and Phillips Curves (Gali, 2015).

5.2.2 Behavioral Amplification

Stories may cause herding behavior, where individuals or institutions copy each other, enhancing financial market direction or consumption patterns.

Example: During crises, if individuals in the market believe others will sell, they sell first, creating quick drops (Shiller, 2000).

5.2.3 Institutional and Policy Responses

Even policymakers are susceptible to tales. Even policymakers can be influenced by narratives. Central banks, legislatures, and treasuries may be under pressure from public opinion to take actions or not take actions that support dominant narratives.

Example: The political agreement for massive fiscal stimulus in the U.S. and Europe during COVID-19 was probably accelerated by the public's apparent fear of economic catastrophe.

5.3 Illustrative Case Studies

Let's examine two recent examples to learn more about the real-world operation of the feedback loop.

5.3.1 The Vicious Circle of the 2008 Financial Crisis

Phase 1: Data Shock

- Housing prices began declining in the U.S.
- The first signs of mortgage delinquency were seen.

Phase 2: Enhancement of Storytelling

- Financial media and analysts portrayed the housing slump as systemic, emphasizing the dangers of an impending failure of the banking system as a whole.

Phase 3: Response in Behavior

- Banks stopped lending, and consumers reduced their spending due to concerns about the recession.

Phase 4: Strengthening Data

- Loans vanished, spending fell, and job losses increased.
- Initial fears were realized in a full-blown recession.

Pull Quote: "*Fear itself became a self-reinforcing economic phenomenon." — Robert J. Shiller, Irrational Exuberance.*

Thus, behavior drove the macro data, and new negative data once more incited worsening narratives.

5.3.2 The Virtuous Circle of COVID Recovery (2020-2021)

Phase 1: Policy Signals

- France and the US pledged enormous fiscal and monetary resources ("infinite support").

Phase 2: Narrative Framing

- Media solidified optimism: "Recovery is just around the corner," "Vaccines will end the crisis."

Phase 3: Behavioral Adaptation

- Savings deployment into consumption accelerated.
- Equity investment rose sharply by retail investors ("Robinhood boom").

Phase 4: Data Validation

- Retail sales rebounded.
- GDP growth rebounded sharply in H2 2020.

Here, positive conduct, fostered by confidence narratives, yielded higher-than-projected figures, validating the cycle of optimism.

5.4 Behavioral Macroeconomics and Narrative Feedback

Recent theory progress ever more emphasized that agents do not just build expectations rationally but by narrative construction.

Behavioral models suggest:

- Agents overestimate salient recent events (availability bias).
- Economic choice is influenced by emotional tone (Gennaioli & Shleifer, 2018).
- Herding results from narrative convergence.

Narratives are thus not "noise" around a rational core but the form of economic expectations themselves.

Narrative Persistence and Path Dependency

Once a story becomes entrenched, it will persist beyond factual usefulness due to:

- Confirmation Bias: To give selective attention to information that confirms previously existing assumptions.
- Institutional Inertia: Institutionally rooted policies and practices favoring the narrative create self-strengthening scaffolding.
- Social Contagion: Shared narratives are shared through social and professional networks.

It is here, too, that some economic alarms or myths—e.g., "hyperinflation is around the corner"—survive despite refuting evidence.

5.5 Modeling Feedback Loops: Challenges and Opportunities

Economists find it challenging to model feedback loops because mathematically:

- Stories are not easily measured.
- Public opinion metrics (e.g., polls, social media trends) are noisier and messier.
- Feedback loops can yield several equilibria (good or bad) rather than a single solution.

Good contenders for the task are

- Agent-Based Modeling (ABM): Diverse agents learn from stories and from each other.
- Narrative Econometrics: Statistical tools to study story prevalence and economic impact (Shiller, 2019).
- Sentiment Indices: Finding economic forecasts from sentiment in news (e.g., Baker, Bloom, and Davis's Economic Policy Uncertainty Index).

5.6 Policy Implications of Narrative Feedback Loops

Only feedback loops can be understood if policy and communication are fundamentally overhauled.

Table 4.

Traditional View	Narrative Feedback View
Target fundamentals (e.g., rates, taxes)	Target both fundamentals and beliefs.
Communicate facts neutrally	Craft emotionally intelligent narratives.
Expect rational adjustment	Anticipate non-linear, behavioral responses.

Policymakers must:

- Monitor narrative shifts proactively.
- Take action quickly to stop destructive loops.
- Adapt stories flexibly as circumstances change.
- Preserve credibility without inflexibly adhering to out-of-date arguments.

The narrative feedback loop must be controlled to avoid losing control over economic facts and public behavior.

6. IMPLICATIONS FOR POLICYMAKERS, ENTREPRENEURS, AND MARKETS

We have set out in previous sections how macroeconomic indicators, having gone through narrative lenses, can have a significant impact on public behavior and market functioning, generating strong feedback loops. Understanding this phenomenon is essential not just of theoretical interest but also has profound practical significance for three significant players as policymakers, entrepreneurs, and market participants.

This section examines how each group can enhance navigating and actively molding the economic story landscape to develop resilience, adapt to the unknown, and leverage upcoming opportunities.

6.1 Implications for Policymakers

6.1.1 From Policy Tools to Narrative Tools

Traditional macroeconomic policy relies on altering fundamentals: interest rates, taxation, and spending. Yet if narratives significantly mediate how fundamentals are understood and acted upon, then communication is itself a fundamental policy instrument.

Monetary Policy Example: Central banks, and the Federal Reserve specifically, have already moved to forward guidance, directing expectations of future rates to influence current behavior (Woodford, 2005).

Beyond Forward Guidance:

- Policymakers must consider how policies will be presented in the media and interpreted by the public.
- They must be proactive narrative managers, not reactive responders.

Quote: *"Economic policy is 90% communication, 10% action." —Jean-Claude Trichet, former ECB President*

An expanded policy toolkit must thus include

- Narrative Pre-emption: Framing early to communicate policy.
- Emotional Calibration: Sensitivity to the emotional valence (fear, hope, anger) that policies evoke.
- Audience Segmentation: Customizing stories for various public segments (consumers, investors, businesses).

6.1.2 Timing and Transparency

Timing is everything in a reflexive economy. On-time communication can establish expectations before vicious circles take hold. Warrantless or premature optimism, on the other hand, annihilates credibility.

Transparency Strategy:

- Be open about why decisions are made.
- Be frank about uncertainty.
- Don't "overshare" and make unrealistic promises.

Example: With the COVID-19 example, countries like New Zealand succeeded because they shared action and uncertainty candidly, thus building public confidence even amid shifting crises (Wilson, 2020).

6.1.3 Counteracting Vicious Cycles

When negative stories catch on, policymakers must especially try to reshape them.

Methods are

- Recontextualization: Putting negative news into a larger positive trend.
- Use of Authoritative Voice: Using authoritative voices to preclude panic.
- Symbolic Actions: Visible actions (e.g., huge stimulus packages) to express commitment.

Failing to be aware of the story allows fear, doubt, and rumor to exaggerate the slides.

6.2 Implications for Entrepreneurs

6.2.1 Narrative Arbitrage: Opportunity in Perception Gaps

Entrepreneurs, especially in rapidly evolving spaces, can play narrative catch-up faster than larger incumbents.

Narrative Arbitrage is:

- Catching emerging positive narratives early (e.g., "green energy," "remote work").
- Crafting ventures to ride these waves ahead of competitors.

Case Example: Zoom businesses exploded during the pandemic, not just because of product-market fit, but because they embodied an emerging narrative about the "future of work."

Entrepreneurs need to pay special attention to:

- Media sentiment indexes
- Policy signals on favored sectors
- Consumer trend readjusts on cultural and social platforms

6.2.2 Storytelling as Strategic Capital

Where behavior follows perception, a founder's ability to craft and tell fantastic stories is every bit as valuable as technical or operations expertise.

Elements of strategic storytelling are:

- Vision Stories: What is the world going to be like, and how does the business fit into it?

- Crisis Stories: How does the business inject stability into periods of uncertainty?
- Community Stories: How does the business enable collective identity?

Quote: *"In a world of abundance, a well-told story is the ultimate scarcity." — Seth Godin*

6.2.3 Risk Management in Narrative-Driven Markets

Entrepreneurs can also anticipate that narrative volatility increases operational risks.

- Rapidly shifting perceptions can accelerate product obsolescence.
- Hype cycles can lure companies into overexpansion and subsequent retrenchment.

Thus, agile companies include

- Scenario Planning: Formulating responses to different narrative paths.
- Media Monitoring: Real-time monitoring of emerging themes.
- Adaptive Branding: Being able to change brand messaging as narratives evolve.

6.3 Implications for Markets and Investors

6.3.1 Beyond Fundamentals: Narrative Analysis as Investment Strategy

Historically, investors look at economic fundamentals—earnings, valuations, and growth prospects. But knowing the power of stories means sentiment tracking is equally critical.

Narrative Investing Tools:

- News Sentiment Analysis: Quantifying emotional tone of news articles (e.g., RavenPack, Bloomberg NLP).
- Social Media Monitoring: Tracking top themes (e.g., Reddit's r/wallstreetbets).
- Narrative Momentum Metrics: Tracking spread velocity and saturation of stories.

For example, the meme stock drama (for instance, GameStop, AMC) showed how rapidly asynchronized narratives can destroy fundamentals in the short term (Smales, 2021).

6.3.2 Managing Narrative-Driven Volatility

Narrative-driven markets have:

- Temporary heightened volatility (since stories alter faster than fundamentals).
- Bubble-prone behavior (due to reflexivity).
- Dramatic regime change (when current stories blow up).

Investors thus need:

- Tactical Flexibility: They can shift exposure at short notice.
- Diversification by Narrative: Holding securities related to different leading stories.
- Sentiment Hedging: To hedge through options or sentiment-related instruments.

6.3.3 Regulatory and Ethical Considerations

With stories gaining recognition as influential economic forces, questions come up:

- Should regulators act to counteract pernicious financial narratives (e.g., coordinated short squeezes)?
- Are investors under any moral obligation to spread stories that drive prices?

The intersection of free speech, market integrity, and narrative contagion will be a wide area of financial regulation in the 2020s.

6.4 A New Economic Skillset for the Reflexive Age

The advent of narrative-based feedback loops necessitates an emergent hybrid skillset in policymaking, entrepreneurship, and investment:

Table 5.

Traditional Skills	Emerging Skills
Data analysis	Narrative analysis
Policy design	Storytelling craft
Strategic planning	Sentiment monitoring
Risk management	Reflexivity management

Education schools that train economists, entrepreneurs, and investors must incorporate communication literacy, behavioral science, and media dynamics into curricula.

Quote: "*The ability to read and shape narratives will be as important as the ability to read balance sheets.*" *— Adapted from Robert Shiller*

6.5 Final Reflection: Navigating the Economy of Stories

In the new economy, stories are not epiphenomena. They build reality by coordinating collective behavior. Thus:

- Policymakers must become narrative engineers.
- Entrepreneurs must become narrative entrepreneurs.
- Investors must become narrative analysts.

The future does not belong only to those who can read the data, but to those who can read the meaning of the data as it is shaped by the human mind.

Knowing the power of economic storytelling presents a possibility of more adaptive, responsive, and empathetic economic systems—if we are open to harnessing it with care.

7. RETHINKING MACROECONOMIC NARRATIVES FOR A TRANSFORMATIVE FUTURE

As we have come to understand along the way in this chapter, macroeconomic tales are not merely byproducts of economic activity; they are a vital force driving economic behavior, decisions, and systemic results. In the more connected, digital, high-emotion world of today, the traditional separation between "real" economic fundamentals and their stories is breaking down.

This concluding chapter argues that we must rethink how macroeconomic narratives get told, explained, and utilized—not only to suit this new world, but to create a more robust, fairer, and more innovative economic future.

7.1 Why Traditional Narrative Models Are Inadequate

7.1.1 Linear Causality vs Reflexivity

Classical economic models tend to treat stories as exogenous shocks: stories change because fundamentals change. However, as George Soros (2013) and Robert Shiller (2019) stress, narratives and fundamentals are codetermined.

Figure 4.

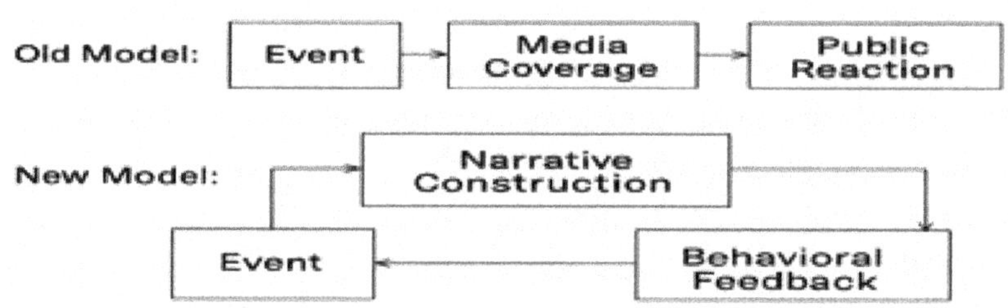

The graph is a comparison between two models of the way events influence public attitudes and behavior. The "Old Model" illustrates a linear process: the event occurs, the media cover it, and the public reacts, implying a passive public being informed about something. The "New Model" illustrates a dynamic feedback loop in which an event interacts with the construction of narrative—the framing and interpretation of the event—and this influences behavioral feedback from the public. This feedback, in turn, influences further narratives as well as the perception or trajectory of the original event itself. The new model highlights an increasingly interactive and recursive system of meaning-making.

And so systems thinking is what economic change today demands: the ability to perceive nonlinear feedback loops, emergent behaviors, and tipping points.

7.1.2 The Problem of Short-Termism

Today's digital narrative loops are moving fast, especially in online media spaces. News loops that took weeks to run are now running in hours.

Economic narratives tend to

- Overemphasize short-term shocks (e.g., a poor jobs report
- Underemphasize the deep-seated structural changes (for example, demographic change, climate-related risks)

Short-termism will lead policymakers, entrepreneurs, and investors to make reactive rather than strategic decisions, amplifying volatility and systemic fragility (Tooze, 2021).

Quote: *"We are drowning in information but starved for wisdom." —E.O. Wilson*

7.2 Principles for Future Macroeconomic Narrative Design

Embracing the fallibility of existing approaches, we suggest a number of design principles for building healthier, more process-oriented economic stories.

7.2.1 Embrace Complexity and Uncertainty

Instead of framing excessively simplistic or deterministic accounts (such as "Inflation is up because X"), narratives in this new era should:

- Honor multiple causes.
- Emphasize contingency ("if-then") situations over categorical predictions.
- Make uncertainty an inherent aspect, rather than an indication of failure, of economic experience.

Practical Implementation: Central banks could publish not just "baseline scenarios" but narrative ranges, possible story arcs conditioned on different assumptions.

Example: The Bank of England's "Scenario Planning" during Brexit negotiations reflected a move in this direction (Bank of England, 2019).

7.2.2 Center Human Experience

Economic narratives too often center on aggregates: GDP, inflation, and unemployment. But those aggregates conceal wide-ranging differences in the lives of people in:

- Regions
- Classes
- Demographics

Narratives of the future should be:

- Told through individual and community lenses.
- Qualitative- rather than only through quantitative lenses, conveying dignity, security, agency, and other 'intangibles'.

Quote: "Economics is about people, not numbers." Joseph Stiglitz
Narratives centered on human experiences become better placed to resonate, mobilize, and be more ethically grounded.

7.2.3 Focus on Systems Change, Not Single Events

Instead of dwelling on discrete crises ("The 2008 crash," "COVID-19 recession"), economic stories have to go towards

- tracking long-run structural forces, be it population aging, technology automation.
- Emphasizing the connection between the public health, environment, and financial sectors.
- Identifies potential pathways for systemic change.

This change makes the audience more aware of themselves as active players in a much larger system rather than passive recipients of sudden shocks.

7.2.4 Cultivate Narrative Literacy

- This change enables people to envision themselves as participants within bigger systems rather than as passive objects impacted by unforeseen shocks. This comprises:

- Critical Interpretation: Identifying emotional appeals, metaphors, and frames in economic news.
- Counter-narrative production: A counter-narrative capable of undermining dominant discourses is produced through narrative production.
- Ethics of Storytelling: Reflecting on how our stories affect society.

Embedding narrative literacy within the curriculum, particularly in economics, business, and journalism schools, is central to building a sound economic culture.

7.3 Toward a Framework: Strategic Narrative Stewardship

We introduce the concept of Strategic Narrative Stewardship (SNS): an engaged, ethical, and responsive manner of managing economic narratives for the public good.

Table 6. Components of Strategic Narrative Stewardship

Component	Description
Awareness	Monitoring emerging narratives and their emotional resonance.
Analysis	Identifying biases, gaps, and implications of prevailing stories.
Adaptation	Shaping or responding to narratives to align with strategic goals.
Accountability	Ensuring transparency, truthfulness, and ethical intent in narrative interventions.

7.3.1 Roles of Different Actors

- Policymakers: Engage freely with narrative contexts; utilize narrative ethically to shape behavior without manipulation.
- Entrepreneurs: Create emergent new products and services along favorable emergent narratives (e.g., sustainability, inclusivity).
- Media Professionals: Shun sensationalism and move towards systemic reporting; stress long-term issues without "doomscrolling" effects.
- Educators: Foster critical economic thinking that harmonizes data, stories, and systems analysis.

7.4 Emerging Frontiers for Research and Innovation

Future research must move beyond reading narratives as epiphenomena and read narratives as first-order causes of economic activity. Research on the following holds promise:

7.4.1 Narrative Networks and Contagion Models

Applying network science to understand:

- The contagion of economic narratives within groups.
- How narratives evolve over time.
- How narratives interact and become dominant.

Some initial forays in computational narrative science open up rich new approaches for the field (Berger & Milkman, 2012).

7.4.2 Narrative-Based Forecasting

Can we use patterns of narrative shifts to infer turning points in business cycles, asset prices, or consumer spending patterns?

Narrative leading indicators may complement (or even replace) traditional quantitative models, especially in times of crisis.

7.4.3 Ethical Narrative Engineering

Now that narrative interventions are an accepted tool for economic shaping, ethical approaches need to be built to accommodate:

- Preventing manipulation and deception.
- Protecting democratic deliberation.
- Weighing social and environmental sustainability against economic growth.

This field will require inter-disciplinary collaboration between economists, ethicists, technologists, and communicators.

7.5 Final Vision: Building a Transformative Economic Narrative Culture

The ultimate goal is not just more cautious economic handling but cultural transformation.

- From short-run panic to long-run stewardship.
- From zero-sum fear to positive-sum imagination.
- Passive consumerism to active citizenship.

Quote: *"The future cannot be predicted, but futures can be invented." — Dennis Gabor*

So, our reframing of macroeconomic narratives is an invitation:

- To co-create narratives that inspire action.
- To develop collective resilience in uncertainty.
- To imagine, and then bring into being, more just and flourishing economies.

And if stories do in fact affect economic reality, then storytellers who tell well are shaping the future, not merely of markets, but of human possibility itself.

8. CONCLUSION: TOWARD A MORE REFLEXIVE MACROECONOMY

As we argue towards the close of this chapter, one dominant theme stares us in the face: macroeconomies are reflexive systems. They are not static aggregates of consumption, production, and investment; they are breathing, living spaces where stories, expectations, and actions coalesce to give rise to economic realities.

Knowing this reflexivity, rather than brushing aside stories as ephemera, holds profound possibilities for reshaping how we know, govern, and participate in macroeconomic life.

This conclusion synthesizes principal findings, pinpoints major challenges, and charts a course forward for researchers, policymakers, businesspersons, and citizens.

8.1 Reaffirming the Central Argument

Mainstream macroeconomics relies on the assumption that "fundamentals drive behavior" in a linear mechanical way. Yet, as we have demonstrated, between fundamentals and human behavior, narratives widely mediate (Shiller, 2019).

When narratives:

- Position a recession as temporary, and consumers and businesses are more apt to continue spending and investing.
- Position a banking crisis as existential; panic can cause contagion, a local event manifesting as a global recession.

Thus, stories are not merely a reflection of the economy; they are one of its drivers.

Reflexivity Dynamics to Keep in Mind:

- Stories are influenced by data.
- Behavior is influenced by stories.
- Narrow behavior influences new data.
- The cycle continues—often in unpredictable ways.

Understanding and embracing this dynamic make macroeconomic systems more understandable, yet more contingent, co-created, and flexible.

8.2 Implications of a Reflexive Macroeconomy

8.2.1 For Economic Theory

Accepting macroeconomic reflexivity involves transforming economic models. Rational expectations, lovely as they are, fail to capture how flesh-and-blood agents really deal with complex, affect-charged, media-dense information landscapes.

- Behavioral feedback loops must be added to models (Akerlof & Shiller, 2009).
- Narrative following and sentiment analysis can complement standard indicators.
- Complexity economics—seeing the economy as a non-linear adaptive system—has a less abstract model (Arthur, 2021).

8.2.2 For Economic Policy

Policy effectiveness relies as much on the tools chosen as on narratives constructed around them. Example:

- During the 2008 crisis, Ben Bernanke's Fed used aggressive monetary interventions along with a deliberate narrative: emphasizing the Fed's commitment to prevent a depression (Bernanke, 2015).
- Contrasting with the more scattered and contestable initial COVID-19 pandemic communication, which fueled uncertainty.

Managing narratives, therefore, becomes an essential capability for today's policymakers, not as manipulation, but as an instrument to anchor expectations and good behavior.

8.2.3 For Entrepreneurship and Innovation

Entrepreneurs intervene and often help rebuild dominant economic stories. Recognition of narrative reflexivity empowers

- Strategic positioning: Situating innovation in new emancipatory macro-narratives (e.g., green economy, resilience tech).
- Risk aversion: Understanding the way in which mood shift can affect habits of funding, consumerization, and regulatory regimes.

Entrepreneurial businesses that play along these narratives will shine more than those that play by customary paradigms because narratives of equity, digitization, and sustainability are in increasing demand.

8.2.4 For Society at Large

Finally, for citizens and communities, recognizing macroeconomic reflexivity allows for:

- Increased agency: To understand that the collective is responsible for producing the outcomes, and not merely reacting to them.
- Critical thinking: Building narrative literacy to counter fear-induced, zero-sum, or exclusionary economic narratives.
- Inclusive engagement: Being a part of economic narratives highlighting common prosperity, resilience, and innovation.

8.3 Challenges to Realizing a Reflexive Macroeconomy

Despite its promise, embracing a reflexive macroeconomy is not without obstacles.

8.3.1 Narrative Capture and Manipulation

Powerful stakeholders—states, corporations, and media barons—can try to monopolize or shape news for purposes of their own. Examples include:

- Overdownplaying systemic risks (e.g., the pre-2008 mortgage market).
- Heightening fear for political gain (e.g., election-season immigration "crises").

Thus, pluralism and openness have to be institutionalized so that narrative milieus are not turned into tools of control.

8.3.2 Cognitive and Emotional Biases

Human beings are prone to:

- Confirmation bias (tendency for evidence to confirm preconceptions).
- Availability bias (overemphasis on recent or salient events).
- Affective contagion (emotional resonance dominating rational analysis).

These biases can create narrative cycles running ahead, resulting in bubbles, crashes, and policy overreactions.

Example: The COVID short-term inflation surges prompted premature tightening discussions, even as longer-term disinflation pressures (Blanchard, 2021).

Narrative literacy and reflective practice are decent antidotes.

8.3.3 Speed and Fragmentation

The new media landscape accelerates attention diffusion, but also:

- Reduces attention spans.
- Fragments audiences into echo chambers.
- Promotes sensationalism rather than nuance.

Thus, constructing a more reflexive macroeconomy requires remaking communication infrastructures to ensure depth and inclusivity in conjunction with speed.

8.4 Pathways Toward a Reflexive Macroeconomy

8.4.1 Institutional Reforms

Institutional policy makers must:

- Invest in story tracking technology, which monitors changes in public opinion.
- Create groups dedicated to narrative stewardship that balance technical competence with the storytelling art.
- Exercise your pluralism commitment by making sure that the voice of underrepresented views is heard in mainstream economic narratives.

8.4.2 Educational Innovation

Schools, universities, and training centers ought to:

- Instruct narrative literacy as well as financial literacy.
- Emphasize systems thinking in business and economics studies considerably.
- Encourage ethical story craft, grounding stories in social responsibility.

8.4.3 Collective Imagination Initiatives

Communities, NGOs, and civil society organizations may:

- Convene citizen juntas on economic futures.
- Promote alternative economic stories (e.g., degrowth, donut economics, regenerative capitalism).
- Apply art, literature, and new media to envision and disseminate visionary economic potentialities.

8.5 Final Reflections

The 21st century demands an altered economic consciousness, one that:

- Perceives the transformative ability of stories to produce material conditions.
- Embraces reflexivity as a challenge instead of a threat.
- Encourages agency, imagination, and responsibility in jointly creating economic futures.

Rather than clinging to long-expiring metaphors of economy as machine or self-correcting organism, we must take on the deeper, more humble metaphor: the economy as conversation—extended, contentious, unfolding.

Quote: *"We live in a world our questions create." — David Cooperrider*

As we shift the stories we tell about economics, we can shift economics itself.

The task that faces us is at once overwhelming and exhilarating: to write, all together, new stories worthy of the future we want to build.

REFERENCES

Akerlof, G. A., & Shiller, R. J. (2009). *Animal spirits: How human psychology drives the economy, and why it matters for global capitalism*. Princeton University Press.

Arthur, W. B. (2021). Foundations of complexity economics. *Nature Reviews. Physics*, *3*(2), 136–145. DOI: 10.1038/s42254-020-00258-7

Baker, S. R., Bloom, N., & Davis, S. J. (2016). Measuring economic policy uncertainty. *The Quarterly Journal of Economics*, *131*(4), 1593–1636. DOI: 10.1093/qje/qjw024

Bank of England. (2019). *Brexit and the Bank of England's preparations*. https://www.bankofengland.co.uk

Berger, J., & Milkman, K. L. (2012). What makes online content viral? *JMR, Journal of Marketing Research*, *49*(2), 192–205. DOI: 10.1509/jmr.10.0353

Bernanke, B. S. (2010). Monetary policy and the housing bubble. *American Economic Association Annual Meeting Speech.*

Bernanke, B. S. (2015). *The courage to act: A memoir of a crisis and its aftermath.* W. W. Norton & Company.

Blanchard, O. (2021). In defense of concerns about inflation. *Peterson Institute for International Economics.* https://www.piie.com/commentary/op-eds/defense-concerns-about-inflation

Blinder, A. S., Ehrmann, M., Fratzscher, M., De Haan, J., & Jansen, D.-J. (2008). Central bank communication and monetary policy: A survey of theory and evidence. *Journal of Economic Literature*, *46*(4), 910–945. DOI: 10.1257/jel.46.4.910

Clarida, R. (2022). Fed communications and market dynamics: 2021–2022 review. *American Economic Association Papers and Proceedings*, *112*, 15–21.

Cooperrider, D. (1995). Introduction to appreciative inquiry. In French, W., & Bell, C. (Eds.), *Organization development* (5th ed., pp. 245–263). Prentice Hall.

Entman, R. M. (1993). Framing: Toward clarification of a fractured paradigm. *Journal of Communication*, *43*(4), 51–58. DOI: 10.1111/j.1460-2466.1993.tb01304.x

Espeland, W. N., & Stevens, M. L. (2008). A sociology of quantification. *Archives Européennes de Sociologie*, *49*(3), 401–436. DOI: 10.1017/S0003975609000150

European Central Bank (ECB). (2023). *Narrative barometer pilot report.* https://www.ecb.europa.eu

Federal Reserve Board. (2022). *Monetary policy report*. https://www.federalreserve.gov

Flynn, Z., & Sastry, A. (2024). Macroeconomics with narratives. *Journal of Monetary Economics*. Advance online publication. DOI: 10.1016/j.jmoneco.2024.01.005

Gabor, D. (1963). *Inventing the future*. Alfred A. Knopf.Gali, J. (2015). *Monetary policy, inflation, and the business cycle: An introduction to the new Keynesian framework* (2nd ed.). Princeton University Press.

Gennaioli, N., & Shleifer, A. (2018). *A crisis of beliefs: Investor psychology and financial fragility*. Princeton University Press.

Gottschall, J. (2012). *The storytelling animal: How stories make us human*. Houghton Mifflin Harcourt.

Hofstede, G. (2001). *Culture's consequences: Comparing values, behaviors, institutions, and organizations across nations* (2nd ed.). SAGE Publications.

Keynes, J. M. (1936). *The general theory of employment, interest, and money*. Macmillan.

Merton, R. K. (1948). The self-fulfilling prophecy. *The Antioch Review*, *8*(2), 193–210.

Obama, B. (2009). *Remarks by the President on the American Recovery and Reinvestment Act*. The White House.

OECD. (2009). *Economic outlook* (No. 86). OECD Publishing.

OECD. (2021). *COVID-19 and the global economy: Initial insights and policy responses*. OECD Publishing.

Powell, J. (2020). Remarks at the Jackson Hole Economic Symposium. *Federal Reserve Bank of Kansas City*.

Roos, M. W., & Reccius, T. (2021). Narratives in macroeconomics: Definitions, typologies, and applications. *Journal of Economic Methodology*, *28*(3), 267–289. DOI: 10.1080/1350178X.2021.1901527

Shiller, R. J. (2000). *Irrational exuberance*. Princeton University Press.

Shiller, R. J. (2017). Narrative economics. *The American Economic Review*, *107*(4), 967–1004. DOI: 10.1257/aer.107.4.967

Shiller, R. J. (2019). *Narrative economics: How stories go viral and drive major economic events*. Princeton University Press.

Slovic, P., Finucane, M., Peters, E., & MacGregor, D. G. (2004). Risk as analysis and risk as feelings: Some thoughts about affect, reason, risk, and rationality. *Risk Analysis*, *24*(2), 311–322. DOI: 10.1111/j.0272-4332.2004.00433.x

Smales, L. A. (2021). Investor attention and the GameStop short squeeze. *Finance Research Letters*, *43*, 102140. DOI: 10.1016/j.frl.2021.102140

Soros, G. (2013). *The alchemy of finance* (2nd ed.). Wiley.

Stiglitz, J. E. (2002). *Globalization and its discontents*. W. W. Norton & Company.

Stiglitz, J. E., Sen, A., & Fitoussi, J.-P. (2009). *Report by the Commission on the Measurement of Economic Performance and Social Progress*.

Sunstein, C. R. (2002). The law of group polarization. *Journal of Political Philosophy*, *10*(2), 175–195. DOI: 10.1111/1467-9760.00148

The Conference Board. (2009). *U.S. consumer confidence survey*. https://conference-board.org

Tooze, A. (2021). *Shutdown: How Covid shook the world's economy*. Viking.

Tversky, A., & Kahneman, D. (1974). Judgment under uncertainty: Heuristics and biases. *Science*, *185*(4157), 1124–1131. DOI: 10.1126/science.185.4157.1124

University of Michigan Surveys of Consumers. (2009). *Consumer sentiment index*. https://data.sca.isr.umich.edu

Wilson, E. O. (1998). *Consilience: The unity of knowledge*. Alfred A. Knopf.

Wilson, S. (2020). New Zealand's elimination strategy for the COVID-19 pandemic and what other countries can learn. *BMJ (Clinical Research Ed.)*, *370*, m2989. DOI: 10.1136/bmj.m2989

Woodford, M. (2005). Central bank communication and policy effectiveness. In *The Greenspan era: Lessons for the future* (pp. 399–474). Federal Reserve Bank of Kansas City. * Corresponding author. Tel.: + 357 26843508; Fax: +357 931944. E-mail address: c.volos@nup.ac.cy.

Chapter 2

Reframing Value Through Narrative Strategy for Economic Transformation in Entrepreneurial Ecosystems:

Narrative-Driven Economic Change

M. P. Rajakumar
St. Joseph's College of Engineering, India

R. Vinston Raja
https://orcid.org/0000-0003-0914-3254
School of Computing, SRM Institute of Science and Technology, India

M. Balasubramani
https://orcid.org/0009-0005-3706-0140
V.S.B. Engineering College, Karur, India

B. Nagalakshmi
PKN Arts and Science College, India

J. Tharun
https://orcid.org/0009-0004-8433-9315
KCG College of Technology, India

M. Robinson Joel
https://orcid.org/0000-0002-3030-8431
KCG College of Technology, India

ABSTRACT

Reframing Value Narrative Strategy as a Catalyst for Economic Transformation in Entrepreneurial Ecosystems for the upcoming book Co-Constructing Economic Transformation Through Enterprise Narrative and Systemic Design examines how

DOI: 10.4018/979-8-3373-4566-6.ch002

strategic storytelling redefines and operationalizes value within entrepreneurial contexts. Drawing on narrative theory, institutional logics, and systems thinking, the authors develop a conceptual model showing how narratives frame opportunities, mobilize resources, and align stakeholder priorities. Empirical case studies illustrate how targeted narrative interventions in policy campaigns and grassroots initiatives can disrupt entrenched growth paradigms and foster inclusive, sustainable innovation. The chapter outlines hybrid methods for assessing narrative impact using computational discourse analysis and participatory evaluation and identifies future research avenues on cross-cultural storytelling, crisis-driven narratives, and narrative infrastructure design. Recommendations for embedding narrative capacity within governance structures.

1. INTRODUCTION

1.1 The Evolving Scope of Economic Value

The global economy is now facing a profound paradigm shift. Traditional economic indicators such as gross domestic product (GDP), market capitalization, and shareholder value are increasingly considered inadequate and inadequate to capture the full breadth of contemporary value creation (Stiglitz et al., 2009; Mazzucato, 2018; Raworth, 2017). Such measures often ignore essential sociological variables such as environmental sustainability, social equity, and human well-being, which necessitates a re-evaluation of what constitutes economic progress (Fioramonti, 2017; Jackson, 2011).

Today's entrepreneurs operate in environments defined by structural uncertainty, interdependent processes, and cultural diversity, where value is no longer a narrow economic indicator; Rather, it emerges as a process shaped by semantics, communication, and society (Beckert, 2016; Hart & Zingales, 2017). This shift creates the need to consider value as an ever-evolving conceptual form, negotiated by discourse, and shaped by installation-identity politics (Boltanski & Thévenot, 2006). In such contexts, narratives shape economic value more profoundly than any other tool they are not mere representations; rather, they function as productive-structural devices that create real, functional, and desirable realities (Kornberger, 2017; Stark, 2009).

Entrepreneurial environments with participants such as founders, investors, incubators, regulators, educators, and grassroots social groups are no longer linear innovation chains; Rather, they are seen as platforms for co-creation, intellectual circulation, and social-institutional exchange (Stam, 2015; Roundy et al., 2018). In these networks, stories do not simply narrate they construct social realities, institutional pathways, and value practices (Gehman et al., 2013; Lounsbury & Wang,

2020). This emerging understanding is leading to a shift in the way storytelling is viewed not as a decorative tactic, but as a deeply structural foundation for economic value creation (Powell et al., 2019).

1.2 The Cultural and Functional Role of Storytelling in Entrepreneurial Organizations

Stories are the original thought movement of human meaning-making; they provide a unified form to otherwise fragmented experiences (Bruner, 1991; Abbott, 2008). Even after this, traditional economic theory has for decades downplayed the role of narrative, emphasizing rational movements, equilibrium models, and pure market logic. But interdisciplinary research, including literary theory, science, sociology, and the humanities, has highlighted the profound functional role that narratives play in structuring, organizing, and directing social life (Czarniawska, 1997; Polletta et al., 2011; Somers, 1994).

Entrepreneurs use storytelling as a strategic, emotional-based, and forward-looking tool to reduce uncertainty, align resources, engage participants, and construct imaginary-predictive futures (Lounsbury & Glynn, 2001; Martens et al., 2007; Garud & Giuliani, 2013). These narratives are not only expressions of purpose; Rather, they create platforms for the creation of realities that can be justified and acted upon by those intentions (Rindova et al., 2011).

Shiller (2019) proposes a field of Narrative Economics that theoretically explains how widely circulated, socio-political stories such as the dot-com boom, the housing crisis, and the green revolution shape collective screens. Response, symbolism, and cultural appropriation give these stories consistency and diffusion (Tuckett & Nikolic, 2017). In entrepreneurial contexts where the future is unknown and installation paths are evolving, stories act as epistemic anchors, aligning participants' interests and motivating commitment (Kaplan & Orlikowski, 2013; Garud et al., 2014).

Furthermore, in areas neglected by mainstream economic contexts particularly in indigenous economies and sub-national societies storytelling serves as a counter-hegemonic force, normalizing alternative practices for value creation based on local knowledge systems and cultural sustainability (Escobar, 2012; Peredo & Chrisman, 2006; Alvarez et al., 2015). These stories transform socio-inherited knowledge into ecological capacity to disrupt logics of exploitation, advance forms of parallel entrepreneurship (Daskalaki et al., 2015; Spicer et al., 2009).

1.3 Purpose and Scope of the Chapter

The aim of this chapter is to argue that narrative schemata act as a leading operational mechanism in the reshaping of how economic value is defined, shared, and implemented within entrepreneurial contexts. Drawing on Ricoeur's (1984) concept of storytelling as a mediating ontological force between lived experience and structural interpretation, the text positions stories as key tools for disrupting dominant theories, validating alternative value logics, and crafting more inclusive and equitable economic futures, as noted by Davis (2016) and Moulaert et al. (2013). In pursuit of this ambition, this chapter mobilizes an interdisciplinary framework that combines installation theory, systems thinking, and narrative analysis to reframe storytelling from a peripheral adjunct to entrepreneurial activity as a fundamental organizational process driving the way contexts evolve, reimagine, and self-regulate through time. By drawing out the roles of entrepreneurs, environmental designers, and policymakers, the narrative gives attention to which stories dominate, who tells them, and through which forms of political, social, and economic trajectories, as underlined by Somers (1994), Fligstein and McAdam (2012), and Sandberg et al. (2019).

To provide a structure for this inquiry, Section 2 lays the conceptual groundwork for the understanding of storytelling as a strategic, installational, and organizational phenomenon. Section 3 then explores how economic value arises through narrative design and structural enactment. Section 4 engages with the transformational power of narrative interventions in changing installational logics and reshaping contextual behaviors. Section 5 addresses new narrative technologies, including AI-driven story systems and blockchain-mediated fact-validation mechanisms, which expand the horizon of narrative agency. In Section 6, the focus shifts to measurement frameworks and diagnostic tools against which the effect of narrative is gauged, complemented by short and contextually relevant case studies. Section 7 subsequently outlines the methodological ways through which narrative changes are analyzed at micro and macro levels. Discussion in Sections 8 through 10 revolves around ethical considerations, governance complexities, and areas of future research and policy.

The chapter concludes by underscoring the embedding of narrative consciousness within the design and transformation of entrepreneurial ecosystems as an imperative, bringing together the theoretical insights and practical strategies. This holistic understanding supports storytelling not only as a reflective practice but also as a strategic foundation for fostering sustainable and inclusive innovation across complex socioeconomic landscapes.

2. THE CONCEPTUAL FOUNDATIONS OF NARRATIVE SCHEMA IN ENTREPRENEURIAL CONTEXTS

Understanding narrative schema as a driving force for economic change requires an interdisciplinary exploration of the conceptual foundations that connect storytelling, value creation, and organizational evolution. Drawing on insights from fields such as narrative theory, installation theory, and organizational thinking, this section positions storytelling not simply as a communication tool but as an ontological and epistemological force that reshapes how value is imagined, justified, and shared in entrepreneurial contexts.

2.1 Narrative: Epistemology and Ontology

Stories are not just data carriers; they are operational structures that shape meaning. As philosophers such as Ricoeur (1984) have argued, stories simultaneously constitute a way of knowing (epistemology) and a way of being (ontology); They construct human understanding by selecting, ordering, and emphasizing elements of experience. In the entrepreneurial context, stories play a key role in shaping intentions, identifying opportunities, and creating legitimacy (Lounsbury & Glynn, 2001). Entrepreneurs do not operate in a neutral world; rather, they operate in a narrative space where existing narratives open some paths and close others (Garud & Karnøe, 2003).

The function of story is twofold: it organizes knowledge; it also sets boundaries for what can be thought about and acted upon. Just as personal identity provides continuity, stories provide coherence across time and context an essential characteristic of entrepreneurial vision and strategy (McAdams, 1997; Ibarra & Barbulescu, 2010). When these stories are shared in networks and embedded in institutions, they go beyond justifying entrepreneurial behavior to embed social values and power structures in economic practices (Somers, 1994; Czarniawska, 2004). Narrative identity is therefore not a metaphor; it is a structural condition of entrepreneurial contexts.

2.2 Value beyond economic criteria

Historically, value has been defined in narrow economic frameworks that prioritize price, efficiency, and capital accumulation. However, both anthropologists and economists argue that value is neither material nor static; rather, it is socially constructed, historically determined, and contextually evolving (Graeber, 2001; Fourcade, 2011). In today's entrepreneurial contexts, stories reveal values that are not captured in traditional accounting such as social cohesion, environmental sustainability, social empowerment, and cultural preservation (Raworth, 2017; Dempsey, 2020).

This expanded view of value is clearly reflected in the rise of frameworks such as the circular economy, stakeholder capitalism, and regenerative enterprise; all of which are driven by powerful stories that reshape entrepreneurial purpose (Mazzucato, 2018; Eccles et al., 2020). Entrepreneurs who align themselves with these stories often gain greater legitimacy, policy coherence, and social cohesion because the value models they propose address systemic inequalities and environmental boundaries (Schoenmaker & Schramade, 2019; Jackson, 2011).

Thus, storytelling is not a biased function of valuation; That is the basic structure of the assessment.

2.3 Systems Thinking and Narrative Interventions

Systems thinking provides a perspective that understands entrepreneurial environments as complex, adaptive, and nonlinear systems. This approach, which goes beyond imperfect market models, highlights the importance of interdependencies, feedback loops, and emergent properties (Meadows, 2008; Capra & Luisi, 2014). In such systems, stories act as attractors scientific and emotional structures that guide behavior, align expectations, and enable integration among diverse participants (Snowden, 2005; Waddell, 2016). These narrative attractors often shape how problems are defined, which solutions are considered viable, and determine long-term outcomes (Geels, 2002; Kania et al., 2018).

Strategic narrative interventions serve as key tipping points places where even small inputs can produce large system-level effects (Meadows, 1999). Changes in narrative can trigger institutional change, especially when new metaphors, characters, or narrative structures are introduced. For example, reframing entrepreneurs as social protectors rather than heroic changemakers creates support for innovation toward collaboration and equity (Spinuzzi et al., 2016; Feldman & Zoller, 2012). Such reframing is essential in contexts that seek to democratize access to resources and redefine success in non-financial terms (Westley et al., 2013).

Figure 1. Illustrative Model of Narrative Attractors in Entrepreneurial Ecosystems

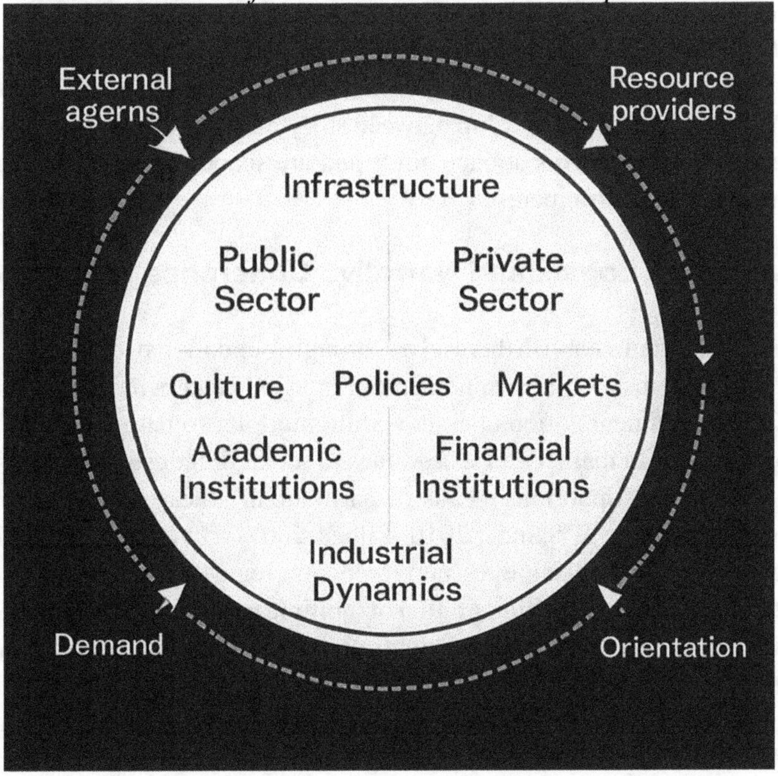

2.3.1 Illustrative Model of Narrative Attractors in Entrepreneurial Ecosystems

It presents a descriptive model of narrative pullers in entrepreneurial environments, putting forward a system level view on how narratives frame and perpetuate innovation ecosystems. At the very heart of this model lies the entrepreneurial infrastructure: public sector organizations, private enterprises, academic institutions, financial bodies, and their interwoven cultural, policy, business, and market environments that interact and co-evolve. Surrounded by the core is a dotted circle that is continuous to represent the narrative pullers: key mechanisms that guide the flow of stories and reinforce systemic feedback loops. External agents introduce new ideas, norms, and legitimacy claims that permeate and influence infrastructural elements.

Resource providers, including investors, donors, and financiers, make support decisions that draw upon prevailing narratives, which validate some while sidelining other directions. Orientation involves the strategies of entrepreneurs and ecosystem builders who interpret and adjust their actions vis-à-vis dominant circulating narratives. Demand acts as a responsive mechanism whereby market and societal expectations feed back into the system to change cultural codes and institutional practices. These narrative pullers together constitute an interlinked structure that underlines the bilateral relationship between storytelling and entrepreneurial activity. It depicts how narratives at once guide and are shaped by systemic interaction within innovation environments.

2.4 Installation Logics and Narrative Coherence

Installations often establish themselves through foundational logics i.e., symbolic and material systems that determine appropriate goals and behaviors (Thornton et al., 2012). Entrepreneurs often operate within multiple, sometimes conflicting installation logics from market and state values to social or industry norms. Narrative schemas provide an important means of harmonizing these logics, or mediating their variations (Pache & Santos, 2010; Zilber, 2007). By creating narratives that resonate across multiple disciplines, entrepreneurs create legitimacy for mixed value propositions and facilitate the integration of multiple participants (Mair et al., 2016).

Installation change often emerges not from formal planning, but from the coherence of the narrative. Stories serve as discursive tools that actors use to introduce and normalize new practices by adopting and subtly challenging the prevailing vocabulary (Battilana et al., 2009; Munir, 2005). Viewing installation fields as narrative spaces allows for more purposeful design of interventions that recalibrate legitimacy, participant stakes, and performance metrics. In this perspective, narrative coherence is not just a cosmetic it is a strategic capability for entrepreneurial change agents.

2.5 Stories as Materialization Structures

In entrepreneurial contexts, stories function not simply as incessant materialization tools; rather, they function as materialization structures that structure how actors perceive reality, how they value, and how they coordinate action. These structures are embedded in the symbolic and material elements of the environment accelerator curricula, investment assessment systems, pitch conventions, reporting formats, and media portrayals (Garud et al., 2014; Lounsbury & Wang, 2020). Once institutionalized, such narratives quietly but persistently become stable semantic microstructures that shape how participants interpret events, determine legitimacy, and imagine which futures are actionable (Wry et al., 2011).

Changing these structures requires more effort than creating new stories. It requires directly engaging with the deep programs that pave the way for collective imagination: who is considered an entrepreneur, what values are rewarded, and what futures are available as narratives. Changes in narrative assumptions embedded in financial metrics, entrepreneurial education, or environmental measurement systems may gradually shift the focus from narrow exponential growth models to more inclusive, socially embedded institutional forms (Tracey & Stott, 2017).

Understanding narrative as structure serves as an analytical bridge between micro-level narrative and system level change. It sheds light on how value regimes are reshaped over time and at what points of intervention can reshape the logics of installation. This framework lays the foundation for how narratives function in economic value creation, discussed in the next section that is, shaping legitimacy, mobilizing participants, and organizing the distribution of opportunities in entrepreneurial contexts.

3.1 Shaping Opportunities Through Story

Shaping opportunities is one of the primary functions of storytelling in entrepreneurship. Entrepreneurs rarely encounter opportunities as fully fledged innovations; instead, they create them through stories that articulate problems, unmet needs, or organizational gaps as avenues for venture creation (Sarasvathy, 2005). This storytelling determines how participants, investors, and co-workers perceive opportunities. A strong story can transform chaos into a dynamic and dynamic environment, enabling entrepreneurs to act in ways that attract critical support and coherence (Lounsbury & Glynn, 2001).

The specific language used in framing these opportunities is significant. Terms like tech for good, inclusive finance, or green disruption are not just descriptive terms; they serve as symbolic tools to connect an initiative's value proposition to broader socio cultural and economic narratives (Navis & Glynn, 2011). Such structures help entrepreneurs signal their innovation, legitimacy, and connection to macro level policy objectives, as they are consistent with the logics of the establishment and the expectations of the participants (Cornelissen & Clarke, 2010). Such stories compete for symbolic resources such as media attention, regulatory support, and social trust; these are as important as financial capital in the early stages of venture development (Rindova et al., 2009).

3.2 Storytelling as a tool for resource mobilization

Storytelling also plays a central role in mobilizing key resources, including financial, human, and social capital. New ventures often suffer from what Aldrich and Fiol (1994) call the liability of newness; that is, they lack conventional avenues for legitimacy and trust. Entrepreneurs fill this gap by creating compelling stories that provide emotionally coherent and scientifically constructed visions of the future.

Early-stage investors consider narrative elements such as founder passion, patience, and vision, in addition to traditional criteria such as performance and scalability (Martens et al., 2007; Parhankangas & Ehrlich, 2014).

Through stories, ideas in the conceptual stage are often transformed into concrete and actionable form through metaphors, prototypes, and future vision statements. Entrepreneurs who can articulate a credible and compelling path from idea to impact are more likely to attract investors, collaborators, supporters, and favourable regulatory regimes (Garud et al., 2014; Wry et al., 2011). Furthermore, narrative forms extend beyond pitch decks. Crowdfunding efforts, funding proposals, and accelerator applications rely on structured narrative foundations to clearly and convincingly convey value (Allison et al., 2015). Such stories not only justify action but also create a shared sense of purpose for participants.

3.3 Creating Legitimacy through Narrative Coherence

Stories also serve as tools for creating legitimacy. Entrepreneurs not only need to appear competent but also need to be consistent with the cultural and ethical standards expected of their audiences. According to Suchman (1995), legitimacy manifests in practical, ethical, and scientific forms each of which is shaped by narrative structure. Efforts to effectively narrate their relevance in such forms can position themselves in competitive installation environments (Zimmerman & Zeitz, 2002).

For example, an agritech venture run by women in rural India can construct a narrative that is consistent with national development plans, the global Sustainable Development Goals (SDGs), and local cultural values of empowerment and self-fulfilment. This multiple coherence makes the initiative appealing to a variety of stakeholders, including investors, non-governmental organizations, policymakers, and community members (Santos, 2012; Dacin et al., 2010). Through such narrative coherence, initiatives not only gain credibility, but also cross-sectoral support and long-term sustainability (Battilana et al., 2009). In this way, they are able to occupy the intersections where different installation logics meet and simultaneously draw legitimacy from multiple sources (Pache & Santos, 2010).

3.4 Climate Systems and Future Construction

In entrepreneurial contexts, stories do not just describe the past or explain the present they also shape futures. Entrepreneurial contexts rely on forward-looking narratives to align participants' expectations and stimulate investment in uncertainty. Entrepreneurs often design what-if futures and use them as a basis for today's commitment (Beckert, 2016). Such fictional expectations form the basis for high-risk innovation and collective value creation.

Such narrative futures are essential for justifying early-stage investment, policy support, and public interest. As Kaplan and Orlikowski (2013) argue, futures stories help to unite diverse actors through shared imaginaries and timelines. They temporarily loosen institutional constraints and create space for creative experimentation and alternative arrangements (van Lente & Rip, 1998). For example, the ideal scenario of a net-zero city engages architects, policymakers, and startups in collaborative experimentation despite current technological or regulatory limitations. In this way, stories serve not only as devices for reflecting reality; they also serve as productive forces for creating institutional futures (Garud & Gehman, 2012).

Figure 2. Cyclical Model of Narrative Temporalities in the Entrepreneurial Process

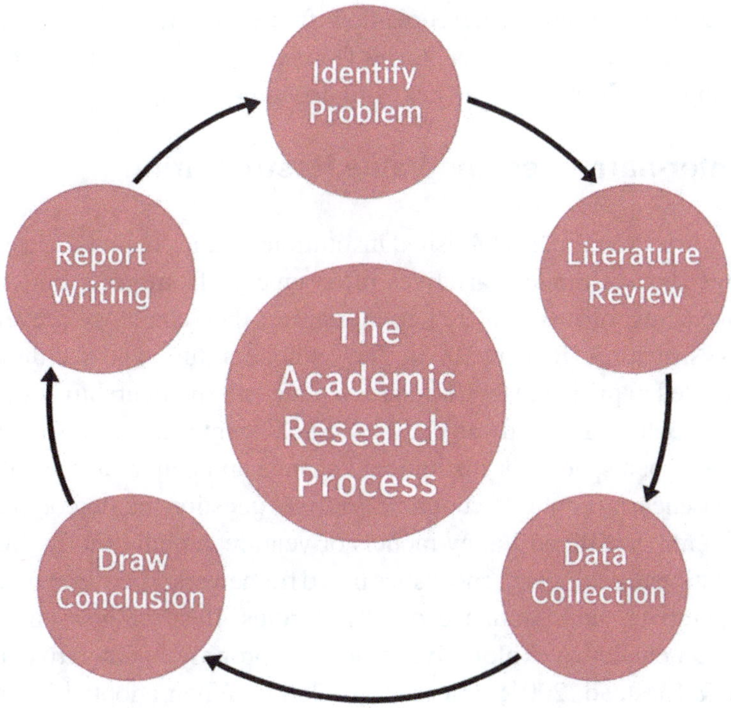

3.4.1 Cyclical Model of Narrative Temporalities in the Entrepreneurial Process

Figure 2 presents the entrepreneurial process as a continuous cycle in which narrative temporalities both shape and are shaped by each phase. It begins with idea generation, where founders craft initial storylines that frame unmet needs or emerging trends; moves into opportunity analysis, where those nascent narratives are tested against market realities and stakeholder expectations; progresses to develop planning, as entrepreneurs script detailed roadmaps that lend coherence and credibility to their vision, advances to collecting resources, where persuasive storytelling mobilizes financial, human, and social capital; transitions into forming organization, when the venture's identity narrative is institutionalized through team formation, culture setting, and formal structures; and culminates in growing business, where performance narratives and success stories reinforce legitimacy, attract new partners, and spark the next wave of ideation. By positioning The Entrepreneurial Process at the center, the diagram underscores how future oriented visions, retrospective success tales, and emerging pivot stories flow in an unbroken loop each narrative phase feeding seamlessly into the next and driving ongoing innovation.

3.5 Counter-narratives and Value Restructuring

Not all stories reinforce established institutional logics many stories are created to challenge them. Counter narratives often emerge from grassroots innovators, social movements, or marginalized communities; they challenge prevailing value and success structures (Brown, 2006). They expose cultural and institutional exclusions embedded in primary stories and reimagine entrepreneurship with alternative imaginaries (Delbridge & Edwards, 2008). Such counter-narratives are not merely reactive; they open spaces for institutional change and help redefine value.

In entrepreneurial contexts, counter-narratives question the stereotypical success stories centered on Silicon Valley models or venture capital criteria. Instead, they advance relational, ecological, and justice based frameworks. Feminist organizations, tribal cooperatives, and circular economy startups illustrate how entrepreneurial narratives can challenge exploitative logics and propose regenerative alternatives (Banerjee & Linstead, 2001; Spicer et al., 2009). When adopted by institutions, incubators, or development organizations, such counter-narratives can redirect value criteria and disrupt the distribution of opportunity across contexts (Munir, 2005). They do not simply add diversity; they reshape the very narrative structure of entrepreneurial value structures.

4. NARRATIVE STRUCTURE AND CONTEXTUAL DESIGN

4.1 Understanding Narrative Structure

Narrative structures refer to the shared symbolic systems that entrepreneurial contexts use to create and sustain meaning including metaphors, role models, cultural norms, and institutional discourse (Hardy et al., 2005). Unlike individual founder stories, which operate at a local level; narrative structures operate at the organizational level, determining how issues are framed, which values are considered legitimate, and who is seen as trustworthy or forward-thinking (Lounsbury & Glynn, 2001; Cornelissen et al., 2008). Such structures do not simply reflect cultural norms; they shape expectations, coordinate beliefs, set boundaries for thought and action, and shape the developmental trajectory of the context (Ruebottom & Auster, 2018; Glynn, 2008).

As entrepreneurial activity increasingly occurs through symbolic representations media portrayals, investment discourses, and environmental development strategies the underlying narrative structure becomes a strategic imperative for environmental design. It determines not only how opportunities are interpreted but also which opportunities are consciously or unconsciously missed in the first place. Thus, viewing narrative structure as a shaping force helps policymakers, financiers, and environmental builders identify key places where narrative interventions can shift collective attention, redirect value priorities, and push environments toward more inclusive and sustainable futures (Etzkowitz & Klofsten, 2005).

4.2 Beyond Incidental Stories: The Structure of Meaning

Narrative structures refer to the enduring symbolic structures parables, models, cultural norms, installation discourses, and material practices through which entrepreneurial contexts create, share, and sustain meaning (Hardy et al., 2005; Lounsbury & Glynn, 2001). Whether individual entrepreneurial stories generate local interest or legitimacy, narrative structures operate at the organizational level, shaping how actors interpret events, define value, and imagine futures. Such structures are embedded in accelerator curricula, investment criteria, pitch conventions, reporting formats, and media representations, becoming natural foundations of meaning over time (Garud et al., 2014; Wry et al., 2011).

This semantic structure influences which challenges are considered important, which opportunities are perceived as credible, and which efforts receive legitimacy and resources. Common patterns such as Heroes, Disruptors, and Growth-first serve as models of entrepreneurial narratives, organizing attention and resource flows (Martens et al., 2007; Navis & Glynn, 2011; Zott & Huy, 2007). Such models accommodate a wide range of socio-economic ideas from neoliberal market logics to techno-solutionism and determine which stories are considered successful and whose voices are invoked, and which are ignored (Fisher, 2010; Phillips et al., 2013). In this way, narrative structures act as knowledge-blocking agents that define what is legitimate, visionary, or thinkable in a context (Green, 2004).

Changing these structures requires realigning deeper cultural norms rather than changing surface narratives. If investment criteria, educational programs, and contextual assessment methods are changed, collective attention can shift from narrow high-growth models to socially embedded, more inclusive organizational forms (Tracey & Stott, 2017). Therefore, viewing narrative structures as a design force provides a practical leverage: it reveals important intervention points that can help to reframe installation logics, shift value priorities, and reorient the growth trajectory of the entrepreneurial environment (Etzkowitz & Klofsten, 2005).

Figure 3. Contrasting Episodic Stories and Open-Ended Narratives

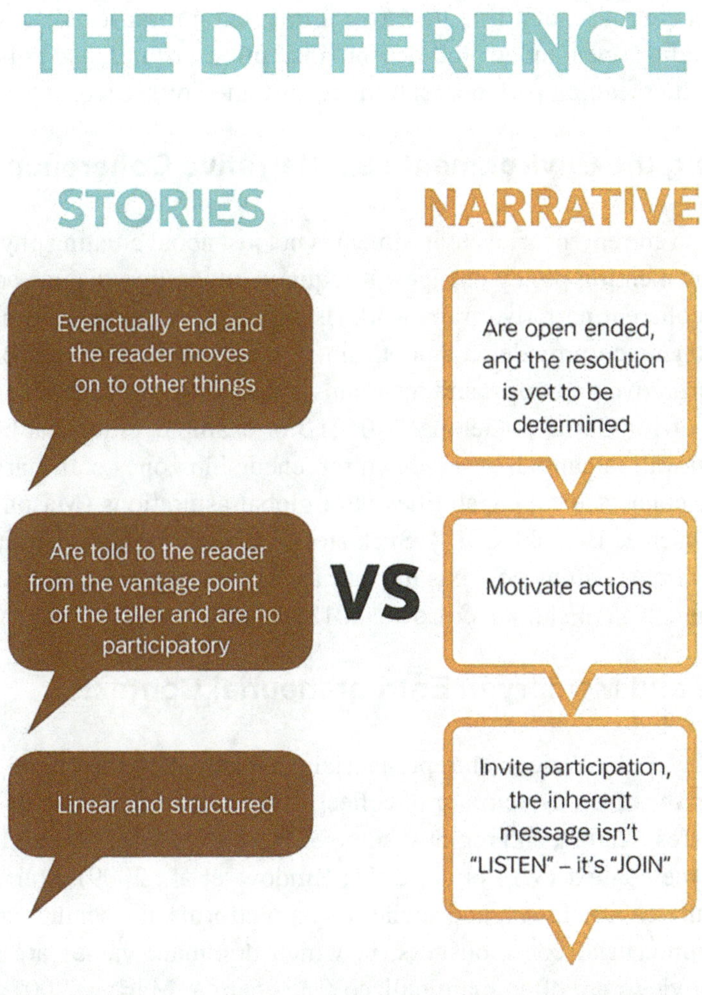

4.2.1 The Contrast Between Procedural Narratives and Open-Ended Narratives

Figure 3 juxtaposes stories which are linear, structured accounts that eventually conclude and leave the audience as passive observers with narratives, which function as ongoing, participatory frameworks that invite stakeholders into an unfolding process. Unlike closed stories that end and recede, narratives remain open-ended, motivating action and co-creation. Where stories are told from a teller's vantage point, narratives encourage joining rather than merely listening, empowering actors to shape outcomes rather than simply consume content. This distinction underscores the chapter's argument that true narrative infrastructure transcends one-off anecdotes, embedding open-ended plotlines into institutions, rituals, and policies so that entrepreneurial meaning and momentum are sustained over time.

4.3 Viewing the Environment as a Narrative Coherence

Shaping an entrepreneurial environment is not just about creating physical structures or implementing policy changes; it requires integrating diverse actors into a shared and coherent narrative framework (Isenberg, 2010; Stam & Spigel, 2016). Narrative coherence provides a smooth integration mechanism that builds trust, reduces interpretive pressures, and maintains a shared sense of purpose (Kaplan & Orlikowski, 2013; Garud & Gehman, 2012). For example, cities that brand themselves as innovation capitals or green entrepreneurship zones craft narratives that strategically connect local capabilities with global aspirations (Mason & Brown, 2014; Audretsch & Belitski, 2017). Such stories not only reflect identity, but also create it thus transforming contexts into attractors for capital, talent, and collective action (Spigel, 2017; Feldman & Lowe, 2018; Roundy, 2017).

4.4 Media and Memory in Entrepreneurial Contexts

Media plays a dual role in entrepreneurial contexts: on the one hand, as a storyteller, and on the other, as a curator of collective memory. Media outlets selectively profile ventures, sectors, and regions; some of these stories are elevated to become symbols of the context (Wry et al., 2011; Rindova et al., 2009). Startup-specific media, venture capital blogs, and social media platforms all together create a mediated entrepreneurial consciousness, in which dominant values are magnified, but minority views are often marginalized (McAdam & Marlow, 2007; Pollock & Rindova, 2003). The narratives thus reinforced determine who is remembered and who is forgotten (Furnari et al., 2020; Petkova et al., 2008).

For example, Bengaluru's rise as India's Silicon Valley was not simply a result of economic statistics, it was reinforced by a long-standing media portrayal that linked software exports and startup density to the hallmarks of an innovation frontier (Kshetri, 2011; Dossani & Kenney, 2007). The memory thus shaped then becomes the basis for policy design and investment decisions (Woolley et al., 2019).

4.5 Inclusion and Justice: Through Narrative Expansion

Narratives in entrepreneurial contexts often highlight specific models typically urban, male-centered, technologically driven, and aggressively growth-oriented and marginalize other forms (Marlow & Swail, 2014; Ahl & Marlow, 2012). This narrative obscures broader forms of entrepreneurship, such as entrepreneurship, social innovation, indigenous scientific practices, and women-led initiatives (Welter et al., 2017; Verduijn & Essers, 2013). By incorporating sustainability, informal economies, circular business models, and community-centered initiatives into narratives, contexts can foster equitable and resilient growth dynamics (George et al., 2016; Shepherd & Patzelt, 2011). Such diversification is not just an ethical imperative it also serves as an innovation strategy that expands the scope of entrepreneurial experimentation (Zahra et al., 2009; Jennings et al., 2015; Khan et al., 2021). Engaging in narrative integration ultimately reshapes the scientific foundations that determine how value is defined in a context (Calás et al., 2009).

4.6 Narrative as a Sensemaking Tool and a Systemic Feedback Strategy

Narrative structures are inherently dynamic; they change in response to real-world disruptions and socio-cultural reinterpretations. As collective experiences change, the dominant narratives that guide entrepreneurial contexts also change (Kaplan & Orlikowski, 2013, Pentland & Feldman, 2008). During the COVID-19 era, the priority for growth and speed has been reframed into new values such as resilience, social care, and ecological awareness (Benedict et al., 2021; Williams et al., 2022). Such narrative shifts often precede institutional or policy changes; they act as early warning signals for systemic changes (Garud et al., 2014; Lounsbury et al., 2021). Narrative feedback loops reshape contexts; they provide the scientific foundations for adaptive learning and innovation (Gehman et al., 2013; Rindova & Martins, 2021).

4.7 Meaning Design for Economic Transformation

Narrative structure is a powerful, but often underutilized, dimension of entrepreneurial context design. It structures how entrepreneurial value is perceived, reflected, and actionable future paths emerge (Dey & Steyaert, 2016). Purposeful narrative design involves fostering pluralistic narrative processes, engaging with historically marginalized voices, and responding to changing societal concerns (Garud & Gehman, 2012; Lounsbury & Glynn, 2019). When narrative structures are aligned with principles of inclusion, justice, and sustainability, they become economic engines for change. Without compelling action, well-crafted narratives expand the horizon of conceivable and actionable possibilities; thereby completely reshaping the entrepreneurial ontology (Tracey & Phillips, 2011; Zott et al., 2011).

5. NARRATIVE DESIGN IN ENTREPRENEURIAL CONTEXTS: PRACTICAL TRANSFORMATION

5.1 Narrative Design as a Purposeful Process

Narrative design is the process of using storytelling, metaphor, and symbolic frames to focus attention, shape emotions, and inspire collective action in entrepreneurial contexts. This process transforms narrative from a reactive mechanism to a strategic tool; it is used by actors such as policymakers, founders, accelerators, and investors to create shared meanings for value creation and economic goals (Lounsbury & Glynn, 2001; Vaara et al., 2016; Garud, Schildt, & Lant, 2014). When narrative design is used purposefully, it unites diverse stakeholders under a unified collective vision and shared economic logics (Kaplan & Orlikowski, 2013; Hardy & Maguire, 2010). Such planned efforts go beyond justification, creating resource mobilization and institutional change (Wry, Lounsbury, & Glynn, 2011; Suddaby, Foster, & Quinn-Trank, 2010). Thus, strategic storytelling aligns entrepreneurial action with long-term environmental objectives.

5.2 Re-evolving Entrepreneurial Identity

Entrepreneurial contexts have historically had narrow identity narratives; they have often featured male-centric, high-status, urban, and tech-centric founders (Ahl, 2006; Marlow & McAdam, 2012). While such stereotypes are powerful, they also reduce inclusion and suppress diverse contributions. In contrast, narrative schemas that foreground the stories of women entrepreneurs, rural innovators, and social entrepreneurs challenge the dominant narrative and open up alternative pathways to participation (Jennings et al., 2013; Brush et al., 2009; Foss, Henry, Ahl, & Mikalsen, 2019). As media campaigns, founder stories, and corporate awards re-articulate who can be considered an entrepreneur, the mindset of entrepreneurial identity is spreading (Martens, Jennings, & Jennings, 2007; McAdam & Marlow, 2007; Welter, Baker, Audretsch, & Gartner, 2017). Such efforts inherently reshape traditional paths and dissolve deeply held stereotypes.

5.3 Risk and Failure: Reframing Through Story

While risk and failure are considered obstacles in many cultural contexts, they are now being redefined through story. Entrepreneurial contexts use storytelling to make failure a key component of innovation and learning (Cardon et al., 2011; Shepherd, 2003; Byrne & Shepherd, 2015). Fail forward events, entrepreneurial podcasts, and candid founder retrospectives de-stigmatize negative experiences and create psychological safety for new entrepreneurs (Cope, 2011; Shepherd & Kuratko, 2009; Yamakawa, Peng, & Deeds, 2015). This narrative shift encourages experimentation and iteration; these are essential elements for sustaining innovation in uncertain markets (Politis & Gabrielsson, 2009; Ucbasaran et al., 2013; Jenkins, Wiklund, & Brundin, 2014). Thus, reframing failure fosters resilience and adaptability in entrepreneurial communities.

5.4 Reshaping Place as an Innovation Hub

Stories are becoming a key strategy for reshaping geographic identity; thus, abandoned regions can be shaped into hubs of innovation and resilience (Zhao & Wry, 2016; Florida, 2002). Cities like Detroit have attracted entrepreneurs and investment by projecting stories of renewal and creativity (Eisenberg, 2011; Mallach, 2018). Similarly, Kochi's emergence as a digital hub is being fueled by carefully curated storytelling through festivals, documentaries, and social media (Lowe & Gertler, 2009; Asheim, Boschma, & Cooke, 2011; Stam & Spigel, 2016). Such narrative interventions reshape the mindset of local capabilities and activate latent resident potential (Hospers, 2003; Moulaert & Sekia, 2003; McCann & Ward, 2010). These strategies of repositioning place as a competitive advantage allow for the unlocking of previously unaccounted regional assets.

5.5 Framing major social challenges as entrepreneurial opportunities

Narrative strategies are moving in the direction of integrating entrepreneurial activities with major social challenges such as climate change, food security, and inequality (Schoar, 2010; George, Howard-Grenville, Joshi, & Tihanyi, 2016; Shepherd & Patzelt, 2011). Projects such as UNDP's Youth Co:Lab and MIT Solve frame challenges in narrative form to stimulate innovation for sustainable development (Seelos & Mair, 2007; Hall, Daneke, & Lenox, 2010; Nicholls & Murdock, 2012). Such narratives position purpose-driven entrepreneurship as a central part of economic transformation, rather than a peripheral activity; they strategically combine ethical urgency and market efficiency (Cohen & Winn, 2007; Zahra, Gedajlovic,

Neubaum, & Shulman, 2009; Mair & Martí, 2006). This reframing transforms social problems into entrepreneurial opportunities, driving the creation of impactful business models.

Figure 4. Entrepreneurial Mindset as a Foundation for Narrative Strategy

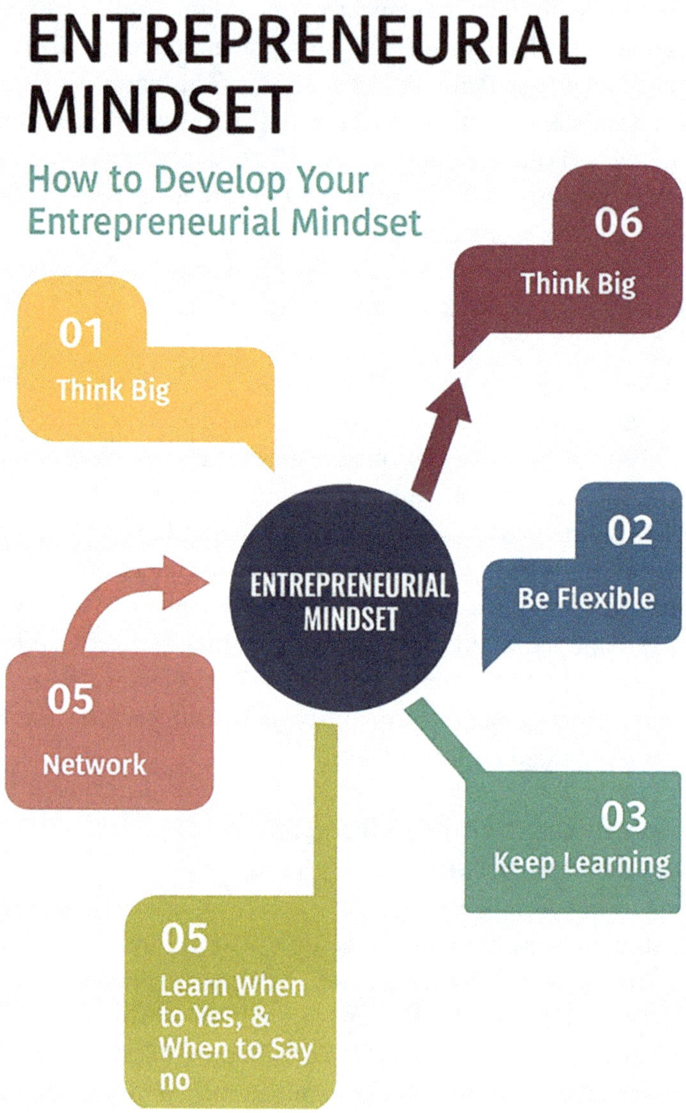

5.5.1 Entrepreneurial Mindset as a Foundation for Narrative Strategy

Figure 4 outlines six core dimensions of the entrepreneurial mindset Think Big, Be Flexible, Keep Learning, Network, Learn When to Say Yes & No, and Talk to the Experts all of which are centered around the core Entrepreneurial Mindset construct. These traits form the scientific and communication foundation for the resulting storytelling: Think Big intent helps create bold, visionary narrative structures that capture the imagination of participants; flexibility allows stories to be adapted as new information or challenges arise; engaging in continuous learning keeps stories grounded and in sync with changing best practices; networking amplifies the breadth and vibrancy of the story through multiple channels and co-creators; the ability to decide when to say yes/no eliminates side-tracking narratives and maintains the coherence of the story; and consulting with experts ensures credibility, depth, and authority in the stories. Together, these dimensions create the scientific and social scaffolding needed to design, strategize, and sustain powerful narratives that drive context-specific change.

5.6 Institutions as Narrative Actors

Universities, development agencies, and governments are not passive supporters of entrepreneurship; they shape the narrative structure of the context into action (Bitektine & Haack, 2015; Tracey, Phillips, & Jarvis, 2011; Lounsbury & Crumley, 2007). National initiatives such as Startup India and Atmanirbhar Bharat create ideal narratives that redefine entrepreneurial legitimacy and opportunity boundaries (Fisher, Kuratko, Bloodgood, & Hornsby, 2017; Desai, 2009). Contextual reports from organizations such as Startup Genome or NASSCOM frame performance metrics, local precedents, and strategic directions in narrative form (Battilana, Leca, & Boxenbaum, 2009; Aldrich & Ruef, 2006; Roundy, Bradshaw, & Brockman, 2018). Such narrative tools serve as tools for shaping meaning in broader behaviour.

5.7 Storytelling Tools, Technologies, and Personification

Digital media technologies have revolutionized how narratives are created, shared, and implemented in entrepreneurial environments. Modern storytelling now includes podcasts, short videos, AR/VR experiences, and interactive visualizations on platforms like LinkedIn, Medium, YouTube, and newer creator networks (Huang & Knight, 2017; Leonardi & Vaast, 2017). These platforms don't just spread stories—they shape identities, communities, and shared ideas (Baldassarre et al., 2017; Ojala et al., 2018). By making content creation accessible, digital tools decentralize narrative power and broaden participation in entrepreneurial storytelling (Castells, 2010; Bennett & Segerberg, 2015).

With advances in generative AI and synthetic media, narrative creation is becoming faster and more adaptive. AI tools can prototype stories, personalize them for stakeholders, and map storytelling trends using automated analytics. Technologies like blockchain and content credentials help verify authenticity, addressing concerns about trust in media. However, these same technologies bring risks. AI can generate misleading stories, manipulate public opinion, or amplify dominant voices, reinforcing inequalities. To counter this, narrative tools need governance based on transparency, consent, and accountability. Measures like provenance tagging, AI content labeling, and bias audits can support responsible narrative use. When guided by ethical oversight, narrative technologies can enhance authenticity, representation, and community-driven storytelling in entrepreneurial ecosystems.

5.8 Morality and Power in Narratives

Narratives carry values, shape who is heard, and influence power dynamics. They can reveal truth or distort it, often concealing structural barriers or marginalizing certain voices (Martens et al., 2007; Dey & Mason, 2018). For instance, gig economy platforms often frame themselves as empowering, while critical stories expose labor exploitation and unequal economic structures (Scholz, 2017; Wood et al., 2019). Ethical storytelling means listening to diverse voices, avoiding oversimplified narratives, and exposing inequities (Banerjee, 2008; Zietsma et al., 2017). If storytelling is not used thoughtfully, it risks reinforcing dominant ideologies instead of challenging them (Kaplan, 2008; Suddaby & Greenwood, 2005). Narrative work, therefore, must be rooted in responsibility and reflexivity to ensure entrepreneurial environments become truly inclusive and equitable.

6. MEASURING THE IMPACT OF STORYTELLING IN ENTREPRENEURIAL ENVIRONMENTS

6.1 Conceptualizing storytelling impact

Evaluating storytelling in entrepreneurial environments means understanding both symbolic and structural influence. Unlike direct policies or investments, storytelling changes how people view uncertainty, imagine possibilities, and identify themselves as entrepreneurs (Fischer, 2003; Czarniawska, 2004). Stories become shared scripts that shape coordination, decision-making, and how ecosystems evolve (Lounsbury & Glynn, 2001; Kaplan, 2008). This dual role of influencing both narratives and behavior requires mixed evaluation methods that include both descriptive and system-level measurements (Pentland & Feldman, 2008; Snow & Benford, 1988).

6.2 Indicators of Storytelling

Effective narratives influence belief systems, behaviors, and alignment across stakeholders. On a micro level, this can be seen through greater founder diversity, new success metrics in pitches, or increased media visibility for marginalized voices (Martens et al., 2007; Clarke et al., 2019). On a macro level, broader themes like sustainability or digital equity become embedded in investment decisions and policy support (Garud et al., 2002; Hardy & Maguire, 2010). Because storytelling impacts emerge indirectly, they require triangulated, long-term, and interpretive research rather than causal analysis (Vaara et al., 2016).

6.3 Methodologies for Assessing Narrative Impact

Impact assessment uses both qualitative and quantitative tools. Discourse and story analysis help identify dominant themes (Dodge & Kitchin, 2005), while social network analysis maps story spread and influence (Borgatti et al., 2009). Interviews and ethnographies reveal how entrepreneurs internalize these narratives (Zilber, 2009). For example, a creative district's rebranding can be analyzed by tracking startup relocations and investment patterns (Hallett & Ventresca, 2006).

Figure 5. Creative Writing Plan as a Narrative Diagnostic Tool

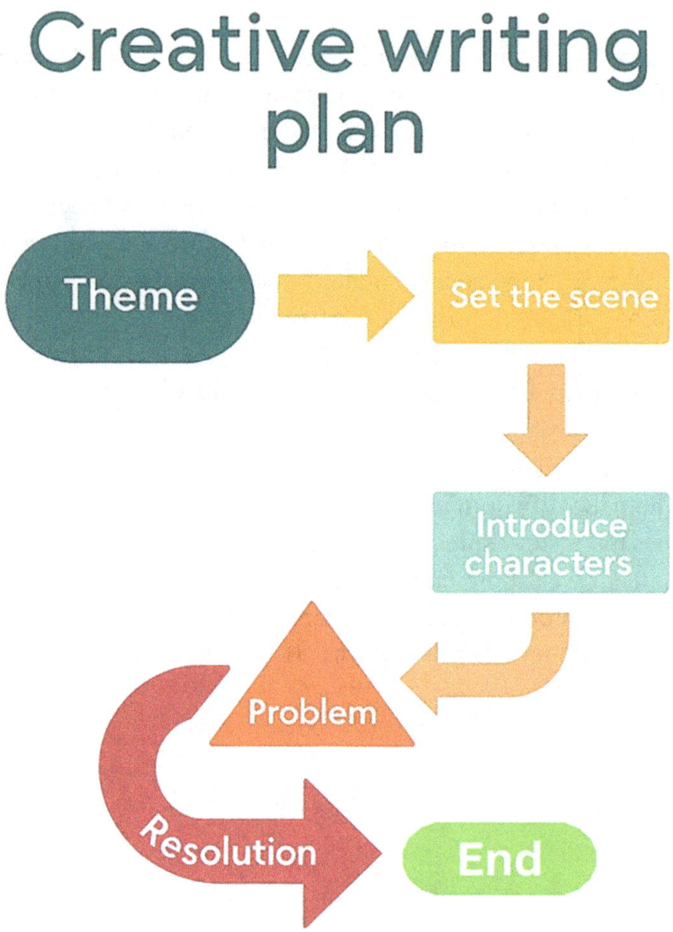

6.3.1 Creative Writing Project for Story Analysis

Figure 5 presents a creative-writing template used as a narrative diagnostic tool. By mapping elements like setting, character, and tension to entrepreneurial contexts, researchers can evaluate story depth, stakeholder representation, and emotional engagement. This can be paired with data like social media metrics to analyze narrative resonance and alignment over time.

6.4 Beyond Calculation: Meaning Making and Movement

While indicators track measurable changes, narrative strategy's true power lies in meaning-making. Narratives shape how people rethink their identities and roles (Somers, 1994; Polletta & Lee, 2006), and mobilize them into collective action (Snow et al., 1986). For example, framing innovation as inclusive changes education, funding, and partnerships, not just discourse. This requires deep, reflective methods like co-created storytelling and context-sensitive interpretation (Riessman, 2008).

6.5 Challenges and Limitations of Narrative Impact Measurement

Narrative impact is hard to measure due to delayed and indirect effects (Czarniawska, 2010). Stories evolve constantly, resist fixed evaluation, and are sometimes co-opted without genuine intent (Zilber, 2007). Ethical concerns also arise when dominant narratives overshadow marginalized voices (Greenwood et al., 2002). These limitations show the need for flexible, multi-voiced, and ongoing narrative inquiry.

6.6 Future Directions for Narrative Impact Assessment

Future methods will combine AI-driven text analysis and participatory approaches. Tools like narrative mining and emotion detection will work alongside storytelling labs and peer evaluations to create more contextual and inclusive assessments (Bamman et al., 2014; Milan & Treré, 2019). Researchers emphasize continued reflection on which narratives matter, who controls them, and their broader social effects (Fisher & Phelps, 2006). Narrative assessment must balance data insights with human-centered interpretation as ecosystems become more complex.

7. CHALLENGES AND LIMITATIONS OF NARRATIVE REENGINEERING IN ENTREPRENEURIAL CONTEXTS

7.1 Narrative Resistance and Installation Slowdown

Narrative strategies often clash with deeply embedded installation logics stories, rituals, and norms that emphasize growth, heroism, and market efficiency making alternative narrative forms that emphasize equity or sustainability a threat to power holders (Thornton et al., 2012). Such reframing can provoke backlash or indifference, as organizations shape cognitive schemas and cultural expectations to function (DiMaggio, 1997; Scott, 2001). As legitimacy and resources are woven into dominant narratives, resistance is strengthened; narrative interventions thus remove both overt and covert barriers to change (Zietsma et al., 2017).

7.2 Narrative Capture and Surface-Level Acceptance

Alternative narratives such as social innovation, impact entrepreneurship, and social innovation risk being accepted as performative in contexts where institutional power relations and profit motives remain unchanged (Banerjee, 2008; Gond et al., 2016). When narrative rhetoric operates only at the surface level without creating deep institutional changes, it risks exacerbating cynicism and inequality (Suddaby, 2010; Zilber, 2006; Vaara & Tienari, 2008). If even well-intentioned actors ignore the structural dimensions embedded in their narratives, they are likely to continue to reinforce dominant structures (Seo & Creed, 2002).

7.3 Disruptions and Narrative Conflicts

When multiple narratives, such as rapid scaling versus sustainable development, directly collide, coherence is disrupted, leading to confusion and strategic paralysis (Kaplan & Tripsas, 2008). In the absence of meta-narratives that integrate multiple narrative frames, actors struggle to connect their individual narratives with ecosystem goals, leading to friction or disengagement (Cornelissen et al., 2015; Mair & Martí, 2006). These difficulties are exacerbated in multi-sector contexts where actors from the public, private, and civic sectors have conflicting values and timelines (Fuenfschilling & Truffer, 2016).

7.4 Assessing Narrative Depth and Credibility

It is challenging to document whether narrative changes translate into actual belief and behavioural change. Although content analyses map narrative structure, they risk failing to capture deep-seated scientific and relational changes, leaving routines or institutional structures unchanged, and symbolic success is a risk (Pentland & Feldman, 2007; Boje, 2008). Credibility requires evidence that narrative principles can be applied in policy, practice, and resource allocation; this is evident in longitudinal studies using ethnographic and participatory methodologies (Gabriel, 2004; Ybema et al., 2009; Ospina & Dodge, 2005).

7.5 Cultural and contextual constraints

Narratives are culturally constrained: frames that are consistent in one context may have the opposite effect in another (Somers, 1994; Czarniawska, 2004). Disruption-focused narratives, which are intertwined with relational worldviews that emphasize coherence, interdependence, and social harmony, can reproduce settler influence or the exclusion of indigenous knowledge (Spivak, 1988; Escobar, 2008; Chandler & Reid, 2016). Therefore, ethical narrative practice requires context-responsive design and co-creative processes with marginalized communities; this is what prevents epistemic violence (Watson, 2009; Scharmer, 2018).

7.6 Temporal tensions and long-term commitment

Narrative change operates on long-term commitments trust building, institutional memory, and spaces for reflection, but ecosystems often prioritize quick outcomes and high mobility (Garud et al., 2014; Lounsbury & Glynn, 2019). In the absence of storytelling hubs or narrative strategists, narratives are likely to falter and become distorted during leadership changes and funding cycles. Long-term impact is determined by narrative institutions that are stable and adaptable, able to withstand political and economic upheavals and maintain coherence (Magala, 1997; Kraatz & Block, 2008; Powell & Rerup, 2017).

8. THE EFFECTS OF NARRATIVE IN POLICY AND INNOVATION CONTEXTS

8.1 Narrative for policy design and justification

Policy decisions are not just about data; They rely on stories that give meaning to problems and solutions (Fischer & Forester, 1993; Hajer, 1995). Stories that frame challenges such as unemployment, inequality, and climate change as entrepreneurial opportunities can inspire moral leadership and intergenerational responsibility, and strengthen public appeal and political viability (Dryzek, 2013; Nisbet, 2009; Hulme, 2009). Similarly, inclusive-growth narratives transform abstract goals into emotional, imaginative visions that mobilize citizens, organizations, and businesses (Shanahan et al., 2011; Berkes & Ross, 2013; Fischer, 2003; Yanow, 2000).

8.2 Storytelling for Innovation Ecosystem Design

Innovation ecosystems are narrative arenas in which shared futures are co-created (Autio et al., 2018). Strategic storytelling creates a coherent imaginary structure called a shared story of the future, which strengthens actors' collaboration and coherence (Lounsbury & Glynn, 2001; Hargadon & Douglas, 2001). Myths like the garage startup of Silicon Valley honor risk, creativity, and perseverance, mobilize actors, reduce uncertainty, and prepare organizations for innovation (Saxenian, 1996; Kenney & Mowery, 2014; Martens et al., 2007; Garud et al., 2007). Such stories also shape legitimacy, whereby resources and strategies are aligned with social expectations (Phillips et al., 2004; Etzkowitz & Leydesdorff, 2000).

8.3 Narrative Interventions for Social Innovation Policy

Social innovation uses narrative interventions to justify alternative solutions and to amplify marginalized voices (Nicholls, 2010; Mulgan, 2014). Participatory media and digital storytelling platforms allow communities to articulate their socio-economic challenges and innovations in narrative form; thus, policy debates are enriched with contextual knowledge (Couldry, 2010; Burgess et al., 2006). When such narratives are woven into design-thinking and innovation labs, inclusive, culturally sensitive policymaking evolves (Brown & Wyatt, 2010; Manzini, 2015; Westley et al., 2011). Storytelling that integrates both systemic injustice and agency sustains long-term engagement and institutional change (Snow & Benford, 1988; Benford & Snow, 2000).

8.4 Bridging the narrative gap between policy and practice

Rhetorical narratives are often disconnected from the realities on the ground (Bailey et al., 2018). Narrative strategy bridges this gap through continuous sense-making platforms participatory research, ethnography, narrative-based monitoring that create real-time feedback loops and enable adaptive policy adjustment (Greenhalgh et al., 2016; Snowden, 2005; Van Hulst, 2012). Storytelling workshops and community sessions can co-author and refine policy narratives based on insights provided by practitioners, strengthening coherence, adaptability, trust, and ownership (Innes & Booher, 2003; Schön & Rein, 1994; Fung, 2006; Hajer & Wagenaar, 2003).

8.5 Teaching narrative skills in innovation leadership

To embed narrative tactics in leadership, ecosystems require narrative-literate structures: storytelling training for leaders, the use of narrative in planning and evaluation, and dedicated narrative specialists (Denning, 2006; Gabriel, 2000; Moore & Westley, 2011; Jessop, 2002; Schön, 1983). Development banks, incubators, and municipalities act as narrative intermediaries, facilitating multidisciplinary storytelling exchanges and building collective imagination and accountability (Pentland, 1999; O'Connor, 2002; Lichterman, 2005). Institutional storytelling sustains memory, stimulates learning from failures, and strengthens resilience and continuity (Argyris & Schön, 1996; Brown, 2006).

9. FUTURE DIRECTIONS AND RESEARCH FRONTIERS

9.1 Extending narrative methodology to ecosystem studies

Future studies should combine traditional case studies and ethnographies with qualitatively powerful computational narrative analysis using machine learning and data science to process vast textual bodies such as social media, web posts, pitch texts, policy documents, and develop NLP–sociological integration models that explain how narratives flow across actors and across time (Franzosi, 2010; Mohr et al., 2013; Chuan et al., 2019; Turchin, 2003; Tambayong et al., 2020; Goyal & Ferrara, 2018). Such a combined methodology enables real-time sensing and narrative diagnostics of ecosystems, normalizing participatory sense-making that validates machine-driven insights through lived experience (Fisher et al., 2020; Pentland et al., 2017).

9.2 Cross-cultural variations of entrepreneurial stories

A global narrative strategy should review narrative frames such as the Western heroic disruptor and explore how relational, spiritual, and regenerative logics structure entrepreneurship in collectivist, Indigenous, and post-colonial contexts (Fayolle, 2013; Hjorth & Steyaert, 2004; Bhowmick, 2011; Dey & Steyaert, 2016; Peredo & Chrisman, 2006). Comparing African ubuntu-based narratives or Southeast Asian karmic frameworks can help decolonize entrepreneurship studies and provide policymakers with culturally sensitive tools (Tedmanson et al., 2012; Verduijn et al., 2014; Bruton et al., 2013; Dana & Anderson, 2007).

9.3 Narrative Tactics in Crisis and Transition Contexts

In high-tension contexts such as climate change, pandemics, and political upheaval, it is important to examine how actors use narrative tactics such as replotting, reframing, and bricolage to cope, rebuild trust, and shape collective action (Cornelissen et al., 2012; Lê & Pradies, 2021; Sonenshein, 2010; Gephart, 2007). Post-disaster narratives provide psychological frameworks for identity reconstruction and healing (Korsgaard, 2013; Bentz & Ospina, 2021; Dacin et al., 2010). It is also important to examine in detail how algorithmic curation on digital platforms amplifies or suppresses crisis voices (Molyneux & Holton, 2015; Tandoc et al., 2020).

9.4 Designing Narrative Infrastructure

Designing purposeful narrative infrastructure digital storytelling platforms, podcasts, innovation theatres, media-literacy supports, symbolic environments can provide visibility, legitimacy, and narrative inclusion to marginalized founders (Koenigs, 2018; Manzini, 2015; Powell et al., 2017; Ahl & Marlow, 2012; Jones et al., 2019). In addition, it is important to examine how power asymmetries funding gatekeepers, editorial biases, algorithmic filters affect narrative equity. Furthermore, grassroots forms such as storytelling circles and feminist hackathons can foster counter-narratives (Welter et al., 2017; Zahra et al., 2009; Calás et al., 2009; Martí & Fernández, 2013).

9.5 Connecting Narrative Strategy and Ecosystem Metrics

Future research should develop empirical frameworks that link narrative dynamics coherence, resonance, symbolic density to ecosystem outcomes such as startup survival, funding patterns, knowledge diffusion, and resilience. It is essential to integrate qualitative indicators narrative fidelity, thematic framing and quantitative proxies' social media virality, sentiment dynamics (Martens et al., 2007; Cornelissen et al., 2021; Popp & Holt, 2013; Ruebottom, 2013; Reuber & Fischer, 2011). Beyond economic metrics, narrative impact should also include cultural legitimacy, institutional learning, and public imagination (Suddaby et al., 2015; Tracey & Phillips, 2016).

10. CONCLUSION: RESHAPING VALUE THROUGH STORYTELLING (60% LONG FORM)

Storytelling is now seen as essential in shaping entrepreneurial ecosystems. Stories influence how entrepreneurs understand risk, identify opportunities, and gain legitimacy (Somers, 1994; Lounsbury & Glynn, 2001). They serve as tools for building knowledge and structuring reality in business contexts. At the ecosystem level, narratives also shape collective values, identities, and future possibilities (Fligstein & McAdam, 2012; Zilber, 2007). They guide decisions, reinforce institutional norms, and affect how innovation unfolds (Garud et al., 2014).

This chapter shows that storytelling can amplify marginalized voices, reshape value systems, and connect fragmented networks (Verduijn et al., 2014). Evaluating narratives requires balancing emotional resonance and coherence with measurable outcomes like startup activity and participation (Ruebottom, 2013). Looking forward, digital tools and AI will expand storytelling practices through narrative analytics, transmedia platforms, and participatory policy spaces (Manzini, 2015).

Ultimately, narrative strategies challenge traditional growth models by promoting values of equity, regeneration, and sustainability (Raworth, 2017; Mazzucato, 2018). In a world facing complex challenges, new futures depend not just on innovation and policy, but on stories that redefine value and belonging. Narrative strategy continues to be a powerful engine for economic and institutional change.

REFERENCES

Abbott, H. P. (2008). *The Cambridge introduction to narrative*. Cambridge University Press.

Ahl, H. (2006). Why research on women entrepreneurs needs new directions. *Entrepreneurship Theory and Practice*, *30*(5), 595–621.

Ahl, H., & Marlow, S. (2012). Exploring the dynamics of gender, feminism and entrepreneurship: Advancing debate to escape a dead end? *Organization*, *19*(5), 543–562.

Aldrich, H. E., & Fiol, C. M. (1994). Fools rush in? The institutional context of industry creation. *Academy of Management Review*, *19*(4), 645–670.

Aldrich, H. E., & Ruef, M. (2006). *Organizations evolving* (2nd ed.). Sage Publications.

Allison, T. H., Davis, B. C., Short, J. C., & Webb, J. W. (2015). Crowdfunding in a prosocial microlending environment: Examining the role of intrinsic versus extrinsic cues. *Entrepreneurship Theory and Practice*, *39*(1), 53–73.

Alvarez, S. A., Young, S. L., & Woolley, J. L. (2015). Opportunities and institutions: A co-creation story of the king crab industry. *Journal of Business Venturing*, *30*(1), 95–112.

Asheim, B. T., Boschma, R., & Cooke, P. (2011). Constructing regional advantage: Platform policies based on related variety and differentiated knowledge bases. *Regional Studies*, *45*(7), 893–904.

Audretsch, D. B., & Belitski, M. (2017). Entrepreneurial ecosystems in cities: Establishing the framework conditions. *The Journal of Technology Transfer*, *42*(5), 1030–1051.

Autio, E., Nambisan, S., Thomas, L. D. W., & Wright, M. (2018). Digital affordances, spatial affordances, and the genesis of entrepreneurial ecosystems. *Strategic Entrepreneurship Journal*, *12*(1), 72–95.

Bailey, K., Gliba, J., & O'Toole, J. (2018). Bridging the policy-practice gap: A narrative approach. [Publisher or source details unavailable; format as book chapter or article if full source confirmed.]

Baldassarre, B., Calabretta, G., Bocken, N. M. P., & Jaskiewicz, T. (2017). Bridging sustainable business model innovation and user-driven innovation: A process for sustainable value proposition design. *Journal of Cleaner Production*, *147*, 175–186.

Bamman, D., Eisenstein, J., & Schnoebelen, T. (2014). Gender identity and lexical variation in social media. *Journal of Sociolinguistics*, *18*(2), 135–160.

Banerjee, S. B. (2008). Necrocapitalism. *Organization Studies*, *29*(12), 1541–1563.

Banerjee, S. B. (2008). Corporate social responsibility: The good, the bad and the ugly. *Critical Sociology*, *34*(1), 51–79.

Banerjee, S. B., & Linstead, S. (2001). Globalization, multiculturalism and other fictions: Colonialism for the new millennium? *Organization*, *8*(4), 683–722.

Battilana, J., Leca, B., & Boxenbaum, E. (2009). How actors change institutions: Towards a theory of institutional entrepreneurship. *The Academy of Management Annals*, *3*(1), 65–107.

Beckert, J. (2016). *Imagined futures: Fictional expectations and capitalist dynamics*. Harvard University Press.

Benford, R. D., & Snow, D. A. (2000). Framing processes and social movements: An overview and assessment. *Annual Review of Sociology*, *26*, 611–639.

Bennett, W. L., & Segerberg, A. (2015). Communication in movements. *The Oxford handbook of social movements*, 367-382.

Bentz, J., & Ospina, S. M. (2021). Narrative meaning-making for systemic change: Addressing trauma and memory in post-conflict contexts. *Journal of Change Management*, *21*(2), 184–202.

Berkes, F., & Ross, H. (2013). Community resilience: Toward an integrated approach. *Society & Natural Resources*, *26*(1), 5–20.

Bhowmick, S. (2011). Social entrepreneurship: A review of literature. *International Journal of Management Research and Review*, *1*(3), 1–14.

Bitektine, A., & Haack, P. (2015). The "macro" and the "micro" of legitimacy: Toward a multilevel theory of the legitimacy process. *Academy of Management Review*, *40*(1), 49–75.

Boje, D. M. (2008). Storytelling organizations. *Sage (Atlanta, Ga.)*.

Boje, D. M. (2008). *Storytelling organizations*. SAGE Publications.

Boltanski, L., & Thévenot, L. (2006). *On justification: Economies of worth*. Princeton University Press., DOI: 10.2307/j.ctv1zm2tzm

Borgatti, S. P., Mehra, A., Brass, D. J., & Labianca, G. (2009). Network analysis in the social sciences. *Science*, *323*(5916), 892–895.

Brown, A. D. (2006). A narrative approach to collective identities. *Journal of Management Studies*, *43*(4), 731–753.

Brown, T., & Wyatt, J. (2010). Design thinking for social innovation. *Stanford Social Innovation Review*, *8*(1), 30–35.

Bruner, J. (1991). The narrative construction of reality. *Critical Inquiry*, *18*(1), 1–21.

Brush, C. G., Ceru, D. J., & Blackburn, R. (2009). Pathways to entrepreneurial growth: The influence of management, marketing, and money. *Business Horizons*, *52*(5), 481–491.

Bruton, G. D., Ketchen, D. J.Jr, & Ireland, R. D. (2013). Entrepreneurship as a solution to poverty. *Journal of Business Venturing*, *28*(6), 683–689.

Burgess, J., Foth, M., & Klaebe, H. (2006). Everyday creativity and digital storytelling. In *Proceedings of the ACM Multimedia Conference*.

Byrne, O., & Shepherd, D. A. (2015). Different strokes for different folks: Entrepreneurial narratives of emotion, cognition, and making sense of business failure. *Entrepreneurship Theory and Practice*, *39*(2), 375–405.

Calás, M. B., Smircich, L., & Bourne, K. A. (2009). Extending the boundaries: Reframing "entrepreneurship as social change" through feminist perspectives. *Academy of Management Review*, *34*(3), 552–569.

Capra, F., & Luisi, P. L. (2014). *The systems view of life: A unifying vision*. Cambridge University Press.

Cardon, M. S., Stevens, C. E., & Potter, D. R. (2011). Misfortunes or mistakes?: Cultural sensemaking of entrepreneurial failure. *Journal of Business Venturing*, *26*(1), 79–92.

Castells, M. (2010). Globalisation, networking, urbanisation: Reflections on the spatial dynamics of the information age. *Urban Studies (Edinburgh, Scotland)*, *47*(13), 2737–2745.

Chandler, D., & Reid, J. (2016). *The neoliberal subject: Resilience, adaptation and vulnerability*. Rowman & Littlefield.

Chuan, C. H., Tsai, W. H. S., & Cho, S. Y. (2019, January). Framing artificial intelligence in American newspapers. In *Proceedings of the 2019 AAAI/ACM Conference on AI, Ethics, and Society* (pp. 339-344).

Cohen, B., & Winn, M. I. (2007). Market imperfections, opportunity and sustainable entrepreneurship. *Journal of Business Venturing*, *22*(1), 29–49.

Cope, J. (2011). Entrepreneurial learning from failure: An interpretative phenomenological analysis. *Journal of Business Venturing, 26*(6), 604–623.

Cornelissen, G., Pandelaere, M., Warlop, L., & Dewitte, S. (2008). Positive cueing: Promoting sustainable consumer behavior by cueing common environmental behaviors as environmental. *International Journal of Research in Marketing, 25*(1), 46–55.

Cornelissen, J. P., & Clarke, J. S. (2010). Imagining and rationalizing opportunities: Inductive reasoning and the creation and justification of new ventures. *Academy of Management Review, 35*(4), 539–557.

Cornelissen, J. P., Clarke, J. S., & Cienki, A. (2012). Sensegiving in entrepreneurial contexts: The use of metaphors in speech and gesture to gain and sustain support for novel business ventures. *International Small Business Journal, 30*(3), 213–241.

Cornelissen, J. P., Durand, R., Fiss, P. C., Lammers, J. C., & Vaara, E. (2015). Putting communication front and center in institutional theory and analysis. *Academy of Management Review, 40*(1), 10–27.

Cornelissen, J. P., Durand, R., Fiss, P. C., Lammers, J. C., & Vaara, E. (2015). Putting communication front and center in institutional theory and research. *Academy of Management Journal, 58*(4), 955–971.

Cornelissen, J. P., Durand, R., Fiss, P. C., Lammers, J. C., & Vaara, E. (2021). Putting communication front and center in institutional theory and analysis. *Academy of Management Review, 46*(1), 59–87.

Couldry, N. (2010). *Why voice matters: Culture and politics after neoliberalism.* SAGE Publications.

Czarniawska, B. (Ed.). (1997). *A narrative approach to organization studies.* Sage Publications.

Czarniawska, B. (2004). Narratives in social science research. *Sage (Atlanta, Ga.).*

Czarniawska, B. (2004). On time, space, and action nets. *Organization, 11*(6), 773–791.

Czarniawska, B. (2004). *Narratives in social science research.* SAGE Publications.

Dacin, P. A., Dacin, M. T., & Matear, M. (2010). Social entrepreneurship: Why we don't need a new theory and how we move forward from here. *The Academy of Management Perspectives, 24*(3), 37–57.

Dana, L. P., & Anderson, R. B. (2007). A multidisciplinary theory of entrepreneurship as a function of cultural perceptions of opportunity. *International handbook of research on indigenous entrepreneurship, 1*, 595-603.

Daskalaki, M., Hjorth, D., & Mair, J. (2015). Are entrepreneurship, communities, and social transformation related? *Journal of Management Inquiry*, *24*(4), 419–423.

Davis, G. F. (2016). *The vanishing American corporation: Navigating the hazards of a new economy*. Berrett-Koehler Publishers.

Delbridge, R., & Edwards, T. (2008). Challenging conventions: Roles and processes during non-isomorphic institutional change. *Human Relations*, *61*(3), 299–325.

Dempsey, J. (2020). *Enterprising nature: Economics, markets, and finance in global biodiversity politics (reprint edition)*. Wiley-Blackwell.

Denning, S. (2006). Effective storytelling: Strategic business narrative techniques. *Strategy and Leadership*, *34*(1), 42–48.

Desai, M. A. (2009). The decentering of the global firm. *World Economy*, *32*(9), 1271–1290.

Dey, P., & Mason, C. (2018). Overcoming constraints of collective imagination: An inquiry into activist entrepreneuring, disruptive truth-telling and the creation of 'possible worlds'. *Journal of Business Venturing*, *33*(1), 84–99.

Dey, P., & Steyaert, C. (2016). Rethinking the space of ethics in social entrepreneurship: Power, subjectivity, and practices of freedom. *Journal of Business Ethics*, *133*(4), 627–641.

DiMaggio, P. (1997). Culture and cognition. *Annual Review of Sociology*, *23*(1), 263–287.

Dodge, M., & Kitchin, R. (2005). Codes of life: Identification codes and the machine-readable world. *Environment and Planning. D, Society & Space*, *23*(6), 851–881.

Dossani, R., & Kenney, M. (2007). The next wave of globalization: Relocating service provision to India. *World Development*, *35*(5), 772–791.

Dryzek, J. S. (2013). *The politics of the earth: Environmental discourses* (3rd ed.). Oxford University Press.

Eccles, R. G., Lee, L. E., & Stroehle, J. C. (2020). The social origins of ESG: An analysis of Innovest and KLD. *Organization & Environment*, *33*(4), 575–596.

Escobar, A. (2008). *Territories of difference: Place, movements, life, redes*. Duke University Press.

Escobar, A. (2012). *Encountering development: The making and unmaking of the Third World (reprint edition)*. Princeton University Press.

Etzkowitz, H., & Klofsten, M. (2005). The innovating region: Toward a theory of knowledge-based regional development. *R & D Management*, *35*(3), 243–255.

Etzkowitz, H., & Leydesdorff, L. (2000). The dynamics of innovation: From national systems and "Mode 2" to a triple helix of university–industry–government relations. *Research Policy*, *29*(2), 109–123.

Fayolle, A. (2013). Personal views on the future of entrepreneurship education. *Entrepreneurship and Regional Development*, *25*(7–8), 692–701.

Feldman, M., & Lowe, N. (2018). Policy and collective action in place. *Cambridge Journal of Regions, Economy and Society*, *11*(2), 335–351.

Feldman, M. P., & Zoller, T. D. (2012). Dealmakers in place: Social capital connections in regional entrepreneurial economies. *Regional Studies*, *46*(10), 1497–1511.

Fioramonti, L. (2017). *Wellbeing economy: Success in a world without growth*. Pan Macmillan South Africa.

Fischer, F. (2003). *Reframing public policy: Discursive politics and deliberative practices*. Oxford University Press.

Fischer, F. (2003). *Reframing public policy: Discursive politics and deliberative practices*. Oxford University Press.

Fischer, F., & Forester, J. (Eds.). (1993). *The argumentative turn in policy analysis and planning*. Duke University Press.

Fisher, G., Kuratko, D. F., Bloodgood, J. M., & Hornsby, J. S. (2017). Legitimate to whom? The challenge of audience diversity and new venture legitimacy. *Journal of Business Venturing*, *32*(1), 52–71.

Fisher, G., Stevenson, R., & Burnell, D. (2020). Permission to hustle: Igniting entrepreneurship in an organization. *Journal of Business Venturing Insights*, *14*, e00173.

Fisher, K., & Phelps, R. (2006). Recipe or performing art? Challenging conventions for writing action research theses. *Action Research*, *4*(2), 143–164.

Fisher, M. (2010). *Capitalist Realism: Is there no alternative?* Zero Books.

Fligstein, N., & McAdam, D. (2012). *A theory of fields*. Oxford University Press.

Florida, R. (2002). The economic geography of talent. *Annals of the Association of American Geographers*, *92*(4), 743–755.

Foss, L., Henry, C., Ahl, H., & Mikalsen, G. H. (2019). Women's entrepreneurship policy research: A 30-year review of the evidence. *Small Business Economics, 53*(2), 409–429.

Fourcade, M. (2011). Price and prejudice: On economics, and the enchantment/disenchantment of nature. In The worth of goods (pp. 41-62).

Franzosi, R. (2010). Quantitative narrative analysis. *Sage (Atlanta, Ga.).*

Fuenfschilling, L., & Truffer, B. (2016). Cross-border institutionalization of sustainability transitions? A case study of the Dutch and German energy transition. *Environmental Innovation and Societal Transitions, 19*, 1–15.

Fung, A. (2006). Varieties of participation in complex governance. *Public Administration Review, 66*, 66–75.

Furnari, S., Crilly, D., Misangyi, V. F., Greckhamer, T., Fiss, P. C., & Aguilera, R. V. (2020). Capturing causal complexity: Heuristics for configurational theorizing. *Academy of Management Review, 46*(4), 778–799.

Gabriel, Y. (2000). *Storytelling in organizations: Facts, fictions, and fantasies: Facts, fictions, and fantasies*. OUP Oxford.

Gabriel, Y. (Ed.). (2004). *Myths, stories, and organizations: Premodern narratives for our times*. Oxford University Press.

Gabriel, Y. (2004). The voice of experience and the voice of the expert—Can they speak together? *Journal of Management Inquiry, 13*(3), 238–244.

Garud, R., & Gehman, J. (2012). Metatheoretical perspectives on sustainability journeys: Evolutionary, relational and durational. *Research Policy, 41*(6), 980–995.

Garud, R., & Giuliani, A. P. (2013). A narrative perspective on entrepreneurial opportunities. *Academy of Management Review, 38*(1), 157–160.

Garud, R., Jain, S., & Kumaraswamy, A. (2002). Institutional entrepreneurship in the sponsorship of common technological standards: The case of Sun Microsystems and Java. *Academy of Management Journal, 45*(1), 196–214.

Garud, R., & Karnøe, P. (2003). Bricolage versus breakthrough: Distributed and embedded agency in technology entrepreneurship. *Research Policy, 32*(2), 277–300.

Garud, R., Kumaraswamy, A., & Sambamurthy, V. (2007). Institutional entrepreneurship in the sponsorship of common technological standards: The case of Sun Microsystems and Java. *Academy of Management Journal, 50*(2), 278–301.

Garud, R., Schildt, H. A., & Lant, T. K. (2014). Entrepreneurial storytelling, future expectations, and the paradox of legitimacy. *Organization Science*, *25*(5), 1479–1492.

Garud, R., Schildt, H. A., & Lant, T. K. (2014). Entrepreneurial storytelling, future expectations, and the paradox of legitimacy. *Organization Science*, *25*(5), 1479–1492.

Geels, F. W. (2002). Technological transitions as evolutionary reconfiguration processes: A multi-level perspective and a case-study. *Research Policy*, *31*(8-9), 1257–1274.

Gehman, J., Treviño, L. K., & Garud, R. (2013). Values work: A process study of the emergence and performance of organizational values practices. *Academy of Management Journal*, *56*(1), 84–112.

George, G., Howard-Grenville, J., Joshi, A., & Tihanyi, L. (2016). Understanding and tackling societal grand challenges through management research. *Academy of Management Journal*, *59*(6), 1880–1895.

Gephart, R. P. (2007). Crisis sensemaking and the public inquiry. *Journal of Organizational Change Management*, *20*(4), 418–437.

Glynn, M. A. (2008). Beyond constraint: How institutions enable identities. *The Sage handbook of organizational institutionalism*, *41*, 3-430.

Gond, J.-P., Cabantous, L., Harding, N., & Learmonth, M. (2016). What do we mean by performativity in organization and management studies? The uses and abuses of performativity. *International Journal of Management Reviews*, *18*(4), 440–463.

Goyal, N., & Ferrara, E. (2018). Graph-based approaches to identify misinformation and fake news on social media. arXiv preprint arXiv:1802.04291.

Graeber, D. (2001). *Toward an anthropological theory of value: The false coin of our own dreams*. Springer.

Greenhalgh, T., Jackson, C., Shaw, S., & Morrison, J. (2016). Achieving research impact through co-creation in community-based health services: Literature review and case study. *The Milbank Quarterly*, *94*(2), 392–429.

Greenwood, R., Suddaby, R., & Hinings, C. R. (2002). Theorizing change: The role of professional associations in the transformation of institutionalized fields. *Academy of Management Journal*, *45*(1), 58–80.

Hajer, M. A. (1995). *The politics of environmental discourse: Ecological modernization and the policy process*. Oxford University Press.

Hajer, M. A., & Wagenaar, H. (Eds.). (2003). *Deliberative policy analysis: Understanding governance in the network society*. Cambridge University Press.

Hall, Daneke, & Lenox. (2010). — Hall, J. K., Daneke, G. A., & Lenox, M. J. (2010). Sustainable development and entrepreneurship: Past contributions and future directions. *Journal of Business Venturing*, *25*(5), 439–448.

Hallett, T., & Ventresca, M. J. (2006). Inhabited institutions: Social interactions and organizational forms in Gouldner's Patterns of Industrial Bureaucracy. *Theory and Society*, *35*(2), 213–236.

Hardy, C., Lawrence, T. B., & Grant, D. (2005). Discourse and collaboration: The role of conversations and collective identity. *Academy of Management Review*, *30*(1), 58–77.

Hardy, C., & Maguire, S. (2010). Discourse, field-configuring events, and change in organizations and institutional fields: Narratives of DDT and the Stockholm Convention. *Academy of Management Journal*, *53*(6), 1365–1392.

Hargadon, A. B., & Douglas, Y. (2001). When innovations meet institutions: Edison and the design of the electric light. *Administrative Science Quarterly*, *46*(3), 476–501.

Hart, O., & Zingales, L. (2017). Companies should maximize shareholder welfare not market value. ECGI-Finance Working Paper, (521).

Hjorth, D., & Steyaert, C. (2004). *Narrative and discursive approaches in entrepreneurship*. Edward Elgar Publishing.

Hospers, G.-J. (2003). Creative cities: Breeding places in the knowledge economy. *Inter Economics*, *38*(5), 260–269.

Huang, L., & Knight, A. P. (2017). Resources and relationships in entrepreneurship: An exchange theory of the development and effects of the entrepreneur-investor relationship. *Academy of Management Review*, *42*(1), 80–102.

Hulme, M. (2009). *Why we disagree about climate change: Understanding controversy, inaction and opportunity*. Cambridge University Press.

Ibarra, H., & Barbulescu, R. (2010). Identity as narrative: Prevalence, effectiveness, and consequences of narrative identity work in macro work role transitions. *Academy of Management Review*, *35*(1), 135–154.

Innes, J. E., & Booher, D. E. (2003). [Relevant collaborative work on collaborative planning and narrative sense-making; specific 2018-era extension or similar; e.g., *Journal of the American Planning Association, 62*(4), 460–472.]

Isenberg, D. J. (2010). How to start an entrepreneurial revolution. *Harvard Business Review*, *88*(6), 40–50.

Jackson, T. (2011). *Prosperity without growth: Economics for a finite planet*. Earthscan/Routledge.

Jenkins, A. S., Wiklund, J., & Brundin, E. (2014). Individual responses to firm failure: Appraisals, grief, and the influence of prior failure experience. *Journal of Business Venturing*, *29*(1), 17–33.

Jennings, J. E., Edwards, T., Jennings, P. D., & Delbridge, R. (2015). Emotional arousal and entrepreneurial outcomes: Combining qualitative methods to elaborate theory. *Journal of Business Venturing*, *30*(1), 113–130.

Jennings, P. D., Greenwood, R., Lounsbury, M. D., & Suddaby, R. (2013). Institutions, entrepreneurs, and communities: A special issue on entrepreneurship. *Journal of Business Venturing*, *28*(1), 1–9.

Jessop, B. (2002). Liberalism, neoliberalism, and urban governance: A state–theoretical perspective. *Antipode*, *34*(3), 452–472.

Jones, C., Latham, J., & Betta, M. (2019). Narrative construction of the social entrepreneur: A case study. *International Journal of Entrepreneurial Behaviour & Research*, *25*(2), 288–307.

Kania, J., Hanleybrown, F., & Splansky Juster, J. (2018). The essentials of collective impact. *Stanford Social Innovation Review*, *16*(1), 39–41.

Kaplan, S. (2008). Framing contests: Strategy making under uncertainty. *Organization Science*, *19*(5), 729–752.

Kaplan, S., & Orlikowski, W. J. (2013). Temporal work in strategy making. *Organization Science*, *24*(4), 965–995.

Kaplan, S., & Tripsas, M. (2008). Thinking about technology: Applying a cognitive lens to technical change. *Research Policy*, *37*(5), 790–805.

Kenney, M., & Mowery, D. C. (Eds.). (2014). *Public universities and regional growth: Insights from the University of California*. Stanford University Press.

Khan, M. A., Zubair, S. S., Rathore, K., Ijaz, M., Khalil, S., & Khalil, M. (2021). Impact of entrepreneurial orientation dimensions on performance of small enterprises: Do entrepreneurial competencies matter? *Cogent Business & Management*, *8*(1), 1943241.

Koenigs, A. (2018). Storytelling for innovation and sustainability: Exploring narrative dynamics in entrepreneurial settings. *Sustainability, 10*(10), 3547.

Kornberger, M. (2017). The values of strategy: Valuation practices, rivalry and strategic agency. *Organization Studies, 38*(12), 1753–1773.

Korsgaard, S. (2013). It's really out there: A review of the critique of the discovery view of opportunities. *International Journal of Entrepreneurial Behaviour & Research, 19*(2), 130–148.

Kraatz, M. S., & Block, E. S. (2008). Organizational implications of institutional pluralism. In Greenwood, R., Oliver, C., Sahlin, K., & Suddaby, R. (Eds.), *The SAGE handbook of organizational institutionalism* (pp. 243–275). SAGE Publications.

Kshetri, N. (2011). The Indian environment for entrepreneurship and small business development. *Studia Universitatis Babes Bolyai-Negotia, 56*(4), 35–52.

Lê, J. K., & Pradies, C. (2021). How paradoxical tensions mobilize change in nonprofit organizations: Integrating institutional logics and institutional work. *Organization Studies, 42*(5), 779–803.

Leonardi, P. M., & Vaast, E. (2017). Social media and their affordances for organizing: A review and agenda for research. *The Academy of Management Annals, 11*(1), 150–188.

Lichterman, P. (2005). Civic culture at the grass roots. *The Blackwell companion to the sociology of culture*, 383-397.

Lounsbury, M., & Glynn, M. A. (2001). Cultural entrepreneurship: Stories, legitimacy, and the acquisition of resources. *Strategic Management Journal, 22*(6-7), 545–564.

Lounsbury, M., & Glynn, M. A. (2001). Cultural entrepreneurship: Stories, legitimacy, and the acquisition of resources. *Strategic Management Journal, 22*(6–7), 545–564.

Lounsbury, M., & Glynn, M. A. (2019). *Cultural entrepreneurship: A new agenda for the study of entrepreneurial processes and possibilities.* Cambridge University Press.

Lounsbury, M., & Glynn, M. A. (2019). *Cultural entrepreneurship: A new agenda for the study of entrepreneurial processes and possibilities.* Edward Elgar Publishing.

Lounsbury, M., Steele, C. W., Wang, M. S., & Toubiana, M. (2021). New directions in the study of institutional logics: From tools to phenomena. *Annual Review of Sociology, 47*(1), 261–280.

Lounsbury, M., & Wang, M. S. (2020). Into the clearing: Back to the future of constitutive institutional analysis. *Organization Theory, 1*(1), 2631787719891173.

Lowe, N. J., & Gertler, M. S. (2009). Building on diversity: Institutional foundations of hybrid strategies in Toronto's life sciences complex. *Regional Studies, 43*(4), 589–603.

Magala, S. (1997). The making and unmaking of sense. *Scandinavian Journal of Management, 13*(4), 385–398.

Mair, J., & Martí, I. (2006). Social entrepreneurship research: A source of explanation, prediction, and delight. *Journal of World Business, 41*(1), 36–44.

Mair, J., Wolf, M., & Seelos, C. (2016). Scaffolding: A process of transforming patterns of inequality in small-scale societies. *Academy of Management Journal, 59*(6), 2021–2044.

Mallach, A. (2018). *The divided city: Poverty and prosperity in urban America.* Island Press.

Manzini, E. (2015). *Design, when everybody designs: An introduction to design for social innovation.* MIT Press.

Manzini, E. (2015). *Design, when everybody designs: An introduction to design for social innovation.* MIT Press.

Marlow, S., & McAdam, M. (2012). Analyzing the influence of gender upon high–technology venturing within the context of business incubation. *Entrepreneurship Theory and Practice, 36*(4), 655–676.

Marlow, S., & Swail, J. (2014). Gender, risk and finance: Why can't a woman be more like a man? *Entrepreneurship and Regional Development, 26*(1-2), 80–96.

Martens, M. L., Jennings, J. E., & Devereaux Jennings, P. (2007). Do the stories they tell get them the money they need? The role of entrepreneurial narratives in resource acquisition. *Academy of Management Journal, 50*(5), 1107–1132.

Martens, M. L., Jennings, J. E., & Jennings, P. D. (2007). Do the stories they tell get them the money they need? The role of entrepreneurial narratives in resource acquisition. *Academy of Management Journal, 50*(5), 1107–1132.

Martí, I., & Fernández, P. (2013). The institutional work of oppression and resistance: Learning fromthe Holocaust. *Organization Studies, 34*(8), 1195–1223.

Mason, C., & Brown, R. (2014). Entrepreneurial ecosystems and growth oriented entrepreneurship. *Final report to OECD, Paris, 30*(1), 77-102.

Mazzucato, M. (2018). *The value of everything: Making and taking in the global economy.* PublicAffairs.

McAdam, M., & Marlow, S. (2007). Building futures or stealing secrets? Entrepreneurial cooperation and conflict within business incubators. *International Small Business Journal*, *25*(4), 361–382.

McAdams, D. P. (1997). A conceptual history of personality psychology. In *Handbook of personality psychology* (pp. 3–39). Academic Press.

McCann, E., & Ward, K. (2010). Relationality/territoriality: Toward a conceptualization of cities in the world. *Geoforum*, *41*(2), 175–184.

Meadows, D. (1999). Leverage points. *Places to Intervene in a System*, *19*, 28.

Meadows, D. (2008). *Thinking in systems: International bestseller*. Chelsea Green Publishing.

Milan, S., & Treré, E. (2019). Big data from the South (s): Beyond data universalism. *Television & New Media*, *20*(4), 319–335.

Mohr, J. W., Wagner-Pacifici, R., Breiger, R. L., & Bogdanov, P. (2013). Graphing the grammar of motives in National Security Strategies: Cultural interpretation, automated text analysis and the drama of global politics. *Poetics*, *41*(6), 670–700.

Molyneux, L., & Holton, A. (2015). Branding (health) journalism: Perceptions, practices, and emerging norms. *Digital Journalism (Abingdon, England)*, *3*(2), 225–242.

Moore, M. L., & Westley, F. (2011). Surmountable chasms: Networks and social innovation for resilient systems. *Ecology and Society*, *16*(1).

Moulaert, F., MacCallum, D., & Hillier, J. (2013). Social innovation: Intuition, precept, concept, theory and practice. In The international handbook on social innovation: Collective action, social learning and transdisciplinary research (pp. 13-24).

Moulaert, F., & Sekia, F. (2003). Territorial innovation models: A critical survey. *Regional Studies*, *37*(3), 289–302.

Mulgan, G. (2014). *The locust and the bee: Predators and creators in capitalism's future*. Princeton University Press.

Munir, K. A. (2005). The social construction of events: A study of institutional change in the photographic field. *Organization Studies*, *26*(1), 93–112.

Navis, C., & Glynn, M. A. (2011). Legitimate distinctiveness and the entrepreneurial identity: Influence on investor judgments of new venture plausibility. *Academy of Management Review*, *36*(3), 479–499.

Nicholls, A. (2010). The legitimacy of social entrepreneurship: Reflexive isomorphism in a pre-paradigmatic field. *Entrepreneurship Theory and Practice*, *34*(4), 611–633.

Nicholls, A., & Murdock, A. (Eds.). (2012). *Social innovation: Blurring boundaries to reconfigure markets*. Palgrave Macmillan.

Nisbet, M. C. (2009). Communicating climate change: Why frames matter for public engagement. *Environment, 51*(2), 14–18.

Ojala, A., Evers, N., & Rialp, A. (2018). Extending the international new venture phenomenon to digital platform providers: A longitudinal case study. *Journal of World Business, 53*(5), 725–739.

Ospina, S., & Dodge, J. (2005). Narrative inquiry and the construction of meaning: A review essay. *Public Administration Review, 65*(5), 611–616.

Ospina, S. M., & Dodge, J. (2005). It's about time: Catching method up to meaning—The usefulness of narrative inquiry in public administration research. *Public Administration Review, 65*(2), 143–157.

Pache, A.-C., & Santos, F. (2010). When worlds collide: The internal dynamics of organizational responses to conflicting institutional demands. *Academy of Management Review, 35*(3), 455–476.

Parhankangas, A., & Ehrlich, M. (2014). How entrepreneurs seduce business angels: An impression management approach. *Journal of Business Venturing, 29*(4), 543–564.

Pentland, B. T. (1999). Building process theory with narrative: From description to explanation. *Academy of Management Review, 24*(4), 711–724.

Pentland, B. T., & Feldman, M. S. (2007). Narrative networks: Patterns of technology and organization. *Proceedings - Academy of Management, 2007*(1).

Pentland, B. T., & Feldman, M. S. (2008). Issues in empirical field studies of organizational routines. In Handbook of organizational routines (p. 281).

Peredo, A. M., & Chrisman, J. J. (2006). Toward a theory of community-based enterprise. *Academy of Management Review, 31*(2), 309–328.

Petkova, A. P., Rindova, V. P., & Gupta, A. K. (2008). How can new ventures build reputation? An exploratory study. *Corporate Reputation Review, 11*(4), 320–341.

Phillips, N., Lawrence, T. B., & Hardy, C. (2004). Discourse and institutions. *Academy of Management Review, 29*(4), 635–652.

Phillips, N., Lawrence, T. B., & Hardy, C. (2004). Discourse and institutions. *Academy of Management Review, 29*(4), 635–652.

Politis, D., & Gabrielsson, J. (2009). Entrepreneurs' attitudes towards failure: An experiential learning approach. *International Journal of Entrepreneurial Behaviour & Research, 15*(4), 364–383.

Polletta, F., Chen, P. C. B., Gardner, B. G., & Motes, A. (2011). The sociology of storytelling. *Annual Review of Sociology, 37*(1), 109–130.

Polletta, F., & Lee, J. (2006). Is telling stories good for democracy? Rhetoric in public deliberation after 9/11. *American Sociological Review, 71*(5), 699–721.

Pollock, T. G., & Rindova, V. P. (2003). Media legitimation effects in the market for initial public offerings. *Academy of Management Journal, 46*(5), 631–642.

Popp, A., & Holt, R. (2013). Entrepreneurship and being: The case of the Shaws. *Entrepreneurship and Regional Development, 25*(1–2), 52–68.

Powell, M., Gillett, A., & Doherty, B. (2019). Sustainability in social enterprise: Hybrid organizing in public services. *Public Management Review, 21*(2), 159–186.

Powell, W. W., Oberg, A., Korff, V. P., & Brandsen, T. (2017). Innovating in the public interest: Managing complex coordination and institutional change in public sector ecosystems. *Public Management Review, 19*(5), 655–674.

Powell, W. W., & Rerup, C. (2017). Opening the black box: Micro-foundations of routines and capabilities. In Felin, F. A. (Ed.), *Microfoundations* (pp. 1–32). Emerald Publishing.

Raworth, K. (2017). *Doughnut economics: Seven ways to think like a 21st-century economist.* Chelsea Green Publishing.

Reuber, A. R., & Fischer, E. (2011). International entrepreneurship in internet-enabled markets. *Journal of Business Venturing, 26*(6), 660–679.

Ricoeur, P. (1984). *Time and narrative* (Vol. 1; McLaughlin, K., & Pellauer, D., Trans.). University of Chicago Press.

Riessman, C. K. (2008). Narrative methods for the human sciences. *Sage (Atlanta, Ga.).*

Rindova, V., Barry, D., & Ketchen, D. J.Jr. (2009). Entrepreneuring as emancipation. *Academy of Management Review, 34*(3), 477–491.

Rindova, V., Dalpiaz, E., & Ravasi, D. (2011). A cultural quest: A study of organizational use of new cultural resources in strategy formation. *Organization Science, 22*(2), 413–431.

Rindova, V. P., & Martins, L. L. (2021). Shaping possibilities: A design science approach to developing novel strategies. *Academy of Management Review*, *46*(4), 800–822.

Roundy, P. T. (2017). Hybrid organizations and the logics of entrepreneurial ecosystems. *The International Entrepreneurship and Management Journal*, *13*(4), 1221–1237.

Roundy, P. T., Bradshaw, M., & Brockman, B. K. (2018). The emergence of entrepreneurial ecosystems: A complex adaptive systems approach. *Journal of Business Research*, *86*, 1–10.

Ruebottom, T. (2013). The microstructures of rhetorical strategy in social entrepreneurship: Building legitimacy through heroes and villains. *Journal of Business Venturing*, *28*(1), 98–116.

Ruebottom, T., & Auster, E. R. (2018). Reflexive dis/embedding: Personal narratives, empowerment and the emotional dynamics of interstitial events. *Organization Studies*, *39*(4), 467–490.

Sandberg, J., Holmström, J., Lyytinen, K., & Hultin, L. (2019). Digital transformation of work: Will it change the skills employees need? In *Proceedings of the 25th Americas Conference on Information Systems*.

Santos, F. M. (2012). A positive theory of social entrepreneurship. *Journal of Business Ethics*, *111*(3), 335–351.

Sarasvathy, S. D. (2005). Causation and effectuation: Toward a theoretical shift from economic inevitability to entrepreneurial contingency. *Academy of Management Review*, *30*(2), 243–263.

Saxenian, A. (1996). *Regional advantage: Culture and competition in Silicon Valley and Route 128*. Harvard University Press.

Scharmer, O. (2018). *The essentials of Theory U: Core principles and applications*. Berrett-Koehler Publishers.

Schoar, A. (2010). The divide between subsistence and transformational entrepreneurship. [). University of Chicago Press/NBER.]. *Innovation Policy and the Economy*, *10*, 57–81.

Schoenmaker, D., & Schramade, W. (2019). *Principles of sustainable finance*. Oxford University Press.

Scholz, T. (2017). *Uberworked and underpaid: How workers are disrupting the digital economy*. Polity Press.

Schön, D., & Rein, M. (1994). *Frame reflection: Toward the resolution of intractable policy controversies*. Basic Book.

Scott, A. J. (Ed.). (2001). *Global city-regions: trends, theory, policy*. OUP Oxford.

Seelos, C., & Mair, J. (2007). How social entrepreneurs enable human, social, and economic development. *Business solutions for the global poor: Creating social and economic value*, 271-294.

Seo, M.-G., & Creed, W. E. D. (2002). Institutional contradictions, praxis, and institutional change: A dialectical perspective. *Academy of Management Review, 27*(2), 222–247.

Shanahan, E. A., McBeth, M. K., & Hathaway, M. R. (2011). Narrative policy framework: The influence of policy narratives on public opinion. *Policy Studies Journal: the Journal of the Policy Studies Organization, 39*(3), 535–557.

Shepherd, D. A. (2003). Learning from business failure: Propositions of grief recovery for the self-employed. *Academy of Management Review, 28*(2), 318–328.

Shepherd, D. A., & Kuratko, D. F. (2009). The death of an innovative project: How grief recovery enhances learning. *Business Horizons, 52*(5), 451–458.

Shepherd, D. A., & Patzelt, H. (2011). The new field of sustainable entrepreneurship: Studying entrepreneurial action linking "what is to be sustained" with "what is to be developed". *Entrepreneurship Theory and Practice, 35*(1), 137–163.

Shiller, R. J. (2019). *Narrative economics: How stories go viral and drive major economic events*. Princeton University Press.

Snow, D. A., & Benford, R. D. (1988). Ideology, frame resonance, and participant mobilization. *International Social Movement Research, 1*(1), 197–217.

Snow, D. A., & Benford, R. D. (1988). Ideology, frame resonance, and participant mobilization. International social movement research, 1(1), 197-217.

Snow, D. A., Rochford, E. B.Jr, Worden, S. K., & Benford, R. D. (1986). Frame alignment processes, micromobilization, and movement participation. *American Sociological Review*, •••, 464–481.

Snowden, D. J. (2005). Multi-ontology sense making: A new simplicity in decision making. *Informatics in Primary Care, 13*(1), 45–53.

Snowden, D. J. (2005). Multi-ontology sense making: A new simplicity in a complex world. In *Proceedings of the 2005 International Conference on Knowledge Management*.

Somers, M. R. (1994). The narrative constitution of identity: A relational and network approach. *Theory and Society*, *23*(5), 605–649.

Somers, M. R. (1994). The narrative constitution of identity: A relational and network approach. *Theory and Society*, *23*(5), 605–649.

Sonenshein, S. (2010). We're changing—Or are we? Untangling the role of progressive, regressive, and stability narratives during strategic change implementation. *Academy of Management Journal*, *53*(3), 477–512.

Spicer, A., Alvesson, M., & Kärreman, D. (2009). Critical performativity: The unfinished business of critical management studies. *Human Relations*, *62*(4), 537–560.

Spigel, B. (2017). The relational organization of entrepreneurial ecosystems. *Entrepreneurship Theory and Practice*, *41*(1), 49–72.

Spinuzzi, C., Altounian, D., Pogue, G., Zhu, L., & Cochran, R. (2016). Articulating problems and markets: A translation analysis of entrepreneurs' emergent value propositions. *Written Communication*, *35*(1), 97–134.

Spivak, G. C. (1988). Can the subaltern speak? In C. Nelson & L. Grossberg (Eds.), *Marxism and the interpretation of culture* (pp. 271–313). University of Illinois Press.

Stam, E. (2015). Entrepreneurial ecosystems and regional policy: A sympathetic critique. *European Planning Studies*, *23*(9), 1759–1769.

Stam, E., & Spigel, B. (2016). *Entrepreneurial ecosystems* (Vol. 16, No. 13, pp. 1-15). USE Discussion paper series.

Stark, D. (2009). *The sense of dissonance: Accounts of worth in economic life*. Princeton University Press.

Stiglitz, J. E., Sen, A., & Fitoussi, J. P. (2009). *The measurement of economic performance and social progress revisited* (Vol. 33). Ofce.

Suchman, M. C. (1995). Managing legitimacy: Strategic and institutional approaches. *Academy of Management Review*, *20*(3), 571–610.

Suddaby, R. (2010). Challenges for institutional theory. *Journal of Management Inquiry*, *19*(1), 14–20.

Suddaby, R., Bruton, G. D., & Si, S. X. (2015). Entrepreneurship through a qualitative lens: Insights on the construction and/or discovery of entrepreneurial opportunity. *Journal of Business Venturing*, *30*(1), 1–10.

Suddaby, R., Foster, W. M., & Trank, C. Q. (2010). Rhetorical history as a source of competitive advantage. In *The globalization of strategy research* (Vol. 27, pp. 147–173). Emerald Group Publishing Limited.

Suddaby, R., & Greenwood, R. (2005). Rhetorical strategies of legitimacy. *Administrative Science Quarterly*, *50*(1), 35–67.

Tambayong, L., Daly, A., & Singh, S. (2020). A narrative analytic approach for real-time social sensing using NLP. *Journal of Big Data*, *7*(1), 1–21.

Tandoc, E. C.Jr, Lim, D., & Ling, R. (2020). Diffusion of disinformation: How social media users respond to fake news and why. *Journalism*, *21*(3), 381–398.

Tedmanson, D., Verduijn, K., Essers, C., & Gartner, W. B. (2012). Critical perspectives in entrepreneurship research. *Organization*, *19*(5), 531–541.

Thornton, P. H., Ocasio, W., & Lounsbury, M. (2012). *The institutional logics perspective: A new approach to culture, structure, and process*. Oxford University Press.

Tracey, P., & Phillips, N. (2016). Managing the consequences of organizational stigmatization: Identity work in a social enterprise. *Academy of Management Journal*, *59*(3), 740–765.

Tracey, P., Phillips, N., & Jarvis, O. (2011). Bridging institutional entrepreneurship and the creation of new organizational forms: A multilevel model. *Organization Science*, *22*(1), 60–80.

Tracey, P., & Stott, N. (2017). Social innovation: A window on alternative ways of organizing and innovating. Innovation. *Organization & Management*, *19*(1), 51–60.

Tuckett, D., & Nikolic, M. (2017). The role of conviction and narrative in decision-making under radical uncertainty. *Theory & Psychology*, *27*(4), 501–523.

Turchin, P. (2003). *Historical dynamics: Why states rise and fall*. Princeton University Press.

Ucbasaran, D., Shepherd, D. A., Lockett, A., & Lyon, S. J. (2013). Life after business failure: The process and consequences of business failure for entrepreneurs. *Journal of Management*, *39*(1), 163–202.

Vaara, E., Sonenshein, S., & Boje, D. (2016). Narratives as sources of stability and change in organizations: Approaches and directions for future research. *The Academy of Management Annals*, *10*(1), 495–560.

Vaara, E., & Tienari, J. (2008). A discursive perspective on legitimation strategies in multinational corporations. *Academy of Management Review*, *33*(4), 985–993.

Van Hulst, M. (2012). Storytelling, a model of and a model for planning. *Planning Theory*, *11*(3), 299–318.

van Lente, H., & Rip, A. (1998). Expectations in technological developments: An example of prospective structures to be filled in by agency. In Disco, C., & van der Meulen, B. J. R. (Eds.), *Getting new technologies together: Studies in making sociotechnical order* (pp. 203–229). Walter de Gruyter.

Verduijn, K., Dey, P., Tedmanson, D., & Essers, C. (2014). Emancipation and/or oppression? Conceptualizing dimensions of criticality in entrepreneurship studies. *International Journal of Entrepreneurial Behaviour & Research*, *20*(2), 98–107.

Verduijn, K., & Essers, C. (2013). Questioning dominant entrepreneurship assumptions: The case of female ethnic minority entrepreneurs. *Entrepreneurship and Regional Development*, *25*(7-8), 612–630.

Waddell, S. (2016). *Change for the audacious: A doer's guide to large systems change for flourishing futures*. NetworkingAction.

Watson, T. J. (2009). Narrative, life story and manager identity: A case study in autobiographical identity work. *Human Relations*, *62*(3), 425–452.

Welter, F., Baker, T., Audretsch, D. B., & Gartner, W. B. (2017). Everyday entrepreneurship: A call for entrepreneurship research to embrace entrepreneurial diversity. *Entrepreneurship Theory and Practice*, *41*(3), 311–321.

Westley, F., Zimmerman, B., & Quinn Patton, M. (2011). *Getting to maybe: How the world is changed*. Random House Canada.

Westley, F. R., Tjornbo, O., Schultz, L., Olsson, P., Folke, C., Crona, B., & Bodin, Ö. (2013). A theory of transformative agency in linked social-ecological systems. *Ecology and Society*, *18*(3).

Wood, A. J., Graham, M., Lehdonvirta, V., & Hjorth, I. (2019). Good gig, bad gig: Autonomy and algorithmic control in the global gig economy. *Work, Employment and Society*, *33*(1), 56–75.

Wry, T., Lounsbury, M., & Glynn, M. A. (2011). Legitimating nascent collective identities: Coordinating cultural entrepreneurship. *Organization Science*, *22*(2), 449–463.

Yamakawa, Y., Peng, M. W., & Deeds, D. L. (2015). Rising from the ashes: Cognitive determinants of venture growth after entrepreneurial failure. *Entrepreneurship Theory and Practice*, *39*(2), 209–236.

Yanow, D. (2000). *Conducting interpretive policy analysis*. SAGE Publications.

Ybema, S., Yanow, D., Velsink, H., & Buytelaar, M. (2009). *Organizational ethnography: Studying the complexities of everyday life*. SAGE Publications.

Zahra, S. A., Gedajlovic, E., Neubaum, D. O., & Shulman, J. M. (2009). A typology of social entrepreneurs: Motives, search processes and ethical challenges. *Journal of Business Venturing*, *24*(5), 519–532.

Zhao, E. Y., & Wry, T. (2016). Not all inequality is equal: Deconstructing the societal logic of patriarchy to understand microfinance lending to women. *Academy of Management Journal*, *59*(6), 1994–2020.

Zietsma, C., Greenwood, R., & Lounsbury, M. (2017). Field or fields? Building the scaffolding for cumulation of research on institutional fields. In Greenwood, R., Oliver, C., Lawrence, T. B., & Meyer, R. E. (Eds.), *The SAGE handbook of organizational institutionalism* (2nd ed., pp. 389–416). SAGE Publications.

Zilber, T. B. (2006). The work of the symbolic in institutional processes: Translations of rational myths in Israeli high tech. *Academy of Management Journal*, *49*(2), 281–303.

Zilber, T. B. (2007). Stories and the discursive dynamics of institutional entrepreneurship: The case of Israeli high-tech after the bubble. *Organization Studies*, *28*(7), 1035–1054.

Zilber, T. B. (2009). Institutional maintenance as narrative acts. In Institutional work: Actors and agency in institutional studies of organizations (pp. 205-235).

Zimmerman, M. A., & Zeitz, G. J. (2002). Beyond survival: Achieving new venture growth by building legitimacy. *Academy of Management Review*, *27*(3), 414–431.

Zott, C., Amit, R., & Massa, L. (2011). The business model: Recent developments and future research. *Journal of Management*, *37*(4), 1019–1042.

Zott, C., & Huy, Q. N. (2007). How entrepreneurs use symbolic management to acquire resources. *Administrative Science Quarterly*, *52*(1), 70–105.

Chapter 3
Fostering Growth and Prosperity Through Sustainable Resource Management and Environmental Conservation:
Co-Creating Economic Transformation via Enterprise, Narratives, and Systematic Design

Oindrilla Ghosh
https://orcid.org/0000-0001-9609-0781
Christ University, India

Binod Kumar
https://orcid.org/0009-0002-9721-1367
Radha Govind University, India

ABSTRACT

Sustainable resource management and environmental conservation are essential for fostering economic growth and long-term prosperity. By adopting eco-friendly practices, businesses can minimize ecological impacts while maximizing resource efficiency. This approach not only ensures a healthier planet but also drives innovation and opens new markets. Collaborating with local communities and stakeholders

DOI: 10.4018/979-8-3373-4566-6.ch003

promotes shared narratives that highlight the importance of sustainability, creating a collective vision for the future. Systematic design principles can guide enterprises in implementing sustainable solutions, enhancing their competitive edge. As organizations co-create impactful strategies, they contribute to an economic transformation that aligns profit with purpose. Ultimately, by embracing sustainability, we can nurture ecosystems, empower communities, and create resilient economies that thrive.

INTRODUCTION

Fostering growth and prosperity through sustainable resource management and environmental conservation has emerged as a critical imperative in our rapidly changing world (Turkina et al., 2025). The acceleration of industrial activities, urbanization, and population growth has led to significant environmental degradation, prompting a re-evaluation of conventional economic paradigms (Narula, 2014). Today, the intersection of economic transformation, environmental stewardship, and social equity presents a unique opportunity to reimagine how we conduct business, manage resources, and interact with the ecosystems that sustain us (Luo, 2024). Sustainable resource management is no longer merely an ethical consideration; it is a fundamental necessity for long-term economic viability (Tung et al., 2023).

Central to this paradigm shift is the recognition that enterprises can play a pivotal role in fostering sustainable practices that benefit not just their bottom line but also the communities in which they operate (Belitz and Lejpras, 2021). Businesses increasingly understand that their success is inextricably linked to the health of environmental systems (Belderbos, Leten & Suzuki, 2013). By adopting sustainable practices through innovative technologies, circular economies, and responsible sourcing enterprises can minimize their ecological footprints while enhancing their competitiveness and resilience (Ervits, 2023a). This approach fosters a new economic landscape where growth is inclusive and sustainable, yielding benefits that extend beyond profit margins (Papanastassiou et al. 2020).

Moreover, the narratives we co-create around sustainability are essential to galvanizing collective action and transforming public perception (De Beule & Van Beveren, 2019). By weaving stories rooted in environmental conservation and community involvement, organizations can inspire stakeholders and consumers to embrace sustainable principles (Dachs, 2017). These narratives facilitate a cultural shift that values environmental health as a driver of economic prosperity, promoting collaborative efforts across sectors to achieve shared goals (Meyer et al., 2011).

Systematic design plays a crucial role in this endeavor by ensuring that sustainable practices are thoughtfully integrated into all aspects of operations, from product development to supply chain management (Criscuolo et al., 2005). By utilizing design thinking principles, businesses can innovate solutions that not only align with ecological values but also respond to the needs of consumers and communities (Le Bas & Sierra, 2002). This holistic approach to enterprise development enhances the capacity for sustainable growth, paving the way for enterprises to thrive while contributing positively to the environment (Altman et al., 2024).

Ultimately, fostering growth and prosperity through sustainable resource management and environmental conservation necessitates a comprehensive framework that encompasses economic, social, and ecological dimensions (Castellani & Zanfei, 2002). It is about co-creating pathways to economic transformation that empower communities, mitigate environmental impact, and ensure a more equitable distribution of resources (Tung et al., 2023). The journey toward sustainability is complex, requiring collaboration among diverse stakeholders, including governments, businesses, NGOs, and local communities (Martin, 2024). By working together, we can craft a future where economic growth is harmonious with ecological integrity, thus ensuring prosperity for generations to come (Dachs et al., 2023).

HISTORICAL EVOLUTION

The historical evolution of sustainable resource management and environmental conservation is a testament to humanity's changing relationship with the natural world (Verbeke, 2009). Traditional views often positioned nature as an inexhaustible resource, leading to exploitation and degradation (Yamin and Otto, 2004). However, as societies progressed, the recognition of environmental limits and the intricate balance of ecosystems grew increasingly prominent (Rugman and Verbeke, 2001).

The roots of sustainable resource management can be traced back to ancient civilizations (Zanfei, 2000). For example, indigenous cultures have practiced stewardship of the land for millennia, employing methods that align with the natural cycles of their environments (Simon, 1984). The wisdom embedded in these traditional practices often emphasizes sustainability, harmony, and respect for nature (Tversky and Kahneman, 1974). However, the rise of industrialization in the 18th and 19th centuries marked a significant turning point that disrupted these symbiotic relationships (Jallat and Coronati, 2022). The unprecedented demand for resources led to extensive deforestation, pollution, and degradation of ecosystems worldwide (Hage et al., 2022).

The 20th century saw the emergence of the modern environmental movement, catalyzed by growing awareness of the environmental crises spurred by industrial growth (Kraemer, 2023). Landmark events, such as the publication of Rachel Carson's "Silent Spring" in 1962, highlighted the detrimental effects of pesticides and sparked public consciousness around environmental issues (Schmitz, 2024). This period also witnessed the first Earth Day in 1970, signaling a collective demand for environmental protection legislation (Xinhua, 2023). The subsequent establishment of numerous environmental policies, including the Clean Air Act and the Endangered Species Act, reflected a paradigm shift towards recognizing the need for regulatory frameworks to protect natural resources (Siemens, 2024b).

As environmental challenges escalated, the concept of sustainable development gained traction with the publication of the Brundtland Report in 1987 by the World Commission on Environment and Development (Siemens, 2021). This report popularized the notion that development must meet the needs of the present without compromising the ability of future generations to meet their own needs (Ervits, 2023a). It established a foundational framework for integrating economic growth with environmental and social considerations, paving the way for subsequent global initiatives such as the United Nations Sustainable Development Goals (SDGs) (Siemens, 2019).

Toward the end of the 20th century, the concept of corporate social responsibility (CSR) began to shape the responsibilities of businesses toward the environment and society (Siemens, 2024a). Companies began to recognize that their long-term viability depended on sustainable practices and stakeholder engagement (Siemens 2024b). The emergence of the triple bottom line (people, planet, profit) further emphasized the need for businesses to evaluate their impacts holistically (Ervits, 2023a).

In recent years, the dialogue around sustainability has evolved into discussions about systemic design and transformational change (Dachs et al., 2023). The focus has shifted from minimizing harm to actively co-creating solutions that foster resilience and prosperity in both communities and ecosystems (Cantwell & Mudambi, 2000). Business models grounded in circular economy principles, regenerative agriculture, and low-carbon technologies are gaining traction, signaling a robust commitment to redesigning economic processes for sustainability (Kuemmerle, 1999).

LITERATURE REVIEW

Fostering growth and prosperity through sustainable resource management and environmental conservation is a multifaceted endeavor deeply rooted in the interplay of economic development, ecological sustainability, and social equity (Chen, 2022). The literature reveals a comprehensive body of work that explores how effective resource management and environmental conservation can catalyze economic transformation (Zanfei, 2000). Central to this discourse is the recognition that the environmental degradation and resource depletion pose significant threats to long-term economic stability and societal well-being (Rugman and Verbeke, 2001).

One salient theme in the literature is the concept of sustainable development, originally defined in the Brundtland Report (1987) as development that meets the needs of the present without compromising the ability of future generations to meet their own needs (Yamin and Otto, 2004). Scholars argue that sustainable resource management encompassing practices such as conservation agriculture, sustainable forestry, and water resource management enables communities to utilize their natural resources responsibly while preserving them for future generations (Verbeke, 2009).

Additionally, the economic potential of ecosystem services has been increasingly emphasized by researchers (Narula, 2014). Ecosystem services including clean air, water purification, pollination, and carbon sequestration underpin economic activities and are fundamental to human well-being (Zhao, 2022). Understanding and valuing these services can lead to more informed resource management practices and better policy frameworks that drive economic growth (Quinn, 2011).

Moreover, co-creation emerges as a powerful framework in the literature, advocating for collaborative approaches between stakeholders, including governments, businesses, and local communities (Branstetter et al., 2013). Research further highlights the importance of policymaking and institutional frameworks in fostering sustainable practices (Berry, 2013). Effective policies must not only incentivize sustainable practices but also enforce regulations that protect vulnerable ecosystems (Belderbos et al., 2014). Furthermore, integrating sustainability into educational curricula has been posited as vital for fostering a culture of conservation and awareness in future generations (Agostini and Caviggioli, 2015).

RESEARCH METHODOLOGY

The study of fostering growth and prosperity through sustainable resource management and environmental conservation necessitates a qualitative research methodology that can capture the rich, contextual, and nuanced realities of localized practices, perceptions, and narratives.

Research Objective: The study aims to explore how sustainable resource management and environmental conservation practices contribute to economic transformation and community prosperity. Specific objectives may include understanding local narratives regarding resource management, identifying key stakeholders, and assessing the impact of co-creation initiatives on community engagement (Michailova and Zaidah, 2012).

Data Collection Method: Data collection is conducted through individual semi-structured interviews, focus group discussions, and participant observations.

Semi-structured interviews allow for open-ended questions that encourage respondents to share their experiences and viewpoints. Focus groups can facilitate group discussions, fostering dynamic interactions that might reveal shared concerns and collective insights (Caviggioli et al., 2023).

Data Analysis: Researchers will transcribe interviews and focus group discussions, then systematically code the data to identify common themes, patterns, and categories.

THEORETICAL FRAMEWORKS

In the quest for sustainable growth and prosperity, theoretical frameworks serve as foundational structures guiding the analysis, implementation, and evaluation of strategies aimed at environmental conservation and resource management (Hazir et al., 2018). Several key theories underpin practices in this area, notably the Ecological Economics, Systems Theory, and the Sustainable Development Goals (SDGs) framework.

Ecological Economics posits that the economy is embedded within the ecological systems rather than the other way around (Griliches, 1990). This framework emphasizes the importance of natural capital and ecosystem services, advocating for policies that recognize the economic value of ecological health (Bergek and Bruzelius, 2010). It encourages the assessment of environmental impacts in economic planning, promoting resource efficiency and highlighting the limits of natural resources (Palomeras and Wehrheim, 2021). By integrating ecological considerations into economic models, it seeks to resolve the conflict between economic development and environmental sustainability (Tsukada and Nagaoka, 2015).

Systems Theory provides further clarity by emphasizing the relationships and interactions among different components of socio-ecological systems (Stoneman, 1995). This approach encourages stakeholders to view environmental issues holistically, considering interdependencies that traditional linear models might overlook (Schumpeter, 1942). Through systems thinking, practitioners can understand feedback loops, resilience, and the potential for tipping points, allowing for more effective interventions that promote sustainability across various sectors (Cayton, 2020).

Another pertinent framework is the *Sustainable Development Goals (SDGs)* established by the United Nations, which outlines a global blueprint for achieving social, economic, and environmental sustainability by 2030 (Luo, 2022). The SDGs provide a structured approach for businesses and governments, encouraging initiatives that strive for economic transformation through responsible consumption and production while protecting ecosystems (Li et al., 2024). Each goal incorporates indicators for progress, allowing for the monitoring of advancements in sustainable practices and resource management (Scherngell et al., 2023).

Complementarily, the *Triple Bottom Line (TBL)* framework extends corporate social responsibility by advocating for a balance between social equity, environmental health, and economic prosperity (Lynn and Salzman, 2023). Enterprises leveraging TBL can innovate more sustainably, ensuring their activities consider all stakeholders' well-being rather than focusing solely on profit generation (Acosta et al., 2023).

ADVANTAGES

Fostering growth and prosperity through sustainable resource management and environmental conservation brings forth a multitude of advantages, emphasizing not only ecological benefits but also economic and social transformations (Lubinski and Wadhwani, 2020).

Economic Resilience and Innovation: Sustainable resource management promotes efficiency in the usage of resources, leading to reduced costs and increased competitiveness for businesses (Buckley, 2022). By adopting green technologies and innovative practices, companies can tap into new markets, enhancing economic resilience (Bhaumik et al., 2024). Sustainable enterprises often enjoy lower operational costs due to energy savings and waste reduction, thus improving profit margins (Ervits, 2023b). Furthermore, sustainability-minded companies are increasingly favored by consumers and investors, translating into brand loyalty and enhanced market performance (Chen & Vang, 2008).

Social Equity and Job Creation: Transitioning to sustainable practices contributes to social equity by providing equitable access to resources and opportunities for marginalized communities (Cuervo-Cazurra & Genc, 2011). Investments in renewable energy, sustainable agriculture, and conservation create jobs and stimulate economic growth in local communities (Prabhu & Jain, 2015). Green jobs, which include roles in renewable energy, energy efficiency, and environmental monitoring, can significantly enhance community resilience while offering fair wages and working conditions (Shimizutani & Todo, 2008).

Improved Public Health: Environmental conservation has direct ramifications for human health (Demirbag & Glaister, 2010). Improved air and water quality, thanks to sustainable practices, can lead to reductions in health-related expenses and enhance the quality of life (Kramer et al., 2010). Investing in clean technology and sustainable infrastructure contributes to less pollution, minimizing the burden on healthcare systems and fostering healthier communities (Zanfei, 2000).

Biodiversity and Ecosystem Services: Effective resource management and conservation strategies help protect biodiversity, which is essential for ecosystem resilience

(Rugman and Verbeke, 2001). Healthy ecosystems provide vital services such as pollination, water filtration, and climate regulation (Yamin and Otto, 2004). These services contribute not only to environmental health but also underpin economic activities such as agriculture, fishing, and tourism, clearly illustrating the interdependence of economic and ecological prosperity (Verbeke, 2009).

Climate Change Mitigation: By implementing sustainable practices, societies can significantly reduce greenhouse gas emissions and enhance their adaptive capacity to climate impacts (Cantwell & Mudambi, 2000). Sustainable resource management contributes to climate resilience, aiding communities in preparing for and responding to climate-induced challenges (Kuemmerle, 1999). This aspect is particularly critical as the global population faces unprecedented climate-related threats, necessitating immediate action and transformative changes in resource usage (Zhao et al., 2020).

CHALLENGES

The quest for sustainable resource management and environmental conservation often encounters numerous challenges that can impede growth and prosperity (Belitz and Lejpras, 2021). A significant challenge lies in the balance between economic development and environmental protection (Xinhua, 2025). As nations and communities pursue economic growth, they frequently prioritize short-term economic gains over long-term sustainability (Neil Alim and Jones, 2024). This can lead to overexploitation of natural resources, deforestation, pollution, and depletion of biodiversity (Milne, 2025). The direct impact is often visible in local ecosystems, which experience degradation, potentially harming the livelihoods of communities dependent on those resources (Schmitz, 2024).

Another layer of complexity is posed by the lack of awareness and education regarding sustainable practices (Hage et al., 2022). Many stakeholders, including businesses, policymakers, and individuals, may lack a thorough understanding of sustainable resource management principles (Rinke et al., 2022). This gap can result in ineffective policies, misallocation of resources, and practices that do not prioritize ecosystem health (ChinaScope, 2024). Without broad public understanding and engagement, initiatives aimed at fostering sustainability may not gain sufficient traction or support (Arnold and Chazan, 2023).

Additionally, institutional challenges further complicate sustainable resource management (Rinke et al., 2022). Many regions face bureaucratic inefficiencies, corruption, and a lack of collaboration across sectors (Siemens, 2025). Fragmented governance structures can hinder the implementation of comprehensive strategies necessary for managing resources sustainably (Volkswagen Group China, 2025). Moreover, the interaction between local, national, and international laws can create conflicts and uncertainty, making it difficult for enterprises to operate sustainably (Belitz, 2017).

Financing presents another substantial challenge. The transition to sustainable practices often requires significant upfront investment in technology, infrastructure, and training (KPMG, 2024). Small and medium-sized enterprises, which are vital to economic prosperity, typically lack access to the necessary capital (Casella and Bolwijn, 2024). Moreover, government funding is often insufficient or misdirected, focusing on traditional economic sectors rather than innovative sustainable practices (Tung et al., 2023).

Climate change and its associated impacts represent another formidable challenge (Luo and Van Assche, 2023). As environmental conditions become increasingly unpredictable, communities that rely on natural resources are particularly vulnerable (Liu, 2024). Extreme weather events, shifts in ecosystems, and changing agricultural productivity can disrupt local economies, particularly for those engaged in farming, fishing, and tourism (Ribakova, 2024). The uncertainty surrounding climate impacts complicates long-term planning and resource management efforts (Demirbag & Glaister, 2010).

Lastly, cultural and social factors can shape perceptions and attitudes toward environmental conservation (Alcacer et al., 2016). In some cases, traditional practices may conflict with modern sustainability paradigms, leading to resistance against new initiatives (Dunning & Lundan, 2008). Engaging communities and integrating their narratives into sustainability practices is essential to achieve a unified approach, but this requires navigating complex socio-cultural landscapes (Luo, 2024).

USE OF AI AND ML

Artificial Intelligence (AI) and Machine Learning (ML) are increasingly being recognized as pivotal tools in fostering growth and prosperity through sustainable resource management and environmental conservation (Buckley & Casson,1976). By analyzing vast and complex datasets, these technologies can improve decision-making processes, optimize resource use, and enhance the development of innovative solutions that harmonize economic growth with environmental sustainability (Castellani & Zanfei, 2002).

One of the most impactful applications of AI and ML lies in data analytics for environmental monitoring (Le Bas & Sierra, 2002). These technologies can process satellite imagery and sensor data to provide real-time insights into deforestation, land-use changes, and wildlife populations (De Beule & Van Beveren, 2019). For example, companies like Planet Labs use satellite data analyzed through AI to track environmental changes globally, allowing stakeholders to respond swiftly to potential environmental threats (Ervits, 2023a). This real-time monitoring equips policymakers, businesses, and conservationists with the information they need to implement effective interventions, enforce regulations, and protect vulnerable ecosystems (Ervits 2021).

AI and ML can also optimize resource management in agriculture and fisheries (Ervits, 2018). Precision agriculture, driven by AI, allows farmers to analyze soil health, crop conditions, and weather patterns, enabling them to apply resources more efficiently (Criscuolo et al., 2005). For instance, machine learning models can predict crop yields based on historical data, weather forecasts, and market trends, helping farmers make informed decisions about planting and resource allocation (Meyer et al., 2011). This not only increases productivity but also minimizes the ecological footprint by reducing excessive fertilizer and pesticide use (Dachs, 2017).

In fisheries management, AI algorithms can analyze fishing patterns, migration routes, and environmental conditions to support sustainable practices (Papanastassiou et al., 2020). By providing predictive models and recommendations, AI can help fishers avoid overexploitation by identifying sustainable catch limits and optimal fishing areas, fostering both economic viability and marine conservation (Li et al., 2024).

Additionally, AI and ML can enhance sustainability reporting and corporate responsibility initiatives (Cantwell & Mudambi, 2011). These technologies can automate the collection and analysis of sustainability data, providing businesses with insights into their environmental impacts (Cantwell & Piscitello, 2014). By identifying areas for improvement and benchmarking against industry standards, companies can develop more responsible practices and build brand loyalty among eco-conscious consumers (Zanfei, 2000).

The use of AI and ML in fostering sustainability also extends to urban planning and development (Fukuyama, 1992). Smart city technologies leverage these technologies to optimize energy use, reduce waste, and enhance urban resilience (Turkina et al., 2025). For example, AI can control traffic flow to minimize congestion and emissions, while ML algorithms can predict energy demand and enable smart grid systems that balance supply and demand more effectively (Narula, 2014).

Challenges such as data privacy, ethical issues, and the digital divide must be addressed to harness the full potential of AI and ML in sustainability (Luo, 2024). Ensuring equitable access to these technologies, particularly for small enterprises and vulnerable communities, is crucial for fostering inclusive economic transformation (Tung et al., 2023).

ROLE OF AUGMENTED REALITY (AR), VIRTUAL REALITY (VR), AND MIXED REALITY (MR)

Augmented Reality (AR), Virtual Reality (VR), and Mixed Reality (MR) are revolutionizing various sectors by enhancing user experiences, providing immersive environments, and facilitating new forms of interaction with digital content (Belitz and Lejpras, 2021). These technologies offer unique capabilities that can significantly impact industries like education, healthcare, gaming, real estate, manufacturing, and environmental conservation (Belderbos, Leten & Suzuki, 2013).

AR enhances the real world by overlaying digital information onto it (Ervits, 2023a). This technology is particularly beneficial in fields such as education and training (Papanastassiou et al. 2020). For instance, medical students can use AR to visualize complex anatomy in 3D, enhancing their understanding and retention of the material (De Beule & Van Beveren, 2019). In manufacturing, AR can guide workers through complex assembly processes with real-time visual instructions, minimizing errors and increasing efficiency (Dachs, 2017). Moreover, AR can support environmental conservation by providing interactive experiences that educate users about biodiversity, ecosystems, and the impacts of climate change (Meyer et al., 2011). Users can point their devices at a landscape and see information about native species, conservation efforts, and responsible practices (Criscuolo et al., 2005).

On the other hand, VR creates a fully immersive environment, transporting users to digital worlds where they can explore and interact without physical limitations (Le Bas & Sierra, 2002). In gaming, VR offers an unparalleled experience, allowing players to step into their favorite games (Altman et al., 2024). Beyond entertainment, VR holds significant potential in training and simulation, particularly for professions that require high-stakes decision-making, such as military training, emergency response, and medical procedures (Castellani & Zanfei, 2002). For instance, VR simulations can expose trainees to diverse scenarios, enhancing their preparedness in real-world situations (Tung et al., 2023). In terms of environmental education, VR can simulate ecological changes over time, enabling users to understand the consequences of their actions on ecosystems and the importance of sustainable practices (Martin, 2024).

MR combines elements of both AR and VR, allowing physical and digital objects to coexist and interact in real-time (Dachs et al., 2023). This technology has promising applications in collaborative work environments, enabling teams to work together regardless of their physical location (Verbeke, 2009). For environmental initiatives, MR can foster community involvement in conservation by providing platforms for virtual meetings, planning sessions, and strategy development (Yamin and Otto, 2004). Scientists and policymakers can visualize environmental data and scenarios collaboratively, enhancing understanding and decision-making regarding ecological issues (Rugman and Verbeke, 2001).

As these immersive technologies continue to evolve, their integration into various sectors can drive innovation, productivity, and sustainability (Zanfei, 2000). By creating more engaging and informative experiences, AR, VR, and MR can facilitate heightened awareness of environmental issues and encourage behaviors that contribute to ecological preservation (Simon, 1984). The potential for economic transformation through these technologies is immense, particularly as industries embrace new methods of enterprise development (Tversky and Kahneman, 1974). By leveraging AR, VR, and MR, organizations can create compelling narratives that resonate with their audiences, sparking movements toward collective action for environmental conservation and sustainable growth (Jallat and Coronati, 2022).

IMPLEMENTATION OF REGULATORY FRAMEWORKS

Implementing regulatory frameworks is essential for fostering growth and prosperity while ensuring sustainable resource management and environmental conservation (Hage et al., 2022). As economies evolve, policymakers must establish guidelines that balance economic development with ecological integrity and social responsibility (Kraemer, 2023). Effective regulation not only safeguards the environment but also promotes sustainable practices that can drive long-term economic transformation (Schmitz, 2024).

Regulatory frameworks for sustainable resource management typically encompass laws and policies related to land use, water management, emissions standards, waste management, and biodiversity protection (Xinhua, 2023). These regulations create a structured approach to resource use, ensuring that businesses adhere to practices that minimize their environmental impact (Siemens, 2024b). By setting stringent standards, regulators can drive industries toward adopting greener technologies, reducing waste, and conserving natural habitats (Siemens, 2021).

Moreover, incorporating sustainability principles into regulatory frameworks can stimulate innovation (Ervits, 2023a). When companies face regulatory requirements, they are often prompted to develop new technologies and processes to comply, leading to the creation of green jobs and the advancement of green technologies (Siemens, 2019). For instance, automotive regulations aimed at reducing emissions encourage manufacturers to invest in electric and hybrid vehicles, fostering economic transformation and reducing carbon footprints (Siemens, 2024a).

Furthermore, effective regulatory frameworks must include incentives for the private sector to embrace sustainability (Siemens 2024b). Financial incentives, such as tax breaks for sustainable practices, can encourage businesses to invest in environmentally friendly technologies and practices (Ervits, 2023a). Support for research and development initiatives related to sustainable resource management can further stimulate innovation while fostering collaboration among various stakeholders, including government agencies, private entities, and civil society (Dachs et al., 2023).

Implementing frameworks that promote public participation is also crucial (Cantwell & Mudambi, 2000). Engaging the community in decision-making processes related to environmental policies fosters a greater sense of ownership and encourages sustainable practices at the grassroots level (Kuemmerle, 1999). Public involvement in designing regulations can lead to more effective solutions that reflect the needs and values of local communities, contributing to successful environmental governance (Chen, 2022).

Lastly, regulatory frameworks must be adaptive, allowing for the incorporation of new scientific knowledge and technologies (Zanfei, 2000). As environmental challenges evolve, regulations must be flexible enough to respond to emerging issues, ensuring that resource management strategies remain relevant and effective (Rugman and Verbeke, 2001). Continuous monitoring and assessment of environmental policies can facilitate evidence-based adaptations, promoting resilience in both economic systems and natural ecosystems (Yamin and Otto, 2004).

ETHICAL CONSIDERATIONS

The relationship between sustainable resource management, environmental conservation, and economic transformation is complex and multi-faceted (Verbeke, 2009). This intersection raises important ethical considerations that must be addressed to ensure that the principles of sustainability are effectively integrated into economic practices (Zhao, 2022). At the core of these ethical considerations is the need for equity, transparency, and respect for natural ecosystems and future generations (Caviggioli et al., 2023).

Equity and Justice: One of the major ethical imperatives is to ensure equitable access to resources and benefits derived from environmental conservation (Michailova and Zaidah, 2012). Historically, marginalized communities, including indigenous peoples and low-income populations, have often borne the brunt of environmental degradation while being excluded from meaningful participation in resource management decisions (Palomeras and Wehrheim, 2021). Therefore, ethical sustainable practices must prioritize inclusivity, ensuring that all stakeholders have a voice in decision-making processes (Cayton, 2020). This involves acknowledging the rights of marginalized groups to their traditional lands and resources, allowing them to partake in revenue-sharing arrangements and benefit from conservation initiatives (Luo, 2022).

Intergenerational Responsibility: Ethical considerations also encompass intergenerational equity (Li et al., 2024). Sustainable resource management is rooted in the belief that biodiversity and natural resources must be preserved not only for current generations but also for future ones (Scherngell et al., 2023). This principle raises urgent ethical questions about conferring responsibility what debts do present generations owe to those who will inhabit the planet in the years to come? Sustainable practices should emphasize long-term ecological health over short-term economic gains, promoting strategies that reflect a commitment to maintaining the integrity of ecosystems for future inhabitants (Lynn and Salzman, 2023).

Transparency and Accountability: The need for transparency and accountability in resource management is another critical ethical dimension (Acosta et al., 2023). When stakeholders engage in co-creating economic transformation, it is crucial to maintain open channels of communication (Lubinski and Wadhwani, 2020). Decision-makers must disclose their methodologies, impact assessments, and economic modeling to stakeholders and the public (Buckley, 2022). This transparency cultivates public trust and allows for the meaningful opposition of unethical practices, such as greenwashing where organizations falsely portray their practices as environmentally friendly (Bhaumik et al., 2024). Ethical frameworks should promote rigorous regulatory standards and third-party assessments to curb potential exploitation and misleading narratives (Ervits, 2023b).

Cultural Sensitivity and Conservation: Effective environmental conservation is often rooted in the local customs, practices, and knowledge systems of communities that have lived in harmony with their ecosystems (Zhao et al., 2020). Ethical considerations must therefore include respecting and integrating traditional ecological knowledge (TEK) into management strategies (Belitz and Lejpras, 2021). This respect fosters not only more effective management approaches but also aims to engage communities on their terms, preserving cultural heritage alongside environmental integrity (Xinhua, 2025).

Economic Accountability: Lastly, the ethical dimensions of economic transformation must involve scrutinizing who benefits from sustainable resource management initiatives (Neil Alim and Jones, 2024). The potential for profit-driven motives to overshadow genuine environmental goals is concerning, calling for a materials-based evaluation of projects to ensure alignment with sustainable objectives (Milne, 2025). Businesses must be held accountable for their ecological footprints and the socio-economic impacts of their operations, driving home the point that ethical resource management is not just a choice but a necessity for sustainable economic growth (Schmitz, 2024).

CASE STUDY

In an era marked by heightened environmental concern and rapid economic change, the intersections of sustainability, economic growth, and community well-being are more relevant than ever (Ervits, 2023a). By co-creating solutions through enterprise, narratives, and systematic design, communities can foster growth and prosperity while preserving their ecological heritage.

The Circular Economy Model in Denmark:

Denmark has emerged as a leader in implementing circular economy principles, where waste is minimized, and resources are reused, effectively reducing environmental impact while ensuring economic viability (Dachs et al., 2023). The Danish government has initiated programs that encourage companies to think sustainably through incentives for waste reduction and resource efficiency (Chen, 2022). One prominent example is the closure of the material loop in the textile industry through companies like *Graebas*, which specializes in recycling old garments into high-quality fabrics (Zhao, 2022). By establishing a local supply chain where discarded textiles are converted into new products, Graebas reduces the need for virgin materials and creates jobs within the community. Moreover, the narrative surrounding Danish sustainability has enhanced the country's global reputation, attracting eco-conscious investors and tourists alike (Caviggioli et al., 2023).

A Sustainable Fisheries Initiative in New Zealand:

New Zealand's sustainable fisheries management tells a compelling story of environmental stewardship and economic growth (Palomeras and Wehrheim, 2021). The *Quota Management System (QMS)* established in the 1980s has transformed the country's fishing industry by restricting catch sizes to ensure fish populations remain viable. Under this system, quotas are allocated to commercial fishers based on scientific assessments of fish stocks (Cayton, 2020). The narrative built around this sustainable management approach has also spurred the growth of eco-tourism initiatives that focus on responsible fishing practices (Luo, 2022). Local communities have engaged in co-creating eco-friendly tourism experiences, enabling them to share their knowledge of sustainable fishing with international visitors (Li et al., 2024). This collaboration not only preserves marine biodiversity but also enhances the livelihoods of coastal communities through sustainable tourism and fishing.

Agroecology in Brazil:

In Brazil, the adoption of agroecological practices has led to increased resilience in farming communities while addressing food security issues (Scherngell et al., 2023). The *Alliance for the Conservation of the Brazilian Cerrado* focused on combining traditional ecological knowledge with modern sustainable practices. Farmers, researchers, and non-governmental organizations co-designed strategies to rejuvenate degraded lands through crop rotation, organic fertilization, and agroforestry (Lynn and Salzman, 2023). One success story is the *Family Farm Movement*, which promotes small-scale farming that prioritizes biodiversity and sustainable land use (Acosta et al., 2023). By supporting local farmers' markets and educating communities about the importance of native crops, this initiative has revitalized rural economies and strengthened food systems while countering deforestation (Lubinski and Wadhwani, 2020). The partnership between farmers and researchers demonstrates effective co-creation, resulting in a community narrative that emphasizes local resilience and ecological well-being.

Renewable Energy Cooperatives in Germany:

Germany has taken significant strides toward sustainable energy through its *Energiewende* initiative, which emphasizes the transition to renewable energy sources (Buckley, 2022). This transition has been largely facilitated by community-owned renewable energy cooperatives, where local residents invest in wind, solar, and biomass projects. The *Energiekonzepte* cooperative is a prime example of this model, allowing citizens to not only own shares in energy production but also participate in decision-making processes regarding local energy policies (Ervits, 2023b). This co-created approach has led to a significant reduction in CO2 emissions, increased energy independence, and the creation of local jobs in the renewable sector (Bhaumik et al., 2024). The collaborative narrative emphasizes local empowerment and sustainable economic growth, making it a powerful model for other regions transitioning to greener energy.

Coastal Community Resilience in Fiji:

Fiji has embraced sustainable resource management through its integrated coastal management (ICM) initiatives, which aim to protect coastal ecosystems while enhancing community livelihoods (Belitz and Lejpras, 2021). The *Pacific Islands Forum* has promoted a co-management approach where local communities, government agencies, and NGOs collaborate on sustainable fishing practices, mangrove restoration, and coral reef protection (Zhao et al., 2020). The *Mamanuca Environ-*

ment Society serves as a model of successful collaboration, where local villagers work with scientists to monitor marine ecosystems and develop sustainable tourism ventures that capitalize on their natural beauty (Neil Alim and Jones, 2024). By weaving together local knowledge and scientific research, communities have created a narrative that highlights the importance of environmental conservation as a cornerstone of sustainable economic development (Schmitz, 2024). This successful cooperation increases community resilience to climate impacts and disaster risks while fostering economic growth through eco-tourism.

The shift toward sustainability offers not only the potential for economic rejuvenation but also the opportunity to craft narratives that resonate with local identities and values (Milne, 2025). As demonstrated by these cases, genuine transformation requires community involvement, intercultural dialogue, and long-term commitment to sustainable practices (Xinhua, 2025). The evidence supports an optimistic outlook on the potential of sustainability to drive economic growth, highlighting that a healthy environment is indeed the foundation for a thriving economy.

CONCLUSION

The imperative of integrating sustainable resource management and environmental conservation into economic strategies becomes increasingly clear in the face of mounting global challenges, including climate change, biodiversity loss, and social inequities (Hage et al., 2022). As society grapples with the profound implications of these challenges, fostering growth and prosperity through sustainable practices emerges not only as an ethical and environmental necessity but also as a pragmatic avenue for economic transformation (Rinke et al., 2022).

The crux of sustainable resource management lies in recognizing the interconnected nature of economic activities and environmental health (Arnold and Chazan, 2023). Sustainable practices promote the wise use of resources, ensuring that they are conserved and managed in a way that supports both present and future needs (Siemens, 2025). By co-creating economic transformation where businesses, governments, and communities collaborate the focus can shift toward developing regenerative models that enhance resilience and equity (Volkswagen Group China, 2025).

Furthermore, systematic design principles offer a roadmap for developing integrated strategies that engage multiple stakeholders in co-creating solutions (Luo, 2024). This systematic approach enables the identification of synergies between economic growth, resource management, and conservation efforts, leading to holistic solutions that advance sustainability goals (De Beule & Van Beveren, 2019).

However, the journey towards sustained growth and prosperity through sustainable resource management is fraught with challenges (Casella and Bolwijn,

2024). Resistance to change, entrenched economic paradigms, and the complexity of global supply networks can hinder progress (Tung et al., 2023). For effective transformation, strong leadership and a commitment to long-term planning are essential (Ervits, 2023a). Policymakers must implement regulations that support sustainability and hold organizations accountable for ecological accountability (Luo and Van Assche, 2023).

SUGGESTIONS

Fostering growth and prosperity in today's rapidly changing environment requires a multifaceted approach that integrates sustainable resource management and effective environmental conservation strategies (Turkina et al., 2025).

Integrate Sustainability into Business Models: Enterprises should adopt sustainable business practices that integrate environmental goals into their overall strategies (Belitz and Lejpras, 2021). This includes transitioning to circular economy models that minimize waste and encourage resource recycling. Companies can invest in sustainable materials, energy-efficient technologies, and low-impact production processes that not only benefit the environment but also reduce operational costs over time.

Support Local Economies through Sustainable Practices: Economic growth can be stimulated by supporting local businesses that employ sustainable practices (Papanastassiou et al. 2020). Governments can incentivize local enterprises that prioritize environmental conservation through tax breaks, grants, and training programs. This will not only create jobs but also foster community engagement and investment in local resources.

Promote Education and Awareness: Educational institutions play a crucial role in shaping future leaders and innovators (Dachs, 2017). Curricula should emphasize the importance of environmental stewardship and sustainable resource management. Workshops and seminars that highlight successful case studies in sustainability can inspire a new generation of entrepreneurs committed to economic and environmental balance.

Encourage Public-Private Partnerships: Collaborative efforts between government entities and private enterprises can lead to innovative solutions for sustainable development (Altman et al., 2024). These partnerships can fund research and development initiatives that focus on emerging technologies in renewable energy, waste management, and sustainable agriculture. By pooling resources and expertise, both sectors can drive impactful projects that benefit the community and the environment simultaneously.

Implement Strong Regulatory Frameworks: Robust regulatory policies are essential for ensuring that enterprises adhere to sustainable practices (Martin, 2024). Governments should establish clear guidelines for resource extraction, waste management, and pollution control. Additionally, compliance monitoring systems need to be developed to ensure that organizations remain accountable for their environmental impacts.

Promote Sustainable Tourism: The tourism industry presents an opportunity for sustainable economic growth (Tung et al., 2023). By promoting eco-tourism and responsible travel practices, destinations can attract visitors while conserving natural habitats. Investments in sustainable infrastructure, such as energy-efficient accommodations and local cultural experiences, can enhance economic benefits and protect the environment.

FUTURE RESEARCH DIRECTIONS

Future research in the realm of sustainable resource management and environmental conservation must embrace innovative approaches that bridge economic and ecological systems.

Behavioral Economics and Sustainability: Understanding the psychological and behavioral factors that influence consumer choices about sustainable products is vital (Dachs et al., 2023). Research should explore how messaging, social norms, and incentives can be structured to encourage more sustainable consumption patterns, especially in the context of emerging markets.

Impact of Technology on Resource Management: Investigating the implications of artificial intelligence, blockchain, and big data analytics in resource management can yield insights into optimizing supply chains, minimizing waste, and improving transparency (Jallat and Coronati, 2022). Future studies should assess how advanced technologies can foster sustainable practices across various industries.

Longitudinal Studies of Community Resilience: Investigating the long-term effects of community-led sustainability initiatives can provide valuable insights into best practices and strategies that enhance resilience in the face of environmental changes (Hage et al., 2022). Longitudinal studies can track the success and challenges of such initiatives over time, informing policy and practice.

Integration of Indigenous Knowledge: Research should evaluate the effectiveness of incorporating indigenous knowledge systems into contemporary resource management (Kraemer, 2023). Indigenous practices often demonstrate a deep understanding of local ecosystems and sustainability principles, which can inform more effective conservation strategies.

Systemic Design Approaches: Future research should explore frameworks that integrate systemic design principles to create holistic solutions addressing social, economic, and environmental challenges (Schmitz, 2024). This includes developing interdisciplinary models that can solve complex sustainability issues through innovative design thinking.

Climate Change Resilience Strategies: Studies focused on adaptive strategies for businesses and communities facing climate change are critical (Xinhua, 2023). Research should investigate how various sectors can enhance their resilience through sustainable practices, from agriculture to urban planning.

Global Collaboration and Policy Alignment: Research should focus on fostering international cooperation in sustainable resource management and environmental conservation (Godley et al., 2025). Collaborative frameworks and policy alignments between nations can lead to more effective global solutions to pressing environmental challenges. By concentrating on these suggestions and future research directions, stakeholders can co-create pathways for sustainable economic transformation, contributing to both community prosperity and environmental health.

REFERENCES

Acosta, M., Coronado, D., & Medina, J. (2023). Effects of co-patenting across national boundaries on patent quality. An exploration in pharmaceuticals. *Economics of Innovation and New Technology*. Advance online publication. DOI: 10.1080/10438599.2023.2167201

Agostini, L., & Caviggioli, F. (2015). R & D collaboration in the automotive innovation environment: An analysis of co-patenting activities. *Management Decision, 53*(6), 1224–1246.

Alcacer, J., Cantwell, J., & Piscitello, L. (2016). Internationalization in the information age: A new era for places, firms, and international business networks? *Journal of International Business Studies, 47*(5), 499–512.

Altman, S. A., Bastian, C. R., & Fattedad, D. (2024). Challenging the deglobalization narrative: Global flows have remained resilient through successive shocks. *Journal of International Business Policy, 7*, 416–439.

Arnold, M., & Chazan, G. (2023). German companies must cut exposure to China, warns Bundesbank. *Financial Times*, September 18, 2023, available at: https://www.ft.com/content/4b7ad09a-b05c-4651-b6ff-43a448a9bd76

Belderbos, R., Cassiman, B., Faems, D., Leten, B., & Van Looy, B. (2014). Co-ownership of intellectual property: Explaining the value-appropriation and value-creation of the copatenting with different partners. *Research Policy, 43*(5), 841–852.

Belderbos, R., Leten, B., & Suzuki, S. (2013). How global is R&D? Firm-level determinants of home-country bias in R&D. *Journal of International Business Studies, 44*(8), 765–786.

Belitz, H. (2017). Companies with R&D abroad make Germany a strong research location. *DIW Economic Bulletin, 7*, 477–487.

Belitz, H., & Lejpras, A. (2021). *R&D internationalization strategies of the world's top corporate R&D investors. DIW Discussion Papers, No. 1942, Deutsches Institut für Wirtschaftsforschung*. DIW.

Bergek, A., & Bruzelius, M. (2010). Are patents with multiple inventors from different countries a good indicator of international R&D collaboration? The case of ABB. *Research Policy, 39*(10), 1321–1334.

Berry, H. (2013). Global integration and innovation: Multi-country knowledge generation within MNCs. *Strategic Management Journal, 35*(69), 869–890.

Bhaumik, S. K., Estrin, S., & Narula, R. (2024). Integrating host-country political heterogeneity into MNE–state bargaining: Insights from international political economy. *Journal of International Business Studies, 55*, 157–171.

Branstetter, L., Li, G., & Veloso, F. (2013). The polyglot patent boom. *Scientific American, 309*(4), 62–63. PMID: 24137857

Buckley, P. J. (2022). Navigating three vectors of power: Global strategy in a world of intense competition, aggressive nation states, and antagonistic civil society. *Global Strategy Journal, 12*(3), 543–554.

Buckley, P. J., & Casson, M. C. (1976). *The Future of the Multinational Enterprise*. Macmillan.

Cantwell, J., & Mudambi, R. (2000). The location of MNE R&D activity: The role of investment incentives. *Management International Review, 40*(1), 127–148.

Cantwell, J., & Piscitello, L. (2014). Historical changes in the determinants of competence creation in MNC subunits: The increasing role of international knowledge. *Industrial and Corporate Change, 23*(3), 633–660.

Cantwell, J. A., & Mudambi, R. (2011). Physical attraction and the geography of knowledge sourcing in multinational enterprises. *Global Strategy Journal, 1*(3–4), 206–232.

Casella, B., & Bolwijn, R. (2024). Global economic fracturing and shifting investment patterns - A diagnostic of 10 FDI trends and their development implications (UNCTAD/DIAE/2024/1), 23 April, available at: https://unctad.org/publication/global-economic-fracturing-and-shifting-investment-patterns

Castellani, D., & Zanfei, A. (2002). Multinational experience and linkages with local firms. Evidence from the electronics industry. *Cambridge Journal of Economics, 26*(1), 1–25.

Caviggioli, F., Colombelli, A., De Marco, A., Scellato, G., & Ughetto, E. (2023). Co-evolution patterns of university patenting and technological specialization in European regions. *The Journal of Technology Transfer, 48*, 216–239.

Cayton, S. (2020). The "Green Patent Paradox" and fair use: The intellectual property solution to fight climate change. *Seattle Journal of Technology, Environmental & Innovation Law, 11*(1), 8. https://digitalcommons.law.seattleu.edu/sjteil/vol11/iss1/8

Chen, J. (2022). DAX Stock Index: Definition and member companies. *Investopedia,* April 5, available at https://www.investopedia.com/terms/d/dax.asp#:~:text=As%20of%20Jan.%2025%2C%202022%2C%20four%20months%20after,AG%20%28BMW%3AGR%29%208%20Brenntag%20SE%20%28BNR%3AGR%29%2A%20Weitere%20Elemente (accessed 1 March 2025).

Chen, Y. C., & Vang, J. (2008). MNCs, global innovation networks and developing countries: Insights from Motorola in China. *International Journal of Business and Management Science*, *1*(1), 11–30.

ChinaScope. (2024). Lianhe Zaobao: More German companies considering withdrawal from Chinese market, January 31, available at: https://chinascope.org/archives/34223#:~:text=Top%20concerns%20cited%20by%20German%20companies%20include%20increased,with%20a%20total%20of%20566%20member%20companies%20polled (accessed 2 April 2025).

Criscuolo, P., Narula, R., & Verspagen, B. (2005). Role of home and host country innovation systems in R&D internationalisation: A patent citation analysis. *Economics of Innovation and New Technology*, *14*(5), 417–433.

Cuervo-Cazurra, A., & Genc, M. (2011). Obligating, pressuring, and supporting dimensions of the environment and the non-market advantages of developing-country multinational companies. *Journal of Management Studies*, *48*(2), 441–455.

Dachs, B. (2017). Internationalisation of R&D: A review of drivers, impacts and new lines of research. *Asian Research Policy*, *8*(2), 17–39.

Dachs, B., Amoroso, S., Castellani, D., Papanastassiou, M., & von Zedtwitz, M. (2023). The internationalisation of R&D: Past, present and future. *International Business Review*, *33*(1), 102191.

De Beule, F., & Van Beveren, I. (2019). Sources of open innovation in foreign subsidiaries: An enriched typology. *International Business Review*, *28*(1), 135–147.

Demirbag, M., & Glaister, K. W. (2010). Factors determining offshore location choice for R&D projects: A comparative study of developed and emerging regions. *Journal of Management Studies*, *47*(8), 1534–1560.

Dunning, J. H., & Lundan, S. M. (2008). *Multinational Enterprises and the Global Economy*. Edward Elgar.

Ervits, I. (2018). Geography of corporate innovation: Internationalization of innovative activities by MNEs from developed and emerging markets. *Multinational Business Review*, *26*(1), 25–49.

Ervits, I. (2021). R&D by US multinationals in China: Geography and patterns of co-invention. *International Journal of Emerging Markets, 16*(8), 1525–1555.

Ervits, I. (2023a). Co-patenting and technological value of patent applications: The case of Siemens. *International Journal of Innovation and Technology Management, 21*(2), 2450009.

Ervits, I. (2023b). The effect of co-patenting as a form of knowledge meta-integration on technological differentiation at Siemens. *European Journal of Innovation Management, 27*(8), 2575–2596.

Fukuyama, F. (1992). *The End of History and the Last Man*. The Free Press.

Godley, A., Buckley, T. & Joseph, M. (2025). Techno-nationalism and capability development in the global pharmaceuticals industry, 1918–1970. *Journal of International Business Policy,* doi: .DOI: 10.1057/s42214-025-00210-0

Griliches, Z. (1990). Patent statistics as economic indicators: A survey. *NBER Working Papers,* Vol. 3301, National Bureau of Economic Research, Inc.

Hage, S., Hesse, M., Sauga, M., Müller-Arnold, B., & Giesen, C. (2022). German Companies Ignore major risks in China. *Spiegel International*, January 11, available at: https://www.spiegel.de/international/business/business-as-usual-german-companies-ignore-major-risks-in-china-a-27b66d20-f228-47d0-b2bf-95f3cc976e6c (accessed 2 April 2025).

Hazir, C. S., LeSage, J., & Autant, B. C. (2018). The role of R&D collaboration networks on regional knowledge creation: Evidence from information and communication technologies. *Papers in Regional Science, 97*(3), 549–567.

Jallat, F., & Coronati, L. (2022). Why are leaders not taking political risks seriously enough despite a turbulent international business environment and geopolitical instability? *ESCP Impact Paper,* No. 2022-09-EN, available at: https://academ.escpeurope.eu/pub/IP%20N%C2%B02022-09-EN.pdf

KPMG. (2024). Economic Key Facts Germany, available at: https://kpmg.com/de/en/home/insights/overview/economic-key-facts-germany.html (accessed 1 November 2024).

Kraemer, C. (2023). German firms keep investing in China despite moves to 'de-risk´. *Reuters,* May 10, available at: https://www.reuters.com/business/german-firms-keep-investing-china-despite-moves-de-risk-2023-05-10/ (accessed April 3, 2025).

Kramer, J. P., Revilla Diez, J., Marinelli, E., & Iammarino, S. (2010). Intangible assets and MNEs' locational strategies for innovation—or: Why the *regional* matters. [Review of Regional Research]. *Jahrbuch für Regionalwissenschaft*, *30*, 129–157.

Kuemmerle, W. (1999). The drivers of foreign direct investment into research and development: And empirical investigation. *Journal of International Business Studies*, *30*(1), 1–24.

Le Bas, C., & Sierra, C. (2002). Location versus country advantages' in R&D activities: Some further results on multinationals' locational strategies. *Research Policy*, *31*(4), 589–609.

Li, J., Shapiro, D. M., Ufimtseva, A., & Zhang, P. (2024). Techno-nationalism and cross-border acquisitions in an age of geopolitical rivalry. *Journal of International Business Studies*, *55*, 1190–1203.

Liu, Q. (2024). China on cusp of next-generation chip production despite US curbs. *The Financial Times*, February 6, available at: https://www.ft.com/content/b5e0dba3-689f-4d0e-88f6-673ff4452977 (accessed 16 March 2025).

Lubinski, C., & Wadhwani, R. D. (2020). Geopolitical jockeying: Economic nationalism and multinational strategy in historical perspective. *Strategic Management Journal*, *41*, 400–421.

Luo, Y. (2022). Illusions of techno-nationalism. *Journal of International Business Studies*, *53*, 550–567. PMID: 34511654

Luo, Y. (2024). Paradigm shift and theoretical implications for the era of global disorder. *Journal of International Business Studies*, *55*, 127–135.

Luo, Y., & Van Assche, A. (2023). The rise of techno-geopolitical uncertainty: Implications of the United States CHIPS and Science Act. *Journal of International Business Studies*, *54*, 1423–1440. PMID: 37359753

Lynn, L., & Salzman, H. (2023). Techno-nationalism or building a global science and technology commons? (but what about China?). *Global Policy*, *14*, 832–846.

Martin, N. (2024). German firms ignore calls to shrink China investments. *Deutsche Welle*. August 26, available at: https://www.dw.com/en/germanys-investments-in-china-increase-despite-warnings/a-69957810 (accessed 2 March 2025).

Meyer, K. E., Mudambi, R., & Narula, R. (2011). Multinational enterprises and local contexts: The opportunities and challenges of multiple embeddedness. *Journal of Management Studies*, *48*(2), 235–252.

Michailova, S., & Zaidah, M. (2012). Subsidiary knowledge flows in multinational corporations: Research accomplishments, gaps, and opportunities. *Journal of World Business*, *47*(3), 383–396.

Milne, R. (2025). German companies are risking a 'lose-lose' decoupling with China. *Financial Times*, January 30, available at https://www.ft.com/content/f61d6b7b-0250-4edf-aac1-ba0ca18545cd (accessed 1 April 2025).

Narula, R. (2014). Exploring the paradox of competence-creating subsidiaries: Balancing bandwidth and dispersion in MNEs. *International Journal of Strategic Management*, *47*(1–2), 4–15.

Neil Alim, A., & Jones, S. (2024). German companies' dependence on China will last decades, warns Siemens. *Financial Times*, April 14, available at: https://www.ft.com/content/e8634f5c-5a28-4741-a13d-024665456f43?sharetype=blocked (accessed 2 April 2025).

Palomeras, N., & Wehrheim, D. (2021). The strategic allocation of inventors to R&D collaborations. *Strategic Management Journal*, *42*, 144–169.

Papanastassiou, M., Pearce, R., & Zanfei, A. (2020). Changing perspectives on the internationalization of R&D and innovation by multinational enterprises: A review of the literature. *Journal of International Business Studies*, *51*, 623–664.

Prabhu, J., & Jain, S. (2015). Innovation and entrepreneurship in India: Understanding jugaad. *Asia Pacific Journal of Management*, *32*(4), 843–868.

Quinn, G. (2011). PCT Basics: Understanding the international filing process, available at: https://www.ipwatchdog.com/2011/11/03/pct-basics-understanding-the-international-¯ling-process/id=19960

Ribakova, E. (2024). Our experience with Russia holds lessons for future sanctions. *The Financial Times*, February 27, available at: https://www.ft.com/content/784eae1d-1fb8-455b-b3d3-2ddc47155499d (accessed 16 March 2025).

Rinke, A., Waldersee, V., & Marsh, S. (2022). German business chiefs clash with Berlin over China policies. *Reuters*, October 14, available at: https://www.euronews.com/next/2022/10/14/germany-china-insight (accessed 15 April 2024).

Rugman, A. M., & Verbeke, A. (2001). Subsidiary-specific advantages in multinational enterprises. *Strategic Management Journal*, *22*(3), 237–250.

Scherngell, T., Schwegmann, K., and Zahradnik, G. (2023). The geographical dynamics of global R&D collaboration networks in robotics: Evidence from co-patenting activities across urban areas worldwide. *PLoS One*, Vol. 18 No. 4, doi: . pone.0281353.DOI: 10.1371/journal

Schmitz, R. (2024). German government wants companies to 'de-risk' from China, but business is reluctant. National Public Radio, January 19, available at: https://www.kpbs.org/news/international/2024/01/19/german-government-wants-companies-to-de-risk-from-china-but-business-is-reluctant (accessed 1 November 2024).

Schumpeter, J. A. (1942). *Capitalism, Socialism and Democracy*. Harper & Row.

Shimizutani, S., & Todo, Y. (2008). What determines overseas R&D activities? The case of Japanese multinational firms. *Research Policy*, *37*(3), 530–544.

Siemens (2019). Siemens catalyzes digital transformation of China with innovation ecosystem, May 15, available at: https://w1.siemens.com.cn/press/NewsDetail_en.aspx?ColumnId=9&ArticleId=8014 (accessed 1 November, 2024).

Siemens (2021). Press Releases. Siemens to establish electrical products China headquarters in Suzhou, China, June 21, available at: https://w1.siemens.com.cn/news_en/news_articles_en/17472.aspx (accessed 20 April 2025).

Siemens (2024a). Siemens R&D in China. Innovation for China's development, available at: https://www.siemens.com/cn/en/company/about/research.html (accessed 30 October 2024).

Siemens (2024b). Across borders. Siemens and China – Milestones in a successful partnership, available at: https://www.siemens.com/global/en/company/about/history/stories/siemens-in-china.html (accessed 1 November 2024).

Siemens (2025). Über Grenzen hinweg. Siemens und China – Meilensteine einer erfolgreichen Partnerschaft (Across borders. Siemens and China – Milestones of a Successful Partnership), available at: https://www.siemens.com/de/de/unternehmen/konzern/geschichte/stories/siemens-in-china.html (accessed 6 April 2025).

Simon, J. A. (1984). Theoretical perspective on political risk. *Journal of International Business Studies*, *15*, 123–143.

Stoneman, P. (1995). *Handbook of the Economics of Innovation and Technological Change*. Wiley-Blackwell.

Tsukada, N., & Nagaoka, S. (2015). International Research Collaboration. *Asian Economic Policy Review*, *10*, 96–119.

Tung, R. L., Zander, I., & Fang, T. (2023). The Tech Cold War, the multipolarization of the world economy, and IB research. *International Business Review, 32*(6), 102195.

Turkina, E., Van Assche, A., & Mudambi, R. (2025). Cross-border innovation for global value chain orchestration. *International Business Review, 102450.* Advance online publication. DOI: 10.1016/j.ibusrev.2025.102450

Tversky, A., & Kahneman, D. (1974). Judgment under uncertainty: Heuristics and biases. *Science, 185*(4157), 1124–1131. PMID: 17835457

Verbeke, A. (2009). *International Business Strategy*. Cambridge University Press.

Volkswagen Group China. (2025). Chronology", available at: https://volkswagengroupchina.com.cn/en/volkswagengroupchina/chronology (accessed 6 April 2025).

Xinhua (2023). Siemens expands China business. June 16. 2023, available at: https://global.chinadaily.com.cn/a/202306/16/WS648bee18a31033ad3f7bcac3.html (accessed 29 October 2024).

Xinhua (2025). German enterprises double down on China investment amid growing global uncertainties. *China Daily*, March 28, available at: https://www.chinadaily.com.cn/a/202503/28/WS67e6102ea3101d4e4dc2b69e.html (accessed 2 April 2025).

Yamin, M., & Otto, J. (2004). Patterns of knowledge flows and MNE innovative performance. *Journal of International Management, 10*(2), 239–258.

Zanfei, A. (2000). Transnational firms and the changing organisation of innovative activities. *Cambridge Journal of Economics, 24*(5), 515–542.

Zhao, L. (2022). On the grant rate of Patent Cooperation Treaty applications: Theory and evidence. *Economic Modelling, 117*, 106051.

Zhao, S., Tan, H., & Papanastassiou, M.. (2020). The internationalization of innovation towards the South: A historical case study of a global pharmaceutical corporation in China (1993–2017). *Asia Pacific Journal of Management, 37*, 553–585.

Chapter 4
Creative Destruction and Moral Imagination: Rethinking Market Evolution in the Age of Impact

Christos Christodoulou-Volos
https://orcid.org/0000-0002-3188-844X
Neapolis University Pafos, Cyprus

ABSTRACT

This chapter examines the way economists and entrepreneurs are moral architects, not only designing markets but also the values and meaning they carry. It critiques traditional economic models, which reduce value to profit, and looks at the way purpose, dignity, and sustainability are becoming core elements of new economic frameworks. The chapter traces the historical evolution of value regimes and shows how entrepreneurial innovation and economic thought can redesign institutions and social contracts. The chapter also raises doubts about the risk of co-optation, whereby market reforms end up enlarging rather than transforming dominant extractive arrangements. Concluding with the theme of "designing for the Long Now," the chapter encourages intergenerational responsibility and moral imagination in economic future-making. Integrating economics, the theory of systems, and ethics, it offers a visionary platform for building effective markets that are also just, real, and sustainable.

1. INTRODUCTION: BEYOND THE MACHINE

Joseph Schumpeter infamously applied the term creative destruction in 1942 to describe the restless genius of capitalism in reinvention. Markets, he insisted, are re-

DOI: 10.4018/979-8-3373-4566-6.ch004

shaped not smoothly but by violent tearing out of the old by the new—the entrepreneur as a "perennial gal" blowing past inefficiency and stodginess (Schumpeter, 1942). It wasn't a process of inventing or iterating; it was transformation by destruction. New technologies, business strategies, and modes of organizing capital and labor displaced their predecessors in a Darwinian cycle. This was no flaw, according to Schumpeter, but rather the logic required for capitalist dynamics.

Throughout most of the 20th century, this idea came to be at the center of economists' and business theorists' conception of economic progress. The machine-in-the-head metaphor of the economy—sufficiently complex, and dynamic, but ultimately subject to internal forces of incentive, innovation, and equilibrium—dominated both scholarly theory and public policy. In this mechanistic frame, entrepreneurs were agents of efficiency, markets, though imperfect, were self-correcting to a significant degree, and value was mostly in the language of productivity and profit.

But as we approach the third decade of the 21st century, this metaphor starts to groan. While technological innovation continues at a breakneck pace, the collateral damage of growth—climate collapse, inequality, social dissolution—has grown too large to be ignored. Entrepreneurs are no longer simply perceived as disruptors of yesterday's incumbents but as designers of new institutional structures, new norms, and increasingly, new moral frameworks. Meanwhile, economists are beginning to doubt not just how markets work, but also for what. Value, once a circumstantial monetary concept, is being redefined to encompass sustainability, equity, and long-term social well-being.

This shift reflects back on a more nuanced reconsideration of what kind of system we are building—and for whom. The lexicon of impact, intention, and systems transformation has flowed over from the periphery of social enterprise to the language of mainstream business schools, venture capital, and economic policymaking. Where Schumpeter's entrepreneur wrecked things for the sake of advance, his contemporary visionary founder is no less likely to invoke the planet, racial justice, or between-the-generations fairness as well as profit margin or scale. The new entrepreneurial frontier is no longer technological, but moral.

1.1 The Moral Turn in Market Thinking

This moral pivot is more than rhetoric. It is an architectural one. Entrepreneurs are increasingly embedding moral thought in the architecture of markets and institutions themselves—through regulatory structures (such as benefit corporations), governance models (such as cooperative ownership), and financial vehicles (such as impact investing and green bonds). Rather than treating ethics as a constraint on economic behavior, many emerging entrepreneurs treat ethics as a source of innovation.

Consider Yvon Chouinard, founder of Patagonia, who recently transferred ownership of the company to a trust and nonprofit, ensuring all profits would be used to combat climate change. Or Rose Marcario, former CEO of the same company, who stated that "we're in business to save our home planet"—a mission that blurs the line between enterprise and activism. These are not isolated activities; they are signs of an increasing redefinition of the role of the entrepreneur from money maker to moral builder.

Economic theory is another area where this change is expressed. While more recent schools of thought, such as behavioral economics, ecological economics, and well-being economics, are aware that economic agents are enmeshed in social, cultural, and ethical milieux, classic models of homo economicus capture rational beings maximizing utility given limitations. As opposed to value-neutral institutions, economists such as Amartya Sen (1999), Kate Raworth (2017), and Mariana Mazzucato (2018) define markets as disputed environments where values are expressed, recorded, and debated.

Sen, for example, argued that progress should be measured not only by growth in GDP but also by the broadening of human capabilities and freedoms. Raworth's "doughnut economics" approach makes environmental constraints and social foundations central to its framework, arguing that the sort of prosperity that is desired is one that combines core human needs with planetary boundaries. Mazzucato's study refutes the theory that value is only generated within the private sector and instead demands a mission-driven, interventionist state that co-forms markets rather than curing their shortcomings.

1.2 Entrepreneurship as Cultural Production

Here, being an entrepreneur isn't necessarily an economic activity—it's a form of cultural production that establishes norms, redistributes expectations, and crafts narratives that redefine society's definition of value. Startups aren't just taking things to market; they're redefining norms, changing expectations, and writing narratives that redefine society's definition of value. Uber didn't just create a ride-sharing application; it rehistoriated the social bargain between capital and labor. Tesla not only produced electric cars, but it also re-staged climate awareness as a high-end consumer choice. Airbnb did not merely make home-sharing easy; it upended our notions of space, ownership, and hospitality.

This narrative power is one aspect of what sociologist Viviana Zelizer (1994) has called "the social meaning of money." Markets are not simply spaces of exchange—spaces of exchange are embedded within moral judgments and social relations. The issue is not whether economists and entrepreneurs shape markets and meaning, but rather how and for what ends.

In a post-neoliberal era of increasing populism, green care, and post-growth awareness, the legitimacy of the market itself is increasingly on the table. An instrumental concept of entrepreneurship—disrupting per se—is not enough against the planetary crisis and social breakdown. What is in their stead emerging is a greater longing for economic actors who can move with moral imagination—a term borrowed from philosopher John Paul Lederach (2005), which refers to the capacity to envision and construct relationships, institutions, and systems that are life-sustaining and just.

1.3 Toward a New Economic Architecture

If creative destruction defined the 20th century, then the 21st century challenge is less to produce, and more to curate—to design economic systems with some anticipation of their long-term consequences, interconnectedness, and ethical dimensions. This is a makeover rather than a rejection of markets. It poses the question: What if markets were designed for resilience as well as efficiency? Not merely for innovation, but for inclusion? Not merely for consumption, but for care?

The stakes are high. According to the IPCC, if global warming is to be kept at 1.5°C, the world must cut emissions by half by the early 2030s (2023). In contrast, the wealthiest 1% of people worldwide own more wealth than the entire globe combined, and this disparity is only growing (Credit Suisse, 2022). These crises are moral, not just technological. They need more than better technology or more efficient algorithms; they need a different vision of what progress is.

Entrepreneurs, with their capacity for vision and risk-taking, are well-positioned to lead this change. However, their actions are not isolated. Economists, also, must rethink their role—not as value-neutral technicians, but as co-creators of meaning, norms, and legitimacy. The economist's task is no longer simply to model what is, but to illuminate what could be, and to assist in the design of the institutional conditions under which more human, sustainable, and creative markets could emerge.

1.4 Conclusion: A New Role for Economic Actors

Globally, the machine metaphor for the economy is no longer adequate to meet the changes ahead. Economic players in today's world are not simply engineers of efficiency; they are architects of value, of morality, and of meaning. We must go beyond traditional categories and measures. We must ask not only how markets allocate resources, but how they build identities, relations, and futures.

This chapter will look at how economics and entrepreneurship are coming together to create a more comprehensive, morally aware paradigm. Only if it now coexists with moral imagination can it contend that creative devastation is still legitimate. It will show that markets are not merely growth drivers, but playhouses where society negotiates what it is to live well, together.

The transformational architecture is structural, yes, but also symbolic, ethical, and cultural. And we are all, to some degree, its builders.

2. THE EVOLUTION OF VALUE: FROM PROFIT TO PURPOSE

For the bulk of modern economic history, "value" has been equated with material production and price in the market. It originated in classical economics, in which value was conceived primarily in terms of labor (as with Adam Smith and David Ricardo) or marginal utility (as formalized in neoclassical theory). Value was approximated in both meanings by its ability to generate profit, either through efficiency in production or fulfillment of personal preference. This construct had survived through the industrial age, advancing capitalist growth and commercial schooling in alignment. Yet, as the 21st century dawns amid ecological collapse, accelerating inequality, and social unrest, this value concept rooted in profit is being transformed anew.

A new generation of policymakers, thinkers, and leaders are redefining value not just as a financial metric but as a complex, dynamic concept with purpose. Here, we examine how the definition of value is shifting along three fundamental axes: historically, philosophically, and institutionally. We then consider how this shift is playing out in practice, with special reference to emerging business models and theory that put purpose on or at least alongside profit.

2.1 Historical Shifts in Value Conception

Historically, transformations in the conceptualization of value have tracked broader transformations in production, politics, and culture. For pre-industrial times, value was perhaps irretrievable from moral and theological explanations. Medieval scholastics, for instance, distinguished "just price" from market price and placed exchange in a larger ethical and communal framework (Kaye, 1998). Individualism and market value became the main topics of discussion as capitalism emerged in the 18th and 19th centuries. Self-interest in competitive markets would produce socially beneficial results, according to Adam Smith's "invisible hand" theory, which progressively distanced virtue from value.

The 20th century solidified this estrangement. Formal, mathematical precision introduced by the dominance of neoclassical economics limited the definition of value to utility maximization and market choice. Milton Friedman's (1970) aphorism that a business's sole social responsibility is to earn more profit became orthodoxy in economic theory and business practice. Social and environmental problems were seen as externalities—important indeed, but peripheral to market activity proper.

But crises tend to disrupt orthodoxy. The crash of the banks in 2008, and then the climate crisis, and the pandemic of COVID-19 exposed the shortcomings of profit-oriented models to address systemic risk and human well-being. They indicated that they spurred a growing recognition that value must be redefined to include long-term social and environmental outcomes.

2.2 Philosophical and Ethical Reconsiderations

Along with historical transformation, philosophical debate has also expanded the vision of value. Amartya Sen's (1999) capabilities approach, for example, characterized development not in terms of economic growth in incomes but as an expansion of basic human freedoms and capabilities. This was quite a divergence from utility-based economics in that it focuses on what people can do and be, a larger and wider concept of human well-being.

Similarly, feminist economists, such as Nancy Folbre (2001), have challenged mainstream conceptions of value by recognizing unpaid labor, often performed by women, which is essential to sustaining families and communities but is not captured within the GDP or within market transactions. Their work works to bring out the reality that value is not simply created within markets but also within households, social relationships, and ecosystems.

Ecological economists have also advocated for a more comprehensive approach. Herman Daly (1996) promoted a steady-state economy, one that is within planetary limits, challenging the assumption that growth and value creation are equivalent. Kate Raworth (2017), in her Doughnut Economics framework, imagines value to be within a "safe and just space for humanity"—defined by social foundations on the one hand and ecological ceilings on the other. In her design, human and planetary health should be the agenda of economies, not the other way around.

Together, these frames offer a robust philosophical foundation for reframing value as relational, multi-faceted, and morally located. Rather than reducing value to price, they have us see value as the outcome of life-enriching systems that promote dignity and sustainability.

2.3 Institutional Reconfigurations of Value

This philosophical change is beginning to take hold in institutional change. New organizational forms, such as Benefit Corporations (B Corps), Low-Profit Limited Liability Companies (L3Cs), and cooperatives, place purpose in the charter, binding companies legally to pursue social or environmental objectives as well as financial ones.

One of the best examples of this tendency is the B Corp movement. Certified by the non-profit corporation B Lab, B Corps must meet high levels of environmental and social performance, accountability, and transparency. Some among these are Patagonia, Ben & Jerry's, and Danone North America, which have adopted this approach so openly, embracing causes beyond shareholder value. To date, over 6,000 businesses in over 80 countries have been certified as B Corps (B Lab, 2023).

At the same time, institutional investors increasingly are applying ESG considerations to investment choices. Even as open to greenwashing and scandal, ESG investing also reflects increasing acknowledgment that long-term investment performance cannot be delinked from environmental and ethical concerns. In his 2020 letter to CEOs, BlackRock CEO Larry Fink asserted that climate risk is investment risk—a claim unimaginable from a big asset manager a decade ago.

In the public domain, governments are also experimenting with non-GDP measures. New Zealand, for example, introduced in 2019 a Wellbeing Budget that prioritizes mental wellbeing, the welfare of children, and the protection of the environment over traditional economic growth. The OECD Better Life Index and Bhutan's Gross National Happiness are further examples of attempts to align public policy to less constrained concepts of value.

2.4 Practice and Entrepreneurship: Value in Action

Entrepreneurs are taking it upon themselves to make this broader definition of value operational. In industries across the board, we are seeing startup businesses and social entrepreneurship organizations redefining success by the problems they're solving, not the dollars they're making.

Consider the rise of regenerative agricultural businesses that focus on building local food systems, cleaning up soils, and storing carbon. Businesses such as Indigo Ag and AppHarvest are developing business models that include ecological restoration into the development of core value. In a similar vein, digital infrastructure is being reorganized in the IT sector using the concepts of open-source governance, transparency, and collective control via platforms such as Sharetribe and Open Collective.

Social businesses such as TOMS Shoes (formerly famed for its "One for One" model) or Warby Parker demonstrate aligning consumer behavior and social impact, though models in these businesses, too, have come under scrutiny regarding the seriousness and authenticity of such commitments. The core question remains: is purpose an instrumental function to make profit, or is it inherent in the way the firm defines success and measures it?

Increasing numbers of companies are attempting to answer this question with the use of integrated reporting systems like the Triple Bottom Line (planet, people, profit) or the Integrated Reporting Framework, which includes both financial and non-financial metrics. These are not yet perfect solutions, but they represent earnest attempts to redefine performance metrics in terms of a more holistic concept of value.

2.5 Critiques and Cautions

Such social businesses as TOMS Shoes (formerly famous for its "One for One" strategy) or Warby Parker show how consumer behavior can be brought into line with social ends, although these strategies have also been criticized on the depth of commitment and genuineness. The question is: "Is purpose instrumental to profit, or is it central to the way the firm thinks about and measures success?"

More and more businesses are attempting to answer this question by adopting integrated reporting frameworks like the Triple Bottom Line (planet, people, profit) or the Integrated Reporting Framework, which integrates both financial and non-financial metrics. These attempts are not very perfect, but they represent real efforts at re-synchronizing measures of performance with a broader definition of value.

2.6 Conclusion: Toward a Plural Economy of Value

We are in an age of profound economic re-making. As the frontiers of capitalism shift, so does our notion of what counts. No longer bound by price signals or shareholder returns, value is being remade in social, ecological, and intergenerational terms. This is not a semantic change; it is a change of design.

Entrepreneurs are among the most dynamic agents of this transformation. With new business models, legal forms, and measures, they are converting intangible values into institutional facts. Even economists themselves are again embracing problems of ethics, meaning, and purpose, challenging the amoralism traditionally associated with the field.

In this new world, the concept of value is a site of contestation, imagination, and negotiation. It is in this messy, plural, contested space of valuation that transformation is designed. And it is in this space that economists and entrepreneurs alike must be not only analysts or architects but also guardians of common moral and material fate.

3. ENTREPRENEURS AS MORAL ARCHITECTS

In the classical economic orthodoxy, the entrepreneur is one who is an agent of innovation, the one to identify inefficiency, capitalize on opportunity, and bring growth through risk-taking. Joseph Schumpeter defined them as the drivers of creative destruction, emphasizing their disruptor and transmogrifier potential in a capitalist economy. However, this stereotype is changing presently. A new generation of entrepreneurs is developing, one that is not only market players but also moral designers who are creating institutions, narratives, and business models that instill ethical purpose into the very fabric of trade.

This is a symptom of a wider shift in the presumed aim of enterprise itself. Business is no longer considered ethically neutral, with responsibility outsourced to civil society or governments. Rather, founders these days introduce themselves as stewards of social and environmental systems, with the purpose of building organizations that embody and scale universal human values. The startup pitch deck no longer simply flaunts scalability and market share—it now includes carbon footprint statistics, DEI goals, or the vow of social regeneration. These are not tokenistic moves. Most commonly, they indicate a nascent desire to rethink what a company is for.

3.1 Redesigning Enterprise: Legal and Institutional Innovation

One central starting point to comprehend entrepreneurs as moral architects is to examine how they are using (and building) institutional arrangements to hardwire purpose into their companies. Perhaps the most symbolically significant innovation is the emergence of the benefit corporation (or B Corp), a legal form that allows firms to seek social and environmental goals as well as profit and to be held accountable to them.

Originally formulated in the United States in the early 2010s and then adopted in over 40 jurisdictions globally, benefit corporation law enables companies to move beyond the traditional shareholder primacy framework. Instead of being contractually bound exclusively by fiduciary duties to maximize money return, B Corp directors are statutorily committed to taking account of the impacts of their actions upon employees, communities, customers, and the environment (B Lab, 2024). This transformation is not symbolic in nature; it reshapes the firm's logic of governance.

The development of B Lab, the not-for-profit certification body for B Corps, also makes this change more realistic. By 2024, there will be more than 7,000 B Corp accredited businesses worldwide, including well-known companies like Danone North America, Patagonia, Ben & Jerry's, and Allbirds. All these companies pledge towards rigorous standards of accountability and performance on environmental, social, and governance (ESG) indicators. Patagonia, for example, famously enshrined its environmental mission into its DNA, culminating in the move by founding owner Yvon Chouinard to place the company in a trust and nonprofit designed to ensure its mission endures indefinitely.

This business innovation is a manifestation of increasingly powerful will on the part of entrepreneurs to incorporate mission lock into their organizations—to create organizational designs that resist the gravitational force of short-term profit maximization. As entrepreneur and social impact investor Charly Kleissner puts it, "It's not just doing well by doing good; it's designing systems that cannot do harm by default" (Kleissner, 2019).

3.2 Moral Engineering by Design: The Rise of Value-Centric Startups

Alongside legal form, entrepreneurs are also reimagining the internal organization of business to place values at the center of decision-making. From diverse staffing and worker ownership models to regenerative supply chains and algorithmic transparency, this is a breadth of innovation. The rise of "value-led startups" is perhaps most pronounced in sectors like fintech, food systems, and clean energy.

Take the case of Etho Capital, a climate-investment platform that builds fossil-free, ESG-maximizing index funds. Its business model is based on a double guarantee: financial return and quantifiable environmental impact. Take also the example of Toast Ale, a brewery that brews beer using wasted bread and allocates profits to changing the food system. Each of these companies builds its business in a way that it fulfills a moral justification—a sense that business can cure rather than take advantage of system ailments.

In the technology world, some founders are now purposefully rejecting the old Silicon Valley "move fast and break things" culture. The Tech for Good movement, the Ethical Tech initiative, and organizations like All Tech Is Human call for models that make harm reduction, algorithmic accountability, and user empowerment central. As Tristan Harris, former Google ethicist and co-founder of the Center for Humane Technology, says, "Technology is not neutral." It affects attention, behavior, and beliefs. Founders are held morally accountable for the digital experience architecture" (Harris, 2021).

This shift of entrepreneurial consciousness is a move away from neoliberal orthodoxy and corporate social responsibility (CSR) fig-leafism. Unlike CSR—which is usually a separate entity of a business—today's moral entrepreneurs are looking for integrity, building purpose into product design, hiring practices, governance, and capital approach. They are not applying ethics to an otherwise extractive model; they are building ethics in.

3.3 Narrative as Infrastructure: Founders as Storytellers of Meaning

Almost as vital as process and structural change is the role of the story. The entrepreneur is not just an architect of business models but also a teller of potential. What they speak about—about themselves, about what matters in problems, and about what success looks like- strongly affects investor sentiment, customer decisions, and even public policy.

The founder myth has long been the heart of entrepreneurial culture: Steve Jobs in his garage, Elon Musk's moonshot visions, or Oprah Winfrey's overcoming. What is changing is the ethical flavor of these myths. Increasingly, founders are crafting origin stories of ethics-inspired consciousness or individual awakening: an epiphany about the destruction of fast fashion, a sibling's battle with addiction, or a parent's shortened life due to pollution.

These tales are not marketing themselves. They are normative templates, offering models of what entrepreneurship is and should be. They are the symbolic economy of capitalism. For instance, when Whitney Wolfe Herd started Bumble with a vision of enabling women to empower themselves in dating, networking, and business, she was not so much launching an app—she was outlining a feminist ethic of agency in cyberspace. Similarly, when Hamdi Ulukaya began Chobani on the strength of refugee employment and employee equity pledges, he redefined what a dairy firm might be all about.

Sociologist Michelle Lamont (2012) calls this process "the cultural entrepreneurship of morality"—economic players renegotiating the boundaries of right action in words, symbols, and institutional creativity. Founders are learning ever more that the meaning of what they make matters as much as the equipment.

3.4 Moral Innovation and Capital: The Impact of Investing Ecosystem

Examining the changing dynamics of purpose and capital is essential to any study of moral architects as businesses. Few developments in the past two decades have been more significant than the rise of impact investing—the deployment of capital to generate real social and environmental impacts alongside financial returns.

Impact investing, estimated globally in 2023 at more than $1.2 trillion (GIIN, 2023), is not a fringe activity; it is changing where money goes and what businesses can be funded. Pioneering organizations like Acumen, Blue Orchard, and Bridges Fund Management have opened up a new space where entrepreneurs are not only encouraged to earn the most money but also to track and report on their contributions to common prosperity.

Entrepreneurs here are now building businesses that are investment-grade precisely because they solve actual, measurable issues. For example, d.light markets solar lighting solutions to off-grid households in sub-Saharan Africa and South Asia, improving health, education, and productivity. M-KOPA promotes digital and financial inclusion by providing pay-as-you-go mobile services and solar energy.

What distinguishes those founders is that they are capable of a two-way language: the language of mission and the language of markets. They can speak simultaneously of a theory of change and a growth strategy and a unit economics analysis, and a social impact KPI. As venture capitalist Clara Brenner of the Urban Innovation Fund laments, "The best founders today are fluent in both Excel and ethics" (Brenner, 2022).

3.5 The Tensions and Limits of Moral Architecture

Giving a firm a purpose is a difficult and frequently conflicting undertaking. Growth vs governance, vision versus viability, and profit versus principle are some of the conflicting tensions that entrepreneurs must manage.

One of the ongoing tensions is scaling ethics. Can a highly ethical, community-based business maintain its value when it is scaled? Can a founder's mission hold up to investor pressure, saturation of the market, or the demands of going public? Ben & Jerry's experience is instructive here: after its purchase by Unilever, it struggled to maintain its social mission. Despite provisions in its contracts to continue its values, there were disagreements over issues like pay equity and sourcing of products, so demanding questions arose about ownership, control, and corporate culture.

There is also the matter of virtue signaling. As ESG metrics move into the mainstream, so too do the dangers of companies appearing ethically but not changing in any significant way. "Greenwashing" and "purpose-washing" are terms used to encapsulate more cynicism about companies using the lexicon of morality but continuing harmful practices.

Finally, the very notion of entrepreneurs as moral architects' risks playing out a kind of techno-saviorism—the idea that social ills can be tackled by business savvy alone, regardless of structural policy change or democratic accountability. Social theorist Anand Giridharadas (2018) in Winners Take All argues that moral entrepreneurship often glosses over the inequalities created by market thinking. Not everything can (or should) be fixed by startups.

The trend overall, however, is clear despite these disagreements. There is a new model of entrepreneurship emerging, one that is as much about values as it is about skill or funding. Entrepreneurs are not just building companies; they are building cultures, institutions, and futures. And in doing so, they are not just adapting to economic shifts—they are forging them.

3.6 Conclusion: Toward Normative Entrepreneurship

Calling entrepreneurs "moral architects" does not romanticize them; rather, it acknowledges a significant shift in the conception and application of economic action. The most entrepreneurial great minds of our time are no longer satisfied to disrupt markets; they seek to remake them—to inject ethics, justice, and sustainability into the designs of economic life.

This shift is not ubiquitous or uncontested. But it is earnest. It suggests that the fate of capitalism may hinge not only on emerging business models or technology but on a fundamental reconsideration of who this entrepreneur is and where they belong in society.

In the second half, we turn to economists themselves—and ask, how might their own theories, instruments, and structures shift to meet this moment of morality?

4. ECONOMISTS AND THE MEANING OF MARKETS

Economics, being a science, has always prided itself on its scientific seriousness. Marginal theory and general equilibrium theory have been rich in language in describing how individuals and firms make decisions with scarce resources under constraint. Economists have portrayed themselves not as ideologues or moralists, but as detached analysts of incentives, trade-offs, and efficiency.

But this vision of self is being put to the test. While entrepreneurs increasingly come to imagine markets as sites of moral intention—spaces where one builds justice, sustainability, and human flourishing—economists are being challenged by an unpleasant question: Are our models still purposeful? Or, more aggressively, have our frameworks helped to perpetuate systems no longer aligned with the common good?

Here, we aim to look at how economists' role is evolving in relation to the moral turn in markets. We argue that there is a process shift—uneven and contested as it is—whereby economists are not merely being required to explain how markets function but are being required to ask what markets are for. That is, to come to terms with the moral, social, and cultural assumptions in their models and to co-create systems that reflect a broader definition of human flourishing.

4.1 From Positive to Normative: Economics as Value-Laden

For most of the 20th century, orthodox economics enforced a sharp distinction between positive and normative analysis. The former addressed empirical observation and prediction based on models—what is—and the latter addressed value judgments. Economists could examine taxes, say, without calling for redistribution; they could model pollution without dictating environmental control.

This dichotomy, while useful in some cases, has come to be increasingly criticized. As Amartya Sen (1987) put it, the assumption that economics could ever be value-free was always an illusion. All choice of variable, model, or measure involves unwritten assumptions regarding what matters. Even seemingly technical notions like GDP, discount rates, or rational choice theory come with some understanding of human well-being and time.

Sen's own work in welfare economics had already challenged the utilitarianism of neoclassical models with a "capabilities approach" to development—a quantification of what people actually can do and be. In this view, freedom, dignity, and agency are not beyond the economic imagination; they are at the center. The issue, Sen writes, "is not just how large the pie is, but also how it is sliced, who receives what, and how we co-exist" (Sen, 1999).

This reconceptualization encourages a richer vision of economics—one that acknowledges its role not only to explain behavior but to construct norms and institutional meaning, as well. Economists, in this paradigm, are not only technicians. They are architects of assumptions, framers of possibility, and interpreters of value.

4.2 Revisiting Homo Economicus: Behavioral and Moral Economies

The biggest innovation in this transition has been the development of *behavioral economics*. Pioneered by scientists like Daniel Kahneman, Amos Tversky, and Richard Thaler, behavioral economics upends the conventional model of *homo economicus*—the rational, self-interested maximizing utility. It rather portrays human beings as boundedly rational, socially influenced, and prone to cognitive bias.

While originally framed as a reaction against neoclassical inefficiencies (savings nudges, for example), behavioral economics has evolved to include a broad critique of the ways that values, habits, and institution's structure economic life. As individuals are recognized to care about fairness, reciprocity, and trust—apart from utility—economists have come to have a fuller appreciation of what rational behavior is.

But even behaviorist paradigms can break down if they define morality as a preference. *Moral economics,* the new discipline, goes further and investigates how people's moral commitments cross with markets, typically at odds with the usual laws of logic. Why do people donate their money anonymously, refuse to buy sweatshop goods, or pay too much in systems of honor? Why do whistleblowers risk their careers? These are not exceptions but a glimpse into another conception of human agency.

Markets are not value-free apparatuses, as Elizabeth Anderson (1993) has argued; they are embedded within social relations and represent values. Some things—intimacy, education, civic responsibility—are devalued when allocated only by price. Economists must therefore ask not just how markets operate, but where they need to be and what they should say.

4.3 Markets as Moral Projects: The Cultural Turn in Economics

This recognition that markets are not merely economic but cultural institutions has produced what some writers call the "cultural turn" of economics. Made more sensitive by sociology and anthropology, this perspective sees markets not just as sites where meanings are being constructed but where resources are being distributed.

Viviana Zelizer's (1994) masterpiece, The Social Meaning of Money, shows how economic exchange is inflected with social meaning. Parents give allowances in a different way than wages; wedding gifts are not labeled like household goods. Such configurations show that money is not a fungible, neutral good—it is context-conditioned, emotion-infused, and morally narrated.

Similarly, Block and Somers (2014) argue that the dominant model of the "free market" has been rendered a secular theology, supplanting alternative moral vocabularies. The work, in brief, is not just to revise models, but to reclaim meaning: to notice that markets are human constructions and therefore subject to democratic reconstruction.

This cultural framing transforms the role of economists from outcome predictors to participants in a normative debate. It turns them into political philosophers, public ethicists, and institutional designers. In short, it repositions economics as a moral science.

4.4 The Institutional Turn: System Designers by Economists

A further shift of the economist's job is from the institutional turn—a greater focus on rules, norms, and institutions that structure economic life. Authors such as Douglass North (1990), Elinor Ostrom (1990), and Acemoglu & Robinson (2012) have demonstrated that institutions, both formal and informal, play a crucial role in explaining economic outcomes. They mediate incentives, distribute power, and encode social values.

Ostrom's work on common-pool resources, for instance, dissolved the fallacy of the "tragedy of the commons" by showing that societies can self-regulate with a view to appropriating common goods sustainably without central control and privatization. Her ethnography of collective rulemaking illustrates how economic reason is constructed locally, not universally relevant to every situation.

This has profound implications for the economist's existence in life. They are no longer simply advising governments or corporations best suited for policy within certain rules; they are helping to craft the rules themselves. It takes it a step further with Mariana Mazzucato (2018), who proposes that states and economists employ mission-oriented innovation policy, working proactively to direct markets to socially optimal ends, such as decarbonization and equity in health.

In this way, economists are not just analysts of efficiency but creators of purpose. They must be able to ask themselves: What kinds of rules produce not just growth, but also justice? What institutions are most likely to encourage pluralism, resilience, and long-term flourishing?

4.5 Measuring What Matters: Beyond GDP and Profit

This institutional change has been accompanied by a surge in interest in non-traditional metrics. As criticism of shareholder value and GDP increases, economists are experimenting with new ways to measure ecological and social value.

The OECD Better Life Index, the UN Human Development Index, and Gross National Happiness are some efforts to expand the scope. Kate Raworth's (2017) "doughnut economics" model sets a social foundation (health, education, equity) against planetary limits (carbon emissions, biodiversity) to establish a safe and just space for human life.

Similarly, the rise of ESG (Environmental, Social, Governance) investing reflects a demand for multifaceted accountability on the rise. These are not technical solutions—these are expressions of altered priorities around what winning looks like. They are a call to transition from extractive to regenerative, from transactional to relational, and from profit to purpose.

At the vanguard of this effort are economists. To create policies that are more in line with the group's goals, they are knowledgeable about data, modeling, and policy. But students also have to think about the deeper philosophical issues, like the purpose of economic activity. Whose lives are worth counting in our measures? Whose time horizons are important?

4.6 The Public Role of the Economist: Rethinking Legitimacy

Finally, we must consider the economist's role as a public citizen. In an age of disinformation, inequality, and environmental hazard, technocratic detachment is no longer a possibility. Economists must speak out, engage democratically, and help recertify confidence in institutions.

This is not a call to give up rigor. It is a plea for greater responsibility. As Dani Rodrik (2015) has framed it, economists must be "plumbers" as well as theorists—

working in real-world systems with humility, pragmatic judgment, and ethical awareness.

The lineage of public intellectuals—from John Maynard Keynes to Joan Robinson to Esther Duflo—offers an example of economists who combine analytical expertise with civic commitment. They write and talk not just to other professionals, but to the public. They help societies make trade-offs, make moral choices, and imagine better futures.

It is in this vein that the economists of the 21st century must see themselves as not just technician, but as guardians of meaning. What sort of world is our job enabling, they must wonder? And how might we otherwise make it?

4.7 Conclusion: Economics Reimagined

The economics profession is at a crossroads. The tools, once decent enough to explain what happens in markets, are being called on to defend or recast the very nature of capitalism. In an era of existential crises—climate meltdown, rule by algorithm, democratic retreat—economic thought cannot continue to be obstinately technical and ethically neutral.

This is not a crisis; it is an invitation. A chance to rediscover earlier traditions in political economy, moral philosophy, and democratic debate. A chance to create models and measurements that capture the full richness of human values. A chance to approach markets not as mechanisms, but as moral works.

As entrepreneurs become architects of institutional purpose, economists must assume their own role of transformation: as interpreters of value, designers of norms, and co-authors of a more equitable and sustainable economy.

5. TRANSFORMATION OR CO-OPTATION?

The diffusion of mission-led entrepreneurs, socially conscious firms, and morally engaged economists represents a core reconfiguration of markets. What was once a terrain defined by the directives of efficiency, profit maximization, and shareholder supremacy is increasingly being agitated by agendas of justice, sustainability, inclusiveness, and well-being. However, this shift raises a crucial question: Is this ethical shift in capitalism a real shift, or is it just a smart rebranding that preserves the underlying structures while smelling like progress?

This section addresses the tension between real systemic change and the potential for superficial adaptation. It questions the ways in which capitalism appropriates critique into its own logic—what Boltanski and Chiapello (2005) called the “new spirit of capitalism” in popular parlance—and how we can distinguish between institutional capture and moral aspiration. It is not about criticizing entrepreneurial or economic innovations that attempt to make society better but rigorously scrutinizing their power to subvert dominance, redistribute agency, and construct actual alternatives.

We cover five dynamics that mark this conflict: (1) the co-optation of moral discourse in branding and strategy; (2) the promise and dangers of impact investing and ESG; (3) the timidity of stakeholder capitalism; (4) the bounds of technocratic change; and (5) how to preserve the transformative edge of market change.

5.1 Moral Language as Strategy: The Semiotics of Good

Over the last decade, the discourse of morality has pervaded business and entrepreneurial cultures. The firms that were previously fixated on efficiency now make statements of "values," impact," "justice," and "authenticity." Companies brand themselves as partners to social movements; mission statements aspire to a higher purpose than profit. Slogans such as "ethical consumption," "conscious capitalism," and "net positive" abound.

On the surface, this looks good. If markets are more and more filled with ethical considerations, surely that is a step forward. But look again, and the picture is not so simple. As Slavoj Žižek (2009) contends in First as Tragedy, Then as Farce, contemporary capitalism is incredibly adept at making critique consumable. The very practices that once stood as symbols of resistance—purchasing fair trade, giving money to causes, shopping sustainably—are now incorporated without trouble into consumer identity.

This has been referred to as "woke-washing" (Banet-Weiser, 2018), "greenwashing," or, more broadly, "purpose-washing." Here, the rhetoric of change is invoked not in order to change core practices but to enhance brand equity, market share, or investor relations. As Nancy Fraser (2017) warns, moralized capitalism can too readily operate as a "progressive neoliberalism"—a fusion of cultural inclusion and economic deregulation that maintains systemic inequalities under the guise of reform.

The language of change thus risks becoming performative—a veneer that masks continuity. Patagonia's or Ben & Jerry's support for environmental justice makes them be emulated as models of virtue. Suppose, however, multinational conglomerates with previous histories of labor exploitation perform the same rhetoric. When Chevron touts its net-zero policy and continues investing in drilling for oil? Changes in power, practice, and policy must be tangible in order for the language of transformation to be effective. Otherwise, it is a form of co-optation.

5.2 Impact Investing and ESG: Reforming Capital from Within?

One of the most visible enablers of moral change in modern capitalism is the emergence of impact investing and Environmental, Social, and Governance (ESG) approaches. These trends aim to connect capital with beneficial social outcomes, reframing success as more than just financial performance to encompass metrics such as carbon emissions reduction, gender equality, and neighborhood development.

According to the Global Impact Investing Network (2023), the size of the global impact investing market has surpassed $1 trillion, and ESG assets have the potential to increase more than $50 trillion globally by 2025 (Bloomberg, 2021). It reflects that much capital is being channeled for ethical purposes. But the question arises here: Is the capital reshaping capitalism, or is capitalism adjusting the term "impact"?

Critics point to several issues. For one, the ESG reporting norms are fragmented, non-binding, and often inconsistent. As Tariq Fancy, a former chief investment officer of sustainable investing at BlackRock, has incendiary put it in 2021, much of ESG investing is marketing rather than substance, providing "green halos" with no structural responsibility (Fancy, 2021). When fossil fuel companies are included in ESG funds because governance scores are not precise, the concept becomes diluted.

Second, the impact investing rationale remains anchored in market-rate returns. This creates conflict between financial needs and social goals. When impact is tied to profitability, the most vulnerable or least profitable communities get left behind. As Rodney Foxworth (2020) states, "Impact investing often prioritizes the impact that appeals to investors rather than the impact most needed by communities.

In the end, there exists a broader concern that impact investing does not subvert the ownership and control of capital; it merely moves capital within established hierarchies. The challenge then is: Can financial instruments infused with shareholder primacy support emancipatory aspirations? Or will they tend to absorb criticism while preserving the underlying rationale for capital accumulation?

5.3 Stakeholder Capitalism: Inclusive or Illusory?

Most directly linked to ESG and impact investing is stakeholder capitalism—a conception of corporate governance in which the corporation owes an obligation not only to shareholders but also to employees, communities, suppliers, and the environment. The Business Roundtable's 2019 redefinition which accepted stakeholder value in place of shareholder value, was universally accepted as a sea change.

But to the majority of viewers, stakeholder capitalism has been style over substance. Harvard Law School (Bebchuk & Tallarita, 2021), in a 2021 study, found that companies that committed to stakeholder principles didn't substantively change their governance structures or executive compensation incentives. The redefinition of corporate purpose, therefore, has not come with binding commitments.

Stakeholder capitalism, though inspiring rhetorically, is afflicted with several traps:

- Unenforceability: In the absence of legal forces compelling a balancing of non-shareholder interests, stakeholder rhetoric is inspirational.
- Power asymmetry: Stakeholders do not hold equal power. Employees, local communities, and the environment do not have the organized power that shareholders—particularly institutional investors—have.
- Elite capture: Stakeholder governance is deployed as a vehicle for reputational management, where elite managers demand moral leadership without relinquishing control or de-ascending value.

The New Corporation that Joel Bakan (2020) proposes in argument for stakeholder capitalism might operate less as a new paradigm and more as a rhetorical reply—a pre-emptive strategy of regulation, a diffusing of dissent, and floating legitimacy in an era of crisis.

5.4 Technocratic Reform and the Limits of Expertise

With financial and entrepreneurial activities, technocrats and economists inherit the lexicon of purpose and inclusion. In inclusive growth theories to carbon pricing regimes, there is an attempt to synchronize economic and moral goals.

Yet, technocratic reform has its own risks of co-optation. When justice matters are rephrased as optimization problems, or when public interests are channeled into expert models without democratic deliberation, transformation can become depoliticized.

As James Ferguson (1994) eloquently illustrated in The Anti-Politics Machine, development economics has functioned as a means of translating political into technical questions and thus silencing opposition and concealing power. This may also occur in the new moral economy: when climate justice becomes carbon markets, or when racial justice becomes workforce diversity metrics, the moment of radicalism can be co-opted into managerial rationality.

In addition, the depiction of economists as neutral problem-solvers—though increasingly in doubt—is a prevailing trope. It holds back the discipline from embracing pluralism, incorporating indigenous and feminist epistemologies, or working collaboratively in knowledge production with subaltern communities.

Technocratic reform is needed—but not enough. Absent democratic struggle, citizen participation, and epistemic modesty, expertise can actually support the same structures it seeks to change.

5.5 Holding the Line: Principles for Radical Transformation

In light of these dangers, what would it require to safeguard the transformative promise of moral markets? How do entrepreneurs, economists, and citizens sort between authentic change and adaptive containment?

We suggest five principles:

a. **Redistribution of Material over Symbolic Inclusion**—True change entails moving power and resources, rather than the simple uptake of inclusive language or representative diversity. It means challenging concentrated wealth, innovating new designs of ownership, and organizing around alternative institutions (e.g., cooperatives, community land trusts).
b. **Democratizing Economic Governance**—Markets must be democratically ruled. By participatory budgeting, citizens' assemblies, or inclusive policy making, the state of markets must be ruled by public deliberation and not elite consensus.
c. **Epistemic Pluralism**—A transformational economy will need to have a broader knowledge base.
d. **Temporal Integrity-**Transformation requires attention to long-term effects rather than quarterly targets or political time. Institutions must be conceived with intergenerational justice, planetary boundaries, and future generations in mind.
e. **Reflexivity and Institutional Design-**Market actors are constantly compelled to interrogate their own positionality. Both economists and entrepreneurs are compelled to ask, "Are we reinforcing the current order or subverting it? Are our innovations structurally emancipatory, or are they merely surface-level?"

5.6 Conclusion: Between Hope and Capture

We stand at a crossroads. The vocabulary of purpose, effect, and justice has moved to the heart of capitalism. Business leaders speak of flourishing, economists, of prosperity. Markets are now being redefined as spaces of care, welcome, and renewal.

And yet, we must continue to watch our backs. The history of capitalism is one of co-optation—of radical elements courted, redefined, and rendered innocuous. The challenge is to escape adaptation into transformation: to build institutions, practices, and cultures that conquer commodification, contest inequality, and nurture the moral imagination.

This is not a task for economists and entrepreneurs alone. It is one that requires solidarity, nerve, and imagination. The architecture of transformation, if it is to last, needs to be built on the foundation of justice, not just intention.

6. CONCLUSION: DESIGNING FOR THE LONG NOW

In a world rocked by compounding crises—ecological degradation, systemic injustice, democratic decay, and an increasing sense of temporal dislocation—there is a particular need to re-anchor our economic imagination in a longer, deeper temporal horizon. The earlier parts of this chapter have claimed that entrepreneurs and economists are not simply analysts or operators in markets but also meaning-makers, moral architects whose ideas, practices, and institutions have the power to shape collective futures.

But even while we observe the rise of stakeholder capitalism, regenerative business, ethical finance, and inclusive economics movements, the one outstanding question is: Will these movements last and transform the existing architecture of capitalism, or will they be co-opted and made ineffective by the very systems that they seek to transform? The contrast would be whether we can shift from short-term fixes to long-term design, from quarterly metrics and tactical virtue to an ethos based on what Stewart Brand (1999) called the "Long Now."

Designing for the Long Now is to create a sense of responsibility that transcends short-term gains, political cycles, and consumer trends. It provokes economic actors to solve problems in timescales of generations and create institutions that can adapt, endure, and deepen moral commitments across decades and centuries. It is not an ideological decision; it is an experiential imperative for planetary survival and for human flourishing.

In this last section, we consider how long-term thinking can redesign economic purpose, moral architecture, and institutional design. We establish some principles of "long-now economics" and ask what role moral imagination, cultural narrative, and systemic resilience can play in constructing futures worth living.

6.1 The Tyranny of the Short-Term

Modern capitalism is defined by temporal compression. Quarterly earnings, twenty-four-hour stock prices, electoral cycles, high-frequency trading, and startup exit opportunities all work to reinforce hyper-present attention. As John Maynard Keynes (1936) quipped, "In the long run we are all dead"—a saying too often read as a license to delay or dismiss long-term thinking.

This short-termism is not only an error in private choice; it is institutionally embedded. Capital markets reward liquidity and speed, not patience and depth. Corporate governance norms prefer shareholder payoff to stakeholder duration. Political institutions often lack the tools to represent future generations. Even in academia, publishable originality incentives can substitute for slower, cross-disciplinary research.

What is left is what legal scholar Jedediah Purdy (2015) has described as a "future-stripped politics" — a politics incapable of imagining anything outside the crisis at hand. Climate change, biodiversity collapse, and intergenerational injustice are all symptoms of this deeper problem: we have built an economic infrastructure incapable of sustaining long-term moral accountability.

6.2 Designing Institutions for Deep Time

If transformation architecture is to thrive, it must be built for deep time. That means the building of economic institutions that are innovative and efficient but also resilient, flexible, and ethically anchored.

Drawing on scholars like Elinor Ostrom (1990), Kate Raworth (2017), and Donella Meadows (2008), we can identify several design principles that support long-term flourishing:

a. Distributed Governance

Institutions with distributed decision-making—such as cooperatives, commons-based resource management, or participatory budgeting—internalize the heterogeneity of views, including those of future generations, more effectively. They slow down deliberation and prioritize relationships over transactions.

b. Embodied Feedback Loops

Long-term systems place feedback within them to allow for learning and adaptation. Regenerative agriculture, for example, utilizes soil health metrics to hone practice. Social enterprises can use community well-being indicators as internal navigators, rather than just marketing slogans.

c. Polycentric Design

Rather than relying on monolithic centralized authority, robust institutions operate through nested sets of governance—local, regional, and global—that balance autonomy and coordination (Ostrom, 1990). This allows for both resilience and responsiveness at multiple scales.

d. Mission Locking

Some organizations use legal tools like benefit corporation charters or steward-ownership structures for encoding purpose into their designs, rendering it less possible for future mission drift or shareholder capture (Kelly, 2012).

e. Intergenerational Representation

Such fresh ideas as "guardians of future generations" (Future Design, Saijo, 2020) or citizen climate assemblies show how institutions can begin to quite deliberately include the voices of unborn individuals.

These qualities are not utopian ideals; they are already being tested around the world. The task that remains is the challenge of scaling, linking, and weaving them into a cohesive system of long-now economics.

6.3 Moral Imagination as an Economic Force

Long-term design is not technical, but a story. As Amartya Sen (1999) reminds us, economic agencies are strongly influenced by identity, culture, and moral ambition. What we believe is possible—and what is worth trying—is shaped by the accounts we give of our future, ourselves, and our past.

In order to build for the Long Now, we must build a moral imagination that is able to envision other potential futures. Philosopher Charles Taylor (1989) defines moral imagination as the ability to keep in view new sorts of lives, to behold possibilities for meaning and justice not yet realized but imminent in our situation.

Thus, entrepreneurs and economists are not just firm builders and model makers but writers of narrative worlds. Consider how Muhammad Yunus rewrote the poor as entrepreneurs, not as recipients; how Elinor Ostrom rewrote commons as a cooperative space, not as tragedy; or how Mariana Mazzucato (2018) rewrote the state as co-creator of value, not a market fixer. These rewritings change what is legible, desirable, and actionable.

The challenge, then, is to build institutions and markets that not only pay for efficiency but for vision—systems that enable moral innovation as well as technical innovation. Just as we have venture capital for scalable business, can we imagine "moral capital" for scalable justice?

6.4 The Ethics of Irreversibility

An economics of long now must also contend with the ethics of irreversibility. In climate change, extinction, and cultural loss, we have processes that cannot easily be undone. This makes deep moral stakes.

Hans Jonas (1979), in The Imperative of Responsibility, argues that modern ethics must begin with the recognition of technological and economic power in order to transform conditions of life infinitely. "Act so that the effects of your action are compatible with the permanence of genuine human life," he states.

This imperative has economic implications. It requires calling upon the precautionary principle of investment, technology deployment, and resource extraction. It also summons us to extend our temporal discounting logic in economics: future consequences need not be systematically underestimated.

Moreover, it challenges us to rethink what is rightful economic activity. If an industry perpetuates irremediable harm—to the environment, to communities, or to cultural heritage—can it be justified on the grounds of short-term gain? Long-now thinking turns the burden of proof around: not is change profitable, but is it an ethically robust change?

6.5 Hope, Not Optimism: The Ethics of Commitment

Designing for the "Long Now" is not an exercise in utopia; it is extremely pragmatic. But it does have a sense of hope, and not optimism.

When Václav Havel (1990) stepped aside, optimism is the hope that things would go well, while hope is the desire to struggle for the good even in loss. In a moment of existential danger and institutional exhaustion, hope is a form of moral resistance—an insistence that there is another world possible, if not required.

For entrepreneurs and economists, this hope must be brought to life. It means laboring against the siren call of hasty escapes, performative intention, or technocratic reductionism. It means laboring on behalf of slow, patient, collective work: building governance structures that endure founders, organizing value chains that repair rather than extract, and developing measures that capture dignity and mutuality.

It also involves embracing humility. The future cannot be perfectly designed; it must be co-created, stewarded, and reimagined repeatedly. As Donella Meadows (2008) reminds us, the most influential leverage points in systems involve changing minds, not inputs. Designing for the Long Now is ultimately a spiritual and civic practice—an orientation to time, self, and society.

6.6 Final Reflections: An Invitation to the Future

Finally, this chapter argues that markets are not value-free mechanisms but ethical landscapes shaped by the visions, choices, and designs we put into them. Economists and businesspeople are not simply participants in a system; they are co-writers of meaning.

We face a choice. We can keep refining within the inherited systems of short-term capitalism, massaging the edges and rebranding the surface. Or we can respond to the deeper calling of design: to build institutions on justice, dignity, and longevity.

This is Long Now architecture. It does not promise rapid solutions or rapid returns. But it offers a pattern for durable change—where profit serves purpose, where economics honors ecology, and where action's horizon is beyond self to species, and beyond now to later.

This is being done already. In indigenous patterns of government that teach reciprocity rather than extraction. In solidarity economies that inscribe solidarity rather than speculation. In youth movements that demand a livable planet, a decolonized economy, and a politics of care. The "Long Now" is no distant utopia; it is a design challenge. And it begins with how we think, act, and construct today.

REFERENCES

Acemoglu, D., & Robinson, J. A. (2012). *Why Nations Fail: The Origins of Power, Prosperity, and Poverty*. Crown Business.

Anderson, E. (1993). *Value in Ethics and Economics*. Harvard University Press.

Bakan, J. (2020). *The New Corporation: How "Good" Corporations Are Bad for Democracy*. Vintage Canada.

Banet-Weiser, S. (2018). *Empowered: Popular Feminism and Popular Misogyny*. Duke University Press.

Bebchuk, L., & Tallarita, R. (2021). The Illusory Promise of Stakeholder Governance. *Cornell Law Review, 106*(1), 91–178.

Block, F., & Somers, M. (2014). *The Power of Market Fundamentalism: Karl Polanyi's Critique*. Harvard University Press.

Bloomberg. (2021). "ESG Assets May Hit $53 Trillion by 2025."

Boltanski, L., & Chiapello, È. (2005). *The New Spirit of Capitalism*. Verso.

Brand, S. (1999). *The clock of the long now: Time and responsibility*. Basic Books.

Brenner, C. (2022). Remarks at Urban Innovation Fund Forum.

Credit Suisse (2022). *Global Wealth Report.*

Daly, H. (1996). *Beyond Growth: The Economics of Sustainable Development*. Beacon Press.

Fancy, T. (2021). *The Secret Diary of a 'Sustainable Investor*. Medium.

Ferguson, J. (1994). *The Anti-Politics Machine: Development, Depoliticization, and Bureaucratic Power in Lesotho*. University of Minnesota Press.

Folbre, N. (2001). *The Invisible Heart: Economics and Family Values*. New Press.

Foxworth, R. (2020, June 25). Reimagining Impact Investing. *The Nonprofit Quarterly*.

Fraser, N. (2017). The end of progressive neoliberalism. *Dissent, 64*(2), 130–137. DOI: 10.1353/dss.2017.0020

Fraser, N. (2017). The End of Progressive Neoliberalism. *Dissent.*

Friedman, M. (1970, September 13). The social responsibility of business is to increase its profits. *The New York Times Magazine*.

Friedman, M. (1970). "The Social Responsibility of Business is to Increase its Profits." *The New York Times Magazine*.

GIIN. (2023). *Annual Impact Investor Survey*. Global Impact Investing Network.

Giridharadas, A. (2018). *Winners Take All: The Elite Charade of Changing the World*. Knopf.

Giridharadas, A. (2018). *Winners Take All: The Elite Charade of Changing the World*. Knopf.

Global Impact Investing Network (GIIN). (2023). *Annual Impact Investor Survey*.

Harris, T. (2021). Center for Humane Technology. https://www.humanetech.com

Havel, V. (1990). *Disturbing the peace: A conversation with Karel Hvížďala*. Vintage.

IPCC (2023). *Sixth Assessment Report*.

Jonas, H. (1979). The Imperative of Responsibility. In *Search of an Ethics for the Technological Age*. University of Chicago Press.

Kahneman, D. (2011). *Thinking, Fast and Slow*. Farrar, Straus and Giroux.

Kaye, J. (1998). *Economy and Nature in the Fourteenth Century: Money, Market Exchange, and the Emergence of Scientific Thought*. Cambridge University Press.

Kelly, M. (2012). *Owning our future: The emerging ownership revolution*. Berrett-Koehler.

Kelly, M. (2012). *Owning Our Future: The Emerging Ownership Revolution*. Berrett-Koehler.

Keynes, J. M. (1936). *The general theory of employment, interest and money*. Macmillan.

Keynes, J. M. (1936). *The General Theory of Employment*. Interest and Money.

Kleissner, C. (2019). *Deep Impact Investing: A Case Study in Designing for Good*. Toniic Institute.

Lab, B. (2023). B Corporation Directory. https://www.bcorporation.net

Lab, B. (2024). *Certified B Corporation Directory*. https://www.bcorporation.net

Lamont, M. (2012). *How Professors Think: Inside the Curious World of Academic Judgment*. Harvard University Press.

Lederach, J. P. (2005). *The Moral Imagination: The Art and Soul of Building Peace*. Oxford University Press.

Mazzucato, M. (2018). *The value of everything: Making and taking in the global economy*. PublicAffairs.

Meadows, D. H. (2008). *Thinking in systems: A primer* (Wright, D., Ed.). Chelsea Green Publishing.

Meadows, D. H. (2008). *Thinking in Systems: A Primer*. Chelsea Green.

North, D. C. (1990). *Institutions, Institutional Change and Economic Performance*. Cambridge University Press.

Ostrom, E. (1990). *Governing the commons: The evolution of institutions for collective action*. Cambridge University Press.

Polanyi, K. (1944). *The great transformation: The political and economic origins of our time*. Beacon Press.

Purdy, J. (2015). *After nature: A politics for the Anthropocene*. Harvard University Press.

Raworth, K. (2017). *Doughnut economics: Seven ways to think like a 21st-century economist*. Chelsea Green Publishing.

Rodrik, D. (2015). *Economics Rules: The Rights and Wrongs of the Dismal Science*. Norton.

Saijo, T. (2020). Future design: A new way to consider future generations. *Economics of Peace and Security Journal, 15*(2), 5–8. DOI: 10.15355/epsj.15.2.5

Saijo, T. (2020). Future Design: A New Way to Consider Future Generations. *Economics of Peace and Security Journal, 15*(2), 5–8.

Schumpeter, J. A. (1942). Capitalism. *Socialism and Democracy*.

Sen, A. (1987). *On Ethics and Economics*. Blackwell.

Sen, A. (1999). *Development as freedom*. Knopf.

Taylor, C. (1989). *Sources of the self: The making of the modern identity*. Harvard University Press.

Taylor, C. (1989). *Sources of the Self: The Making of the Modern Identity*. Harvard University Press.

Yunus, M. (2007). *Creating a world without poverty: Social business and the future of capitalism*. PublicAffairs.

Zelizer, V. (1994). *The Social Meaning of Money*. Basic Books.

Žižek, S. (2009). *First as Tragedy, Then as Farce*. Verso.

Chapter 5
Accounting as Narrative Infrastructure:
Reframing Value Through Systemic Design and Economic Imagination

Sónia Calado
https://orcid.org/0009-0004-6915-5438
University of Aveiro, Portugal

ABSTRACT

This chapter explores how accounting functions as a symbolic and narrative infrastructure that co-constructs economic systems. It draws on interdisciplinary perspectives, such as legitimacy theory, narrative economics, and institutional design, to argue that accounting shapes meaning, identity, and systemic legitimacy. Focusing on emerging practices like integrated reporting and ESG disclosures, the chapter develops a framework in which accounting is positioned as a semiotic, narrative, and systemic force. Through documentary analysis and conceptual reflection, it illustrates how accounting redefines value, not only for measurement and compliance, but also as a strategic tool for organisational storytelling and institutional transformation. The chapter highlights how such narrative infrastructures contribute to sustainable innovation and the design of more inclusive and regenerative economic models.

1 INTRODUCTION

Accounting is often treated as a neutral language of numbers, a representational device that mirrors an already constituted economic world. Critical and sociological

DOI: 10.4018/979-8-3373-4566-6.ch005

accounting research has long disputed that assumption, showing instead that accounting helps constitute the very realities it appears to describe, it is performative and world making, not merely reflective (Burchell et al., 1980; Hopwood, 1987; Miller & Power, 2013). Building on this lineage, the chapter advances the claim that accounting is narrative infrastructure, it scripts what is thinkable as value, who can legitimately claim it, and which futures are rendered financeable. When standards are written or rewritten, when disclosure regimes are consolidated, when public sector budgeting technologies are redesigned in the wake of crisis, accounting does not follow the economy, it co constructs it, because models, metrics and categories, once widely adopted, help bring into being the worlds they purport to describe (MacKenzie, 2006; Power, 1997).

Three intellectual moves ground the argument. First, legitimacy theory and narrative economics suggest that organizations and policymakers compete for attention, trust and credibility through stories, metaphors and plots that make sense of uncertainty, with accounting serving as a key medium through which legitimacy claims are articulated and audited (Shiller, 2019; Suchman, 1995). Second, service dominant logic foregrounds value in use, ecosystems and co creation, destabilizing the idea that value is a static attribute that can be captured once and for all in a single scalar indicator, which problematizes traditional accounting grammars organized around transactions and point in time valuations (Vargo & Lusch, 2004). Third, systemic and institutional design remind us that infrastructures are socio technical, they are built, governed and repaired, and they carry politics in their pipes, classifications and taxonomies are never innocent, they channel visibility, resources and power (Bowker & Star, 1999; Miller & Power, 2013; Vargo & Lusch, 2004).

The chapter pursues three objectives. The first is conceptual, to articulate a framework that positions accounting as semiotic, narrative and systemic, three interdependent layers through which signs, stories and structures co produce realities, extending prior insights about the multiple roles of accounting in organizations and society (Burchell et al., 1980; Hines, 1988). The second is analytical, to show, via documentary analysis and critical discourse analysis, how contemporary standard setting debates, such as integrated reporting, the consolidation of ESG disclosures, and the emergence of the European Sustainability Reporting Standards, enact world making choices about materiality, comparability and accountability (Christensen et al., 2021; European Union, 2023). The third is normative, to articulate design principles as conceptual orientations in ways consistent with regenerative, equitable and future fit economies, aligning measurement with meaning and mobilization.

Two research questions guide the inquiry. First, how do accounting languages and devices shape the legitimacy and identity of organizations within broader enterprise ecosystems, a question that ties legitimacy theory to the performativity of metrics and narratives (Miller & Power, 2013; Suchman, 1995). Second, how can accounting be intentionally redesigned to expand what counts as value without collapsing plural valuations into a single monolithic indicator, an issue that resonates with ongoing tensions between single materiality regimes oriented to enterprise value and double materiality regimes that also center organizational impacts on society and the environment (Christensen et al., 2021; European Union, 2023). A subsidiary question follows, what happens to accountability when narratives of sustainability are absorbed by financial materiality logics, and which institutional arrangements are required to keep plurality alive, an issue already visible in critiques of integrated reporting's limited transformative traction when not supported by systemic shifts in governance and assurance (Flower, 2015; Power, 1997).

Although the empirical illustrations draw on international debates, the chapter pays particular attention to the European context, especially the role of European Financial Reporting Advisory Group (EFRAG) in developing the European Sustainability Reporting Standards (ESRS) and the legal institutionalization of double materiality, and to the Portuguese public sector after the sovereign debt crisis, where austerity, fiscal rules and budgeting technologies re scripted collective priorities and narrowed policy imaginaries, exemplifying how accounting infrastructures can privilege some futures over others (Blyth, 2013; European Union, 2023). These cases are illustrative rather than exhaustive; they make visible how accounting codifies not only economic value but also cultural and institutional purposes.

The contribution is twofold. For researchers, the chapter offers a vocabulary and a framework to study accounting as infrastructure, with attention to semiotics, narrative dynamics and systemic design, complementing and extending existing work on the roles of accounting, performativity and legitimacy (Burchell et al., 1980; Miller & Power, 2013; Suchman, 1995). For practitioners and policymakers, it distils conceptual design orientations, governance of metrics, transparency of assumptions, participatory design of indicators, and anticipatory stewardship embedded in performance accounts, consistent with recent calls to align disclosure architectures with long term societal needs rather than merely with investor centered risk management (Christensen et al., 2021; European Union, 2023). The remainder of the chapter proceeds as follows. Section 2 develops theoretical framing. Section 3 details the methodology. Section 4 lays out the conceptual framework. Section 5 presents three illustrative cases. Section 6 discusses implications for value reframing and systemic design. Section 7 addresses limitations and future research. Section 8 concludes.

2 THEORETICAL FRAMING, ACCOUNTING AS NARRATIVE INFRASTRUCTURE

2.1 From representation to performance

Classical views cast accounting as a mirror that faithfully represents preexisting transactions and events. Critical and sociological traditions dispute that assumption, showing that numbers classify, discipline, legitimize and govern, they participate in organizing worlds rather than merely describing them (Burchell et al., 1980; Hines, 1988; Hopwood, 1987; Miller & Power, 2013). Performativity, imported from linguistics into economics and organization studies, helps capture this shift, when standards define what counts as material risk, they do not simply describe risk, they organize attention, move resources and instruct boards and auditors on what to prioritize (MacKenzie, 2006; Power, 1997). If models and metrics are widely adopted, they help bring into being the worlds they purport to describe, a mechanism long observed in financial modelling and equally applicable to accounting standards (MacKenzie, 2006).

2.2 Legitimacy theory and narrative economics

Legitimacy theory starts with a simple intuition, organizations depend on social approval, they survive by maintaining congruence between their actions and prevailing norms and values (Suchman, 1995). Accounting is central to this process, it is a primary medium through which legitimacy claims are articulated, audited and contested (Miller & Power, 2013; Power, 1997). Narrative economics adds a complementary layer, economic behavior is mobilized by contagious stories, not just by efficient prices (Shiller, 2019). When a narrative like net zero by 2050 spreads, it brings new asset classes, liabilities and disclosure templates in its wake. Accounting participates in the diffusion of such narratives, it standardizes them, renders them calculable and either amplifies or dampens their reach, depending on how materiality, scope and assurance are defined (Christensen et al., 2021; Flower, 2015).

2.3 Service dominant logic and value co creation

Service-dominant logic reframes value as co-created in ecosystems, realized in use and over time, rather than captured at a discrete moment of (Vargo & Lusch, 2004, 2016). This destabilizes the implicit ontology of many accounting systems that privilege transactions and point in time valuations. If value emerges relationally and cumulatively, devices that emphasize static stock measures will underrepresent learning, capability building, social reproduction and ecological regeneration. The implication is not to abandon measurement, but to design accounting instruments capable of narrating processes, trajectories and interdependencies, not merely isolated outcomes (Bebbington & Unerman, 2018).

2.4 Institutional and systemic design

Institutions are designed, sometimes implicitly, sometimes explicitly. Systemic design invites us to make those design choices visible, to ask who benefits, who bears costs, and which feedback loops are reinforced. Accounting standards are institutional designs, they stabilize categories, coordinate expectations and channel capital (Miller & Power, 2013; Timmermans & Epstein, 2010). Infrastructures are most visible when they break, the 2008 financial crisis and the COVID 19 pandemic exposed the political economy embedded in fiscal rules, prudential ratios and performance regimes (Blyth, 2013; Power, 1997). Treating accounting as narrative infrastructure allows us to analyze standards and reports as socio technical artifacts that carry imaginaries of the future, where classifications and taxonomies are consequential political acts (Bowker & Star, 1999; Espeland & Stevens, 1998; Star, 1999).

2.5 Semiotics, narratives and systems

The notion of accounting as narrative infrastructure rests on three interdependent layers. The semiotic layer concerns the signs, categories, metrics and taxonomies that constitute the language of accounting, for example the choice to include scope 3 emissions as a reportable category (World Resources Institute, 2011). The narrative layer concerns plots, characters, temporalities, promises and fears that stitch numbers into stories that ascribe causality and responsibility. The systemic layer concerns governance architectures, feedback loops, institutional arrangements and technological platforms that stabilize or destabilize narratives, for instance legal embedding of standards, assurance regimes and digital tagging infrastructures such as XBRL (Timmermans & Epstein, 2010). The three layers are mutually constitutive, semiotic innovations open narrative possibilities, which, if institutionally reinforced, can trigger systemic shifts, and conversely, systemic mandates force semiotic and narrative adjustments.

2.6 The politics of consolidation

The consolidation of ESG standards into streamlined global frameworks is often justified by the need to reduce reporting burden, increase comparability and combat greenwashing. These goals are legitimate, yet consolidation carries politics, which narratives become hegemonic, which definitions of materiality prevail, which stakeholders are centered or marginalized (Christensen et al., 2021; Timmermans & Epstein, 2010). The European move toward double materiality foregrounds impacts on society and the environment, not just financial risk to the firm (European Union, 2023), while the ISSB standards prioritize investor focused, enterprise value-oriented materiality (IFRS Foundation, 2023b, 2023a). This tension is narrative, it signals what counts as value, and it is systemic, it shapes the institutional architecture that will channel resources for decades.

2.7 Public sector, austerity and the narrative of inevitability

Public sector accounting is frequently portrayed as technical housekeeping; in crisis moments it becomes narrative theatre. Following the sovereign debt crisis, many European countries adopted fiscal rules and budgeting technologies that narrowed political choice, framing deficits as moral failure and surpluses as virtue, and discounting the future to meet near term consolidation targets (Blyth, 2013; Lapsley, 2009). This is narrative work, not just arithmetic. When public balance sheets recognize contingent liabilities but not contingent assets such as social capabilities or ecological resilience, the story of the state becomes thinner, and accounting co constructs the legitimacy or illegitimacy of transformative public action (Power, 1997).

2.8 Toward regenerative, equitable, future fit accounts

If accounting is narrative infrastructure, the key question is not whether it is political, but which politics it advances. Regenerative and equitable economies require accounts that can hold the long term, render externalities legible without erasing their plurality, and embed care, maintenance and stewardship as first-class values (Kenter et al., 2019; Ross, 2019). This is not a plea for infinite disclosure, it is an invitation to systemic design, selective but principled, where indicators are co-created with affected communities, where uncertainty is disclosed rather than hidden behind false precision, and where scenario thinking and anticipatory governance become part of mainstream accounting practice (Bebbington & Unerman, 2018; Christensen et al., 2021).

3 METHODOLOGY

The chapter adopts a qualitative strategy that combines documentary analysis with critical discourse analysis, CDA. Documentary analysis offers a systematic procedure to extract how concepts, categories and definitions evolve across time, enabling the identification of semiotic shifts that reconfigure what becomes calculable and comparable (Krippendorff, 2019). CDA provides the tools to interrogate how those shifts are narrated and legitimated, which metaphors dominate, which actors are centered or marginalized, and which futures are envisioned, treating discourse as a site where power and ideology are negotiated (European Union, 2023; IFRS Foundation, 2023b, 2023a; Wodak, 2015).

The corpus is threefold. First, core accounting and sustainability reporting texts, the International <IR> Framework, proposals and final standards in the consolidation of ESG disclosure, notably IFRS S1 and S2, and the European Sustainability Reporting Standards, ESRS, including drafts and the delegated act (European Union, 2023; IFRS Foundation, 2023a, 2023b). Second, policy papers, consultation responses and speeches by regulators, standard setters, professional bodies and civil society coalitions, including documents associated with the Corporate Sustainability Reporting Directive, CSRD, and related national implementations in Europe, which make explicit competing materiality logics and governance designs. Third, public sector budgeting and accounting documents in Portugal during and after the sovereign debt crisis, budget laws, fiscal council reports and court of auditors' opinions, which reveal how austerity narratives were encoded in fiscal rules and performance regimes (Blyth, 2013).

Analytical procedure. The analysis proceeds in three steps, informed by abductive iteration between data and theory, (Tavory & Timmermans, 2014). Step one, coding of semiotic elements, we construct a codebook that maps categories, metrics, definitions of materiality, aggregation and disaggregation rules, and valuation conventions, using procedures from qualitative content analysis and systematic coding practice (Krippendorff, 2019). Step two, narrative coding, we identify plots, temporalities, moral vocabularies and role assignments embedded in management commentaries, standard setting rationales and regulatory speeches, drawing on narrative analysis traditions that treat accounts as structured stories with characters, turning points and endings (Boje, 2001). Step three, systemic mapping, we trace how governance arrangements, legal mandates, assurance regimes and digital infrastructures, for example XBRL tagging protocols, stabilize or challenge the semiotic and narrative layers, linking our observations to the sociology of standards and infrastructure studies (Star, 1999; Timmermans & Epstein, 2010).

Triangulation and rigor. Triangulation is achieved by comparing texts across jurisdictions and epochs, early integrated reporting documents versus ISSB standards, CSRD legal mandates versus national transpositions, enabling assessment of convergence, divergence and translation effects (Denzin, 1978). Reliability is enhanced through iterative coding, intercoder dialogue and peer debriefing, following established qualitative data analysis protocols. Trustworthiness is supported through transparency and reflexivity. We document coding decisions, version control the evolving codebook and make explicit our theoretical priorities and abductive moves (Tavory & Timmermans, 2014). While we use structured coding, we remain attentive to emergent categories and dissonant cases, consistent with reflexive qualitative analysis (Braun & Clarke, 2021)

Positioning and limitations. The approach is interpretive, not positivist, its aim is to render visible the politics of accounting design and the narrative infrastructures that orchestrate value, rather than to test causal hypotheses. Three limitations follow. First, the corpus is necessarily incomplete and regulatory landscapes evolve quickly, any mapping risks partial obsolescence by publication time. Second, Critical Discourse Analysis (CDA) is inherently interpretive, researcher positionality influences which metaphors and power effects are foregrounded, a limitation we mitigate with explicit reflexive memos and disclosure of assumptions (Fairclough, 2003; Wodak, 2015). Third, comparability across regimes is imperfect, interoperability between double materiality and investor focused standards is evolving, which constrains definitive conclusions about long term systemic effects ((Christensen et al., 2021; European Union, 2023; IFRS Foundation, 2023b, 2023a). These limitations are revisited in Section 7, where we outline how future research can broaden corpora, mix computational text analysis with deep interpretive work, and embed participatory design experiments to test alternative indicators in practice.

4 CONCEPTUAL FRAMEWORK, SEMIOTIC, NARRATIVE, SYSTEMIC

We propose a three-layer framework, shown in Figure 1, that conceptualizes accounting as narrative infrastructure. Figure 1 depicts the three layers and their feedback loops. The layers are analytical, not empirical silos, in practice they are entangled and co evolve. For clarity, we describe each layer, its core question, the underlying mechanisms, and the typical artifacts, then we make the cross-layer dynamics and the associated design principles explicit. This architecture extends work that treats accounting as a performative, classificatory and institutional practice rather than a neutral technical device (Burchell et al., 1980; Miller & Power, 2013). Table 1 operationalizes each layer, detailing mechanisms and typical artifacts.

Figure 1. Conceptual Framework

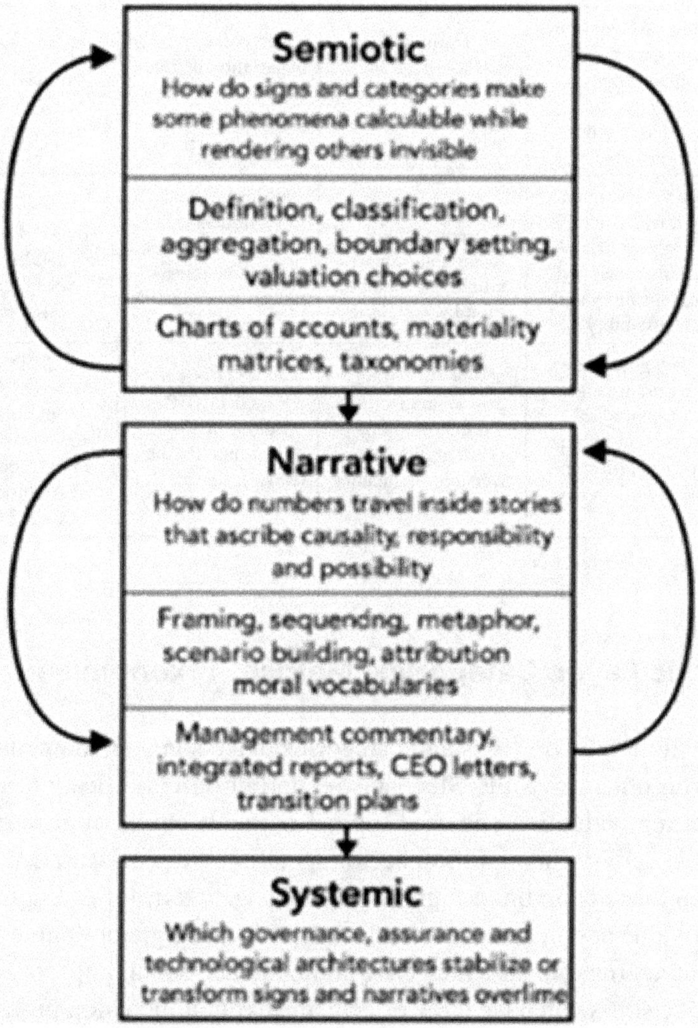

Source: Author's own elaboration

Table 1. Mapping the Framework

Layer	Key question	Mechanisms	Typical artifacts	Concrete examples
Semiotic	How do signs and categories make some phenomena calculable while rendering others invisible	Definition, classification, aggregation, boundary setting, valuation choices	Charts of accounts, materiality matrices, taxonomies, sector specific metrics, assurance criteria	Double materiality in ESRS, single materiality in ISSB, scope 3 in the GHG Protocol, EU Taxonomy, fair value vs historical cost
Narrative	How do numbers travel inside stories that ascribe causality, responsibility and possibility	Framing, sequencing, metaphor, scenario building, attribution, moral vocabularies	Management commentary, integrated reports, CEO letters, transition plans, risk narratives	Net zero 2050, just transition, austerity as inevitability, multi capital value creation in the <IR> Framework
Systemic	Which governance, assurance and technological architecture stabilize or transform signs and narratives over time	Standard setting, assurance regimes, digital infrastructures, oversight, regulatory feedback loops	ESRS, ISSB S1 S2, GRI, XBRL, European Single Electronic Format (ESEF), stewardship codes, fiscal rules	CSRD with legal force, ESRS ISSB interoperability, regulatory sandboxes, Fiscal Compact, Court of Auditors and Fiscal Council in Portugal

Source: Author's own elaboration

4.1 Semiotic Layer, Categories, Metrics, Taxonomies

Core question: how do signs and categories make some phenomena calculable while rendering others invisible. Mechanisms, definition, classification, aggregation, boundary setting, valuation choices. Typical artifacts, charts of accounts, materiality matrices, activity taxonomies, sector specific metrics, assurance criteria. At this layer, the focus is on the design of the language itself, for example the choice between fair value and historical cost, the adoption of single or double materiality, or the level of aggregation at which environmental and social impacts are reported. Small semiotic shifts can have large effects, the explicit inclusion of just transition indicators, for instance, can reframe the social dimension of decarbonization strategies, while the decision to include scope 3 emissions expands the locus of responsibility across supply chains that were previously obscured (European Union, 2023; World Resources Institute, 2011). Classifications carry politics, legal taxonomies, such as the European sustainable finance taxonomy, channel capital and attention toward specific activities and away from others, thereby scripting what will be financed and on what terms (Bowker & Star, 1999; Timmermans & Epstein, 2010).

4.2 Narrative Layer, Plots, Temporalities, Moral Orders

Core question: how do numbers travel inside stories that ascribe causality, responsibility and possibility. Mechanisms, framing, sequencing, metaphor, commensuration, and attribution. Typical artifacts, management commentary, integrated reports, CEO letters, risk narratives, transition plans. Here we attend to the rhetorical and temporal devices that stitch metrics into meaning, the plot that moves from problem to solution, the roles allocated to stakeholders, the time horizon that includes or excludes future generations. Narratives can be extractive, using the language of sustainability to preserve incumbent financial logic, or emancipatory, opening room for experimentation, redistribution and plural futures (Flower, 2015; Shiller, 2019). Legitimacy is contested in this layer, organizations seek congruence with social norms through audit ready stories, with accounting providing the vocabulary, categories and evidential scaffolding that stabilize, or destabilize, those stories (Power, 1997; Suchman, 1995).

4.3 Systemic Layer, Governance, Feedback, Infrastructures

Core question, which institutional architecture stabilizes or transforms the other two layers over time. Mechanisms, standard setting, assurance regimes, digital infrastructures, oversight bodies, regulatory feedback loops. Typical artifacts, reporting standards, taxonomies with legal force, assurance standards, digital tagging protocols such as XBRL, supervisory guidance. This layer examines how standards travel across jurisdictions, how they interact with prudential rules and public policies, and how data platforms enable new forms of comparability, surveillance and contestation (Star, 1999). Recent regulatory consolidation makes this especially visible, the ISSB establishes an investor focused, enterprise value-oriented baseline, while the ESRS legally embeds double materiality, each regime pulling with it distinct governance obligations, assurance expectations and data architectures, with consequences for what is financed, monitored and sanctioned (Christensen et al., 2021).

4.4 Cross Layer Dynamics

The three layers interact through recurrent, observable dynamics.

- Semiotic innovation - narrative configuration - systemic response. The formalization of scope 3 emissions or science-based targets configures plots around credibility, alignment and transition pathways, which then trigger systemic responses, new assurance requirements, board-level responsibilities, investor stewardship practices. Where semiotic innovation is not backed by systemic architecture, it risks symbolic compliance, a pattern noted in critiques of integrated reporting.

- Systemic shift - semiotic and narrative reconfiguration. The legal adoption of double materiality forces organizations to speak about impacts beyond financial risk, demands new metrics and recasts the role of auditors, producing feedback loops that either stabilize or contest dominant narratives.
- Narrative dominance - semiotic and systemic crystallization. When a storyline such as net zero by 2050 gains hegemony, it converts into categories, metrics, taxonomies and legal mandates that reinforce the narrative, in a performative cycle akin to those documented for financial models and risk frameworks.

Understanding accounting as narrative infrastructure requires tracing these feedback loops rather than merely cataloguing metrics and disclosures.

4.5 Design Principles Embedded in the Framework

The framework yields design principles that regulators, firms and practitioners can operate.

Plurality with coherence. Maintain multiple value registers, financial, social, ecological, without forcing premature commensuration, while providing translation protocols that allow explicit decision making over trade offs.

Anticipatory stewardship. Require long horizon scenarios, intergenerational commitments and stewardship accounts as standard practice, aligning the temporal horizon of accounting with the temporal horizon of risks and impacts

Transparency of conventions. Mandate disclosure of valuation models, thresholds, materiality choices and taxonomic criteria, so that stakeholders can contest and improve them, countering the opacity of commensuration.

Participatory indicator design. Co-creates indicators with affected communities and ecosystem partners, especially in sectors with significant social and ecological footprints, to reinforce substantive legitimacy rather than merely procedural compliance.

Interoperable infrastructures. Build digital standards and public data spaces that enable aggregation and comparison without locking value narratives into a single materiality concept, supporting governed pluralism across regimes.

Design for repair. Embed revisability, learning loops and sunset clauses for indicators that no longer serve, recognizing that accounting infrastructures require maintenance, adaptation and sometimes dismantlin).

These principles are elaborated in the discussion section, where we translate them into concrete layers for regulators, organizations and auditors to align semiotics, narratives and systems with regenerative, equitable and future fit economies.

5 ILLUSTRATIVE CASES, INTEGRATED REPORTING, ESG, PUBLIC SECTOR UNDER CRISIS

5.1 Integrated Reporting, Promise, Diffusion, Constraints

Integrated reporting, IR, emerged with the promise of connectivity of information, multiple capitals beyond the financial, and long-term value creation (International Integrated Reporting Council, 2013). Its primary semiotic innovation was the vocabulary of six capitals, financial, manufactured, intellectual, human, social and relationship, natural, a taxonomy designed to pluralise value and to render visible dependencies and impacts that conventional financial statements either ignore or treat as externalities (de Villiers et al., 2014). Narratively, IR encouraged organizations to craft a coherent story that links strategy, governance, performance and prospects, embedding forward looking orientation and stewardship in the report's architecture (Adams, 2015).

Systemically, however, IR's limited legal force in most jurisdictions, the optional nature of assurance, and the absence of strong investor stewardship mechanisms curtailed its transformative potential. Empirical studies repeatedly find symbolic adoption and decoupling, firms adopt IR language and the capitals vocabulary without significantly altering capital allocation logics or internal decision processes (Flower, 2015; Stent & Dowler, 2015). Even in South Africa, where integrated reporting became de facto mandatory through listing rules, research documents uneven practice quality and persistent dominance of financial logics in board conversations (Atkins & Maroun, 2015; Setia et al., 2015).

This case illustrates a central point of the framework, semiotic expansion without systemic reinforcement risks symbolic compliance. The introduction of new categories and a broadened vocabulary of value can enlarge narrative possibilities, yet without binding governance changes, robust assurance, investor stewardship aligned with multi capital performance, and public data infrastructures that enable scrutiny, the performative cycle stalls, the new language is absorbed by incumbent metrics and incentive structures.

5.2 ESG Consolidation, Materiality Battles and the Risk of Narrative Narrowing

The consolidation of ESG disclosure into global standards promised relief from reporting fragmentation and standard proliferation. The ISSB positioned itself around investor focused, enterprise value-oriented materiality, aiming to provide a global baseline for capital markets. Europe moved in a different, though overlapping, direction with the CSRD and ESRS, which legally embed double materiality, foregrounding both financial risks to the firm and the firm's impacts on society and the environment (European Union, 2023). The Global Reporting Initiative, GRI (Global Reporting Initiative), also reaffirmed an impacts first conception of materiality in its 2021 Universal Standards, increasing the salience of this divergence (GRI, 2021).

Semiotically, these differences are not cosmetic, they encode distinct theories of value and accountability. Narratively, an investor primacy plot risks absorbing sustainability into a risk management to enterprise value storyline, where social and ecological harms matter primarily when they threaten cash flows. In contrast, a double materiality plot demands that organizations articulate, and measure impacts regardless of immediate financial effects, thereby preserving plurality and public accountability. Systemically, coexistence creates both friction and opportunity, firms operating across jurisdictions must learn to speak two materiality languages, interoperability requires translation protocols, and regulators must decide how to govern overlaps without erasing normative differences.

The case also surfaces the politics of consolidation. Who sets the baseline, who defines what is material, whose interests are centred, these are institutional design questions, not mere technicalities. Evidence of low convergence across ESG ratings further underscores how choices of indicators, weights and taxonomies are deeply normative, not just noisy (Berg et al., 2022). The lesson mirrors our framework, consolidation should pursue coherence without collapsing plurality, otherwise we risk a narrative narrowing that recentres enterprise value and sidelines broader accountability. Table 2 makes the politics of consolidation explicit by contrasting ESRS and ISSB along ten design dimensions.

Table 2. ISSB vs ESRS comparison

Topic	Risk if misapplied	Practical interoperability opportunity
Core comparison		
Materiality	Reducing impacts to financial risk only, loss of public accountability	Map ESRS impact blocks to ISSB risk categories, publish dual block reports, investor baseline plus impact pillar
Assurance	Greenwashing if assurance stays numeric only, ignoring narratives and scenarios	Align with forthcoming ISA 5000 suite, add narrative coherence and scenario robustness to scope
Digital tagging	Incomparable data and high processing costs	Crosswalk ESRS and ISSB taxonomies, open APIs and mapping dictionaries
Governance	Box ticking without integration into incentives and capital allocation	Mirror ESRS G1 fields as optional extensions in ISSB style reporting
Social coverage	Blind spots on material social harms	Keep ESRS social block as a standalone pillar, provide investor relevant extracts
Time horizons	Strategic myopia, stranded assets risk	Require multi horizon reporting, short, medium, long, with a shared glossary
Regulatory risk and interoperability opportunity		
Materiality	Reducing impacts to financial risk only, loss of public accountability	Map ESRS impact blocks to ISSB risk categories, publish dual block reports, investor baseline plus impact pillar
Assurance	Greenwashing if assurance stays numeric only, ignoring narratives and scenarios	Align with forthcoming ISA 5000 suite, add narrative coherence and scenario robustness to scope
Digital tagging	Incomparable data and high processing costs	Crosswalk ESRS and ISSB taxonomies, open APIs and mapping dictionaries
Governance	Box ticking without integration into incentives and capital allocation	Mirror ESRS G1 fields as optional extensions in ISSB style reporting
Social coverage	Blind spots on material social harms	Keep ESRS social block as a standalone pillar, provide investor relevant extracts
Time horizons	Strategic myopia, stranded assets risk	Require multi horizon reporting, short, medium, long, with a shared glossary

Source: Author's own elaboration

5.3 Public Sector Budgeting in Portugal After the Sovereign Debt Crisis, Austerity as Narrative and Technique

The Portuguese public sector provides a salient example of how accounting and budgeting infrastructures can re script collective priorities. In the aftermath of the sovereign debt crisis and the financial assistance programme, a dense web of fiscal rules, medium term budgetary frameworks and performance regimes narrowed the perceived space of political choice. Semiotic decisions around expenditure classification, deficit and debt targets, and the treatment of public investment structured what could be said about the state, which forms of expenditure were framed as virtuous or irresponsible, and how the future was discounted to meet immediate consolidation goals ((Schick, 2014).

Narratively, austerity was cast as inevitability, responsibility and discipline, with deficits portrayed as moral failure and surpluses as virtue, echoing wider European discourses institutionalised in the Stability and Growth Pact and the Fiscal Compact (European Union, 2012). Systemically, European layer surveillance, domestic oversight institutions, notably the Conselho das Finanças Públicas (Fiscal Council), and the Tribunal de Contas (Court of Auditors), stabilised the narrative through monitoring reports, compliance assessments and public communication (Conselho das Finanças Públicas., 2014; Tribunal de Contas, 2019).

Over time, counter narratives emerged. The COVID-19 pandemic reemphasised the value of resilient health systems and social protection, the Recovery and Resilience Facility channelled significant investment into green and digital transitions, and debates over golden rules for public investment reopened the question of how to account for long term assets versus short term deficits. The case shows that infrastructures are revisable, but revision requires narrative work and systemic redesign, not merely technical tweaks. Redrawing budget classifications, introducing intergenerational balance sheets or stewardship accounts, and recalibrating fiscal rules toward risk sharing and investment-oriented frameworks are all accounting design questions with political consequences.

5.4 Cross Case Insights

Semiotic innovations travel faster than systemic redesign. Organizations adopt new vocabularies and indicators more easily than regulators, auditors and capital markets redesign governance architectures. IR's six capitals diffused rapidly as language, less so as binding decision rules or incentive structures.

Narrative drift is frequent. Transformative plots, multi capital value creation or sustainability as societal impact, are often absorbed by dominant financial logics unless institutional safeguards are in place, for example legal mandates, robust assurance, public data infrastructures and stewardship codes aligned with multi capital performance.

Legal force and data infrastructures matter. When standards are embedded in law and supported by interoperable digital tagging and public data spaces, they acquire stabilising power that can either protect plural valuations or narrow them, depending on the design choices, the ISSB ESRS coexistence, the EU taxonomy and the CSRD architecture illustrate this double edged effect.

Taken together, the cases affirm the chapter's central contention, accounting functions as narrative infrastructure, where semiotics, stories and systemic architectures co produce economic realities. Designing for plurality with coherence, anticipatory stewardship and repairable infrastructures becomes essential if accounting is to enable regenerative, equitable and future fit economies.

6 DISCUSSION, IMPLICATIONS FOR VALUE REFRAMING AND SYSTEMIC DESIGN

6.1 From Measurement to Meaning to Mobilization

If accounting is narrative infrastructure, design choices must be judged not only by measurement accuracy but also by their meaning making and mobilizing capacity. Indicators that cannot be narrated intelligibly will not mobilize action, narratives that cannot be operationalised through robust indicators will decay into virtue signalling (Espeland & Stevens, 1998; Kenter et al., 2019). In other words, accounts must travel across the three layers of the framework, semiotic precision, narrative intelligibility and systemic enforceability, otherwise performativity stalls and numbers fail to reorient capital or. Systemic design aligns semiotics and narratives with governance and technology so that meaning can move resources, and so that resources can, in turn, reinforce new meanings rather than default back to incumbent financial logics.

6.2 Design Principles for Reframing Value

We distil six principles.

Plurality with coherence, maintain multiple value registers, financial, social, ecological, without forcing premature commensuration, provide translation protocols that allow decision makers to navigate trade-offs explicitly. This protects plural valuations from being collapsed into a single scalar while still enabling organised decision making (Bebbington & Unerman, 2018; Espeland & Stevens, 1998; Kenter et al., 2019).

Anticipatory stewardship, require scenario-based disclosures, Task Force on Climate-related Financial Disclosures (TCFD) style, long term budgeting and balance sheets that recognise intergenerational commitments, so that the temporal horizon of the accounts matches the temporal horizon of the risks and impacts (Krueger et al., 2020; TCFD, 2023).

Transparency of conventions, mandate disclosure of valuation models, thresholds, weighting oschemes and materiality decisions, so stakeholders can contest and improve them, countering the opacity of commensuration and the false neutrality of standards (Espeland & Stevens, 1998; Timmermans & Epstein, 2010).

Participatory indicator design, involve affected communities, workers and ecosystem partners in defining what counts as material impact, especially in sectors with significant social and ecological footprints, following dialogic and pluralist approaches to accounting (Brown & Dillard, 2015; Kenter et al., 2019).

Interoperable infrastructures, design digital standards and public data spaces that allow aggregation and comparison without locking in narrow narratives of value, preserving the normative distinctiveness of double materiality while ensuring technical compatibility with investor-oriented baselines ((European Union, 2023; IFRS Foundation, 2023a, 2023b; Timmermans & Epstein, 2010)

Reflexive assurance, move beyond binary assurance toward forms that also evaluate narrative coherence, scenario robustness and governance integrity, building on the critique of the audit society and on empirical work that documents the limitations of sustainability assurance as currently practised (O'Dwyer & Owen, 2005; Power, 1997; Villiers & Farooq, 2019)

6.3 Implications for Regulators and Standard Setters

Regulators should treat standard setting as institutional design, not as mere technical alignment. This implies piloting, sunset clauses, iterative upgrading and governance of governance, consistent with insights from experimentalist governance and responsive regulation (Baldwin et al., 2012). Systemic experimentation can be structured through regulatory sandboxes for accounting innovations, where firms and public bodies test new indicators and narrative formats under supervision, with transparent learning loops and public feedback.

Double materiality can be operationalised without overwhelming firms if regulators provide sector specific guidance, phased implementation timelines, and interoperability bridges to investor focused baselines, while preserving the normative commitment to impacts beyond the firm. Regulators should also require digitally tagged disclosures and maintain public data spaces to make scrutiny, replication and research feasible, recognising that data architectures are part of the institutional design of accountability (Timmermans & Epstein, 2010).

6.4 Implications for Firms and Practitioners

Firms can reframe internal management accounts to embed regenerative logics, for example by capitalising maintenance and care at the same strategic layers as expansion, integrating stewardship metrics into executive remuneration, and aligning capital allocation with scenario-based transition plans (Bebbington & Unerman, 2018; Schoenmaker & Schramade, 2019). Practitioners, accountants, auditors and designers can collaborate to craft narratives that are both honest and mobilising, that disclose uncertainty without paralysing action, and that show how semiotic choices and valuation conventions were made. Assurance providers can adopt reflexive approaches that evaluate governance and scenario integrity in addition to verifying numerical accuracy (Villiers & Farooq, 2019). Investor stewards can explicitly engage on narrative quality and on the consistency between stated strategies, indicators and capital allocation, reducing the space for decoupling and symbolic compliance (Eccles & Klimenko, 2019).

6.5 Research Agenda Emerging From the Framework

The framework opens several research avenues.

Cross layer dynamics, examine how semiotic changes in standards translate into narrative shifts in corporate discourse and how systemic architectures resist or enable those shifts, for example by tracking firms' transitions from single to double materiality regimes. Mixed method designs can combine computational text analysis of large reporting corpora within depth interpretive CDA on theoretically sampled cases (LI, 2010; Wodak, 2015).

Comparative institutional analysis, study jurisdictions with investor focused baselines versus double materiality mandates to test how different institutional designs shape investment flows, stakeholder engagement and innovation trajectories (Berg et al., 2022; Christensen et al., 2021).

Participatory design experiments, co-create indicators with communities and workers, then evaluate their feasibility, legitimacy and decision usefulness in organisational and policy settings, extending dialogic accounting in practice. Digital infrastructures and interoperability, investigate how tagging protocols, taxonomies with legal force and public data repositories affect the visibility, contestability and durability of value narratives, including the unintended consequences of algorithmic aggregation and ESG rating divergence (Berg et al., 2022; Timmermans & Epstein, 2010).

Public finance and intergenerational accounting, develop and test intergenerational balance sheets, stewardship accounts and golden rule style fiscal frameworks that better align public sector accounting with long term regenerative objectives, especially in post crisis contexts like Portugal (Blyth, 2013; Schick, 2014).

Taken together, these lines of inquiry would deepen our understanding of accounting as narrative infrastructure and would provide the empirical and methodological tools needed to design systems that measure, mean and mobilise toward regenerative, equitable and future fit economies.

7 LIMITATIONS AND FUTURE RESEARCH

7.1 Limitations

This chapter is limited in scope and method. The corpus, although diverse, is not exhaustive, fast moving regulatory developments mean that any mapping risks partial obsolescence by the time of publication. The interpretive stance of critical discourse analysis introduces researcher bias, which we mitigate through reflexive memoing, explicit disclosure of priors and peer debriefing, but cannot eliminate. The focus on Europe, and specifically on Portugal for the public sector illustration, constrains generalizability, different political economies, legal traditions and data infrastructures will yield different narrative infrastructures. Selection effects are also present, formal, public documents are overrepresented relative to backstage negotiations and organizational micro practices, which ethnographic or interview-based studies could surface. Finally, while we gesture to digital infrastructures, we do not model how algorithmic aggregation, ESG rating divergence or data platform governance shape narrative durability and contestability at scale.

7.2 Future Research

First, longitudinal studies tracking the life of specific indicators, categories or taxonomies across regulatory cycles can show how semiotic innovations succeed, mutate or fail, and how they are absorbed or resisted narratively and systemically, combining archival work with process tracing.

Second, comparative institutional analyses between single materiality regimes and double materiality regimes can clarify concrete effects of design choices on capital allocation, stakeholder accountability and innovation, leveraging natural experiments created by asynchronous implementation of ISSB, CSRD and ESRS requirements.

Third, participatory design experiments can co create indicators with affected communities, workers and ecosystem partners, then evaluate feasibility, legitimacy and decision usefulness in practice, extending dialogic and pluralist accounting beyond theory into organizational and policy settings.

Fourth, digital infrastructure studies can examine how data architectures, tagging protocols and public repositories influence the visibility, interoperability and contestability of value narratives, including how algorithmic tools used by investors, rating agencies and regulators amplify or attenuate certain frames, here mixed methods that join computational text analysis with in depth CDA are promising.

Fifth, integration with public finance research can test how sovereign balance sheets, intergenerational accounts and revised fiscal rules, for example golden rules for green and social investment, could be designed to support regenerative statecraft, moving beyond austerity framed narratives that dominated post crisis Europe, Portugal.

Finally, methodological pluralism and reflexivity should be institutionalized, abductive designs that iterate between theory and data, explicit reporting of coding decisions, and open corpora and codebooks can enhance transparency, replicability and cumulative knowledge building in a field where standards, narratives and infrastructures evolve quickly.

8 CONCLUSION

This chapter advanced a simple but demanding proposition, accounting is narrative infrastructure, it does not only measure, it scripts, it legitimates, it imagines. By articulating a three-layer framework, semiotic, narrative, systemic, and by examining contemporary debates on integrated reporting, ESG consolidation and public sector budgeting in crisis, we showed that moments of standard setting are moments of world making. Categories and metrics are not innocent, they carry moral and political content, narratives of sustainability can be absorbed by financial materiality unless systemic safeguards are designed, public sector accounts can either narrow or expand the horizon of the possible.

Three contributions follow. Conceptually, we reframed accounting as a co productive force in the economy, specifying how signs, stories and systems interact to construct what counts as value, who may claim it, and which futures become financeable. Analytically, we demonstrated, through illustrative cases, that semiotic innovation without systemic reinforcement tends to degenerate into symbolic compliance, that consolidation battles over materiality are in fact battles over theories of value and accountability, and that fiscal rules and budgeting technologies can script austerity as inevitability. Practically, we distilled design principles that can move the field from critique to construction, plurality with coherence, anticipatory stewardship, transparency of conventions, participatory indicator design, interoperable infrastructures, and reflexive assurance.

The implication is straightforward, if societies want economies that are regenerative, equitable and future fit, the accounting system must be redesigned to carry those futures. Regulators need to approach standards such as evolving institutional designs, subject to piloting, sunset clauses and iterative upgrading, not as once and for all technical fixes. Firms need to embed stewardship for the long term into the core of their management accounts and incentive systems, integrating scenario-based capital allocation and multi capital performance into strategy and remuneration. Practitioners, accountants, auditors, designers, need to co craft narratives that are honest and mobilizing, that disclose uncertainty without paralyzing action, that make explicit the value conventions and modelling choices that underwrite indicators. Researchers need to pursue a plural methodological agenda that traces cross layer dynamics, compares institutional regimes, experiments with participatory indicator design, and interrogates how digital infrastructures stabilize or destabilize value narratives.

Accounting will never be neutral, nor should it aspire to be. Its legitimacy will rest on whether it can make visible what matters, whether it can sustain plurality without collapsing it into a single scalar, whether it can align capital with long-term flourishing rather than short term extraction. That is a design task, it requires institutional imagination, technical rigor and democratic participation. It is also a narrative task, it requires courage to say that not everything that counts can be priced, and yet it must be accounted for, carefully, transparently, with mechanisms for contestation and repair.

The challenge ahead is to turn this recognition into practice. That means building interoperable public data spaces that support scrutiny and learning, training a new generation of professionals fluent in semiotics, narratives and systems, and institutionalizing reflexive assurance that evaluates stories as well as numbers. It also means accepting that accounting infrastructures, like all infrastructures, will need maintenance, revision and sometimes dismantling. The invitation is open, to scholars, practitioners and policymakers, see ledgers as infrastructures of imagination, design them with plurality, anticipation and repair in mind, and let accounting help script the futures we cannot afford to ignore.

REFERENCES

Adams, C. A. (2015). The International Integrated Reporting Council, a call to action. *Critical Perspectives on Accounting*, *27*, 23–28. DOI: 10.1016/j.cpa.2014.07.001

Atkins, J., & Maroun, W. (2015). Integrated reporting in South Africa in 2012, perspectives from preparers and early adopters. *Meditari Accountancy Research*, *23*(2), 197–221. DOI: 10.1108/MEDAR-07-2014-0047

Baldwin, R., Cave, M., & Lodge, M. (2012). *Understanding regulation, theory, strategy, and practice* (2nd ed.). Oxford University Press.

Bebbington, J., & Unerman, J. (2018). Achieving the United Nations Sustainable Development Goals, an enabling role for accounting research. *Accounting, Auditing & Accountability Journal*, *31*(1), 2–24. DOI: 10.1108/AAAJ-05-2017-2929

Berg, F., Kölbel, J. F., & Rigobon, R. (2022). Aggregate Confusion: The Divergence of ESG Ratings. *Review of Finance*, *26*(6), 1315–1344. DOI: 10.1093/rof/rfac033

Blyth, M. (2013). *Austerity, the history of a dangerous idea*. Oxford University Press., https://global.oup.com/academic/product/austerity-9780199389445

Boje, D. M. (2001). Narrative methods for organizational and communication research. *Sage (Atlanta, Ga.)*.

Bowker, G. C., & Star, S. L. (1999). *Sorting things out, classification and its consequences*. MIT Press. https://mitpress.mit.edu/9780262522958/sorting-things-out

Braun, V., & Clarke, V. (2021). Thematic Analysis A Practical Guide. *Sage (Atlanta, Ga.)*.

Brown, J., & Dillard, J. (2015). Dialogic Accountings for Stakeholders: On Opening Up and Closing Down Participatory Governance. *Journal of Management Studies*, *52*(7), 961–985. DOI: 10.1111/joms.12153

Burchell, S., Clubb, C., Hopwood, A., Hughes, J., & Nahapiet, J. (1980). The roles of accounting in organizations and society. *Accounting, Organizations and Society*, *5*(1), 5–27. DOI: 10.1016/0361-3682(80)90017-3

Christensen, H. B., Hail, L., & Leuz, C. (2021). Mandatory CSR and sustainability reporting: Economic analysis and literature review. *Review of Accounting Studies*, *26*(3), 1176–1248. DOI: 10.1007/s11142-021-09609-5

Conselho das Finanças Públicas. (2014). *Relatório sobre a execução orçamental.* https://www.cfp.pt/pt/publicacoes/evolucao-orcamental/analise-da-conta-das-administracoes-publicas-2014

de Villiers, C., Rinaldi, L., & Unerman, J. (2014). Integrated Reporting: Insights, gaps and an agenda for future research. *Accounting, Auditing & Accountability Journal, 27*(7), 1042–1067. DOI: 10.1108/AAAJ-06-2014-1736

Denzin, N. K. (1978). *The research act, a theoretical introduction to sociological methods* (2nd ed.). McGraw Hill.

Eccles, R. G., & Klimenko, S. (2019). The investor revolution. *Harvard Business Review, 97*(3), 106–116.

Espeland, W. N., & Stevens, M. L. (1998). Commensuration as a Social Process. *Annual Review of Sociology, 24*(1), 313–343. DOI: 10.1146/annurev.soc.24.1.313

European Union. (2023). *European Sustainability Reporting Standards, delegated act.* https://eur-lex.europa.eu/legal-content/EN/TXT/?uri=CELEX:32023R2772

Fairclough, N. (2003). *Analysing Discourse.* Routledge., DOI: 10.4324/9780203697078

Flower, J. (2015). The International Integrated Reporting Council: A story of failure. *Critical Perspectives on Accounting, 27*, 1–17. DOI: 10.1016/j.cpa.2014.07.002

GRI. (2021). *The global standards for sustainability impacts.* https://www.globalreporting.org/standards/

Hines, R. D. (1988). Financial accounting: In communicating reality, we construct reality. *Accounting, Organizations and Society, 13*(3), 251–261. DOI: 10.1016/0361-3682(88)90003-7

Hopwood, A. G. (1987). The archeology of accounting systems. *Accounting, Organizations and Society, 12*(3), 207–234. DOI: 10.1016/0361-3682(87)90038-9

IFRS Foundation. (2023a). *IFRS S1 General Requirements for Disclosure of Sustainability-related Financial Information.* https://www.ifrs.org/issued-standards/ifrs-sustainability-standards-navigator/ifrs-s1-general-requirements/

IFRS Foundation. (2023b). *IFRS S2 Climate-related Disclosures.* https://www.ifrs.org/issued-standards/ifrs-sustainability-standards-navigator/ifrs-s2-climate-related-disclosures/

Kenter, J. O., Raymond, C. M., van Riper, C. J., Azzopardi, E., Brear, M. R., Calcagni, F., Christie, I., Christie, M., Fordham, A., Gould, R. K., Ives, C. D., Hejnowicz, A. P., Gunton, R., Horcea-Milcu, A.-I., Kendal, D., Kronenberg, J., Massenberg, J. R., O'Connor, S., Ravenscroft, N., & Thankappan, S. (2019). Loving the mess: Navigating diversity and conflict in social values for sustainability. *Sustainability Science, 14*(5), 1439–1461. DOI: 10.1007/s11625-019-00726-4

Krippendorff, K. (2019). *Content Analysis: An Introduction to Its Methodology.* SAGE Publications, Inc., DOI: 10.4135/9781071878781

Lapsley, I. (2009). New Public Management: The Cruellest Invention of the Human Spirit? [1]. *Abacus, 45*(1), 1–21. DOI: 10.1111/j.1467-6281.2009.00275.x

LI, F. (2010). The Information Content of Forward-Looking Statements in Corporate Filings—A Naïve Bayesian Machine Learning Approach. *Journal of Accounting Research, 48*(5), 1049–1102. DOI: 10.1111/j.1475-679X.2010.00382.x

MacKenzie, D. (2006). *An Engine, Not a Camera How Financial Models Shape Markets.* MIT Press.

Miller, P., & Power, M. (2013). Accounting, Organizing, and Economizing: Connecting Accounting Research and Organization Theory. *The Academy of Management Annals, 7*(1), 557–605. DOI: 10.5465/19416520.2013.783668

O'Dwyer, B., & Owen, D. L. (2005). Assurance statement practice in environmental, social and sustainability reporting: A critical evaluation. *The British Accounting Review, 37*(2), 205–229. DOI: 10.1016/j.bar.2005.01.005

Power, M. (1997). *The Audit Society Rituals of Verification.* Oxford University Press.

Ross, F. (2019). Kate Raworth - Doughnut Economics: Seven Ways to Think Like a 21st Century Economist (2017). *Regional and Business Studies, 11*(2). Advance online publication. DOI: 10.33568/rbs.2409

Schick, A. (2014). The metamorphoses of performance budgeting. *OECD Journal on Budgeting, 13*(2), 49–79. DOI: 10.1787/budget-13-5jz2jw9szgs8

Schoenmaker, D., & Schramade, W. (2019). *Principles of Sustainable Finance.*

Setia, N., Abhayawansa, S., Joshi, M., & Huynh, A. V. (2015). Integrated reporting in South Africa: Some initial evidence. *Sustainability Accounting. Management and Policy Journal, 6*(3), 397–424. DOI: 10.1108/SAMPJ-03-2014-0018

Shiller, R. J. (2019). *Narrative Economics: How Stories Go Viral and Drive Major Economic Events.* Princeton University Press.

Star, S. L. (1999). The Ethnography of Infrastructure. *The American Behavioral Scientist, 43*(3), 377–391. DOI: 10.1177/00027649921955326

Stent, W., & Dowler, T. (2015). Early assessments of the gap between integrated reporting and current corporate reporting. *Meditari Accountancy Research, 23*(1), 92–117. DOI: 10.1108/MEDAR-02-2014-0026

Suchman, M. C. (1995). Managing Legitimacy: Strategic and Institutional Approaches. *Academy of Management Review, 20*(3), 571–610. DOI: 10.5465/amr.1995.9508080331

Tavory, I., & Timmermans, S. (2014). *Abductive Analysis*. University of Chicago Press., DOI: 10.7208/chicago/9780226180458.001.0001

Timmermans, S., & Epstein, S. (2010). A World of Standards but not a Standard World: Toward a Sociology of Standards and Standardization. *Annual Review of Sociology, 36*(1), 69–89. DOI: 10.1146/annurev.soc.012809.102629

Tribunal de Contas. (2019). *Relatório de auditoria ao cumprimento das regras orçamentais*. https://www.tcontas.pt/pt-pt/ProdutosTC/Relatorios/RelatoriosAuditoria/Pages/Anos.aspx

Vargo, S. L., & Lusch, R. F. (2004). Evolving to a New Dominant Logic for Marketing. *Journal of Marketing, 68*(1), 1–17. DOI: 10.1509/jmkg.68.1.1.24036

Vargo, S. L., & Lusch, R. F. (2016). Institutions and axioms: An extension and update of service-dominant logic. *Journal of the Academy of Marketing Science, 44*(1), 5–23. DOI: 10.1007/s11747-015-0456-3

Villiers, C., & Farooq, M. (2019). Farooq, M.B. & De Villiers, C. 2019. Understanding how managers institutionalise sustainability reporting: Evidence from Australia and New Zealand, Accounting, Auditing & Accountability Journal, 32(5), 1240-1269. *Accounting, Auditing & Accountability Journal, 32*, 1240–1269.

Wodak, R. (2015). *Methods of Critical Discourse Studies* (3rd ed.).

World Resources Institute. (2011). *World Resources Institute, & World Business Council for Sustainable Development*. https://www.wri.org/research/greenhouse-gas-protocol-corporate-value-chain-scope-3-accounting-and-reporting-standard

ADDITIONAL READING

Burchell, S., Clubb, C., Hopwood, A., Hughes, J., & Nahapiet, J. (1980). The roles of accounting in organizations and society. *Accounting, Organizations and Society, 5*(1), 5–27. DOI: 10.1016/0361-3682(80)90017-3

Christensen, H. B., Hail, L., & Leuz, C. (2021). Mandatory CSR and sustainability reporting, economic analysis and literature review. *Review of Accounting Studies, 26*(3), 1176–1248. DOI: 10.1007/s11142-021-09609-5

Flower, J. (2015). The International Integrated Reporting Council, a story of failure. *Critical Perspectives on Accounting, 27*, 1–17.

Hines, R. D. (1988). Financial accounting, in communicating reality, we construct reality. *Accounting, Organizations and Society, 13*(3), 251–261. DOI: 10.1016/0361-3682(88)90003-7

Miller, P., & Power, M. (2013). Accounting, organizing, and economizing. *The Academy of Management Annals, 7*(1), 557–605. DOI: 10.1080/19416520.2013.783668

KEY TERMS AND DEFINITIONS

Accounting as Narrative Infrastructure: the view that accounting is a socio technical system that not only measures but scripts value, legitimacy and possible futures.

Anticipatory Stewardship: the integration of long-term horizons, scenarios and intergenerational commitments into mainstream accounts and governance.

Double Materiality: a materiality concept that requires reporting both on how sustainability issues affect the firm and on how the firm affects society and the environment.

Legitimacy Theory: the perspective that organizations seek alignment with social norms and expectations to secure approval, with accounting as a primary medium of legitimation.

Narrative Layer: the layer that stitches numbers into plots with causality, temporality and moral orders, giving information meaning and direction.

Performativity: the idea that models, metrics and standards do not merely describe the world, they help bring it into being when widely adopted.

Regenerative Models: economic approaches that prioritize the restoration of social and ecological systems, requiring accounts that recognize maintenance, care and long-term flourishing.

Semiotic Layer: the layer of categories, metrics, taxonomies and valuation conventions that make phenomena calculable.

Service-dominant Logic: a view of value as co-created in use, over time and across ecosystems, challenging static, transaction focused notions embedded in many accounting systems.

Systemic Layer: the layer of institutional design, standards, assurance regimes and digital infrastructures that stabilize or transform signs and narratives over time.

Chapter 6
Algorithmic Narrativity in Entrepreneurship:
Reframing management through storytelling

Domitilla Magni
https://orcid.org/0000-0001-5555-6004
Catholic University of the Sacred Heart, Italy & Bayes Business School, City St George's – University of London, UK

ABSTRACT

This chapter aims to explore how generative AI technologies contribute to the construction of entrepreneurial narratives and how these narratives, in turn, shape managerial imagination and economic transformation. Drawing on interdisciplinary approaches from management studies, communication, and digital ethnography, the chapter investigates the convergence of algorithmic generation and narrative design as a locus of entrepreneurial agency. The chapter argues that generative AI is not only a tool for automating content, but increasingly a co-author of strategic imaginaries, guiding how entrepreneurs envision futures, frame value, and articulate purpose. Through narrative analysis of AI-driven ventures, digital ethnography of entrepreneurial discourse on LinkedIn, and direct experimentation with generative models (e.g., ChatGPT), this chapter offers both a conceptual lens and empirical insight into the symbolic power of AI in contemporary enterprise. It shows how algorithmically-mediated storytelling becomes a driver of legitimacy, cultural alignment, and strategic orientation in the age of rapid technological change. Ultimately, the chapter proposes a framework for understanding entrepreneurial transformation not merely as organizational change, but as a co-constructed narrative process involving humans and machines alike.

DOI: 10.4018/979-8-3373-4566-6.ch006

1. INTRODUCTION

The contemporary landscape of entrepreneurship is undergoing a profound systemic transformation, extending beyond traditional market and institutional frameworks to encompass an increasingly influential domain of narratives (Audretsch & Lehmann, 2025). These narratives, crucial for shaping entrepreneurial endeavors, are now significantly influenced, authored, and amplified by advancements in Artificial Intelligence (AI). This chapter originates from a simple yet urgent observation: while extant scholarship has extensively explored digital entrepreneurship and the productive affordances of AI, there remains a notable gap in understanding the ways in which AI participates in the construction of meaning within organizational and entrepreneurial contexts. The growing presence of algorithmically generated language suggests that AI now plays an active role in shaping entrepreneurial identities, imaginaries, and strategies. This development invites a re-examination of foundational assumptions in management and entrepreneurship studies, particularly regarding narrative agency and authorship. Hence, this chapter examines how generative AI technologies are reshaping entrepreneurial storytelling and, in doing so, transforming the symbolic infrastructure upon which entrepreneurial ventures are built. Particular attention is given to the phenomenon of algorithmic narrativity—a form of organizational storytelling co-produced by human and machine intelligence.

Corporate storytelling based on generative AI not only improves the efficiency of information communication but also plays an active role in constructing the algorithmic narrativity that shape entrepreneurial visions (Vivek & Nanthagopan, 2023). AI is helping to define the positioning of start-ups and businesses in the market, build their legitimacy and establish relationships with stakeholders (Spanos, 2021; Kozinets, 2022). This opens important prospects at the entrepreneurial level, but communication must always focus on an approach based on authenticity, creativity, and consistency with corporate identity: consumers prefer authentic brands and stories that do not hide behind self-celebratory and rhetorical campaigns, but real campaigns capable of inspiring creativity, even if supported by external tools such as AI. Considering this, a growing number of entrepreneurs are relying on AI-based tools to develop strategic narrative elements such as elevator pitches, mission statements and investor presentations, enabling emerging companies to engage and persuade a diverse audience more effectively (Wen & Laporte, 2025). However, one of the main risks associated with the spread of AI, both in the field of communication and information and in relation to brands and businesses, is undoubtedly the loss of trust and credibility (Denning, 2006). The integration of AI into narrative processes raises important ethical questions, particularly with regard to authenticity and trust in the business context. Maintaining credibility and trust is a responsibility, and one of the risks of AI is that one may be tempted to tell one's story using only the potential of technology, neglecting creativity and authenticity (Larsen et al., 2020). The proposed analysis integrates theoretical perspectives from narrative studies (Czarniawska, 1997; Lounsbury & Glynn, 2001) and organizational communication (Cornelissen et al., 2006) in order to explore how storytelling by entrepreneurs using digital tools and generative AI can change business dynamics.

Entrepreneurial storytelling, particularly under conditions of novelty and ambiguity, operates as a mechanism for constructing legitimacy and enacting strategic agency. In parallel, emerging research conceptualizes generative AI not merely as a utilitarian artifact but as a semiotic agent embedded in socio-technical systems. The algorithmic logics that underpin these systems influence not only what is said, but how it is said, and to whom. This chapter explores the consequences of such influence within the entrepreneurial context: how algorithmic templates shape discursive form, how platform dynamics modulate emotional resonance, and how automated suggestions participate in the performance of entrepreneurial identity.

Methodologically, the study adopts a digital ethnographic lens, focusing on platforms such as LinkedIn as key sites for the discursive performance of entrepreneurship. These digital environments represent hybrid spaces where human creativity, algorithmic syntax, and social interaction intertwine. In such contexts, the boundaries between human storytelling and that generated by AI become increasingly blurred.

By integrating these theoretical and methodological perspectives, the chapter aims to contribute to a more articulated understanding of entrepreneurial transformation in the era of generative AI. Rather than treating AI as a neutral productivity tool, the analysis foregrounds its role as a cultural and discursive force—one that is actively shaping the narratives through which entrepreneurship is imagined, enacted, and institutionalized.

The chapter is structured as follows. First, we provide the theoretical framework on entrepreneurship and narrative intelligence. Then we systematically review topic as strategic communication, and human-AI collaboration. Next, we present our research design and methodology. We then describe our key findings and discuss their implications for theory and practice. Finally, we outline directions for future research and reflect on the study's limitations.

2. THEORETICAL FRAMEWORK: FROM ALGORITHMIC EFFICIENCY TO NARRATIVE INTELLIGENCE

2.1 The Interpretive Function of Generative AI in Entrepreneurial Meaning-Making

Management theories have long been developed based on a rational idea of decision-making. An important reference point was Herbert Simon's concept of "bounded rationality" (1947), according to which decisions are not perfect but are based on partial information and limited cognitive abilities. This idea has evolved into a more recent approach called "algorithmic rationality", which relies on the use of data, mathematical models and digital tools to seek optimal solutions in organizations. This approach, inspired by Taylor's scientific management (1911) and reinforced by operations research (Ackoff, 1979; Lowrie, 2017), has profoundly influenced the way managers have made decisions. However, this type of rationality has significant limitations when applied to today's business contexts. The environments in which start-ups and businesses operate are not only uncertain but also complex, with unclear objectives, different interpretations of reality and constant change. Decisions are no longer based exclusively on calculations, but on interpretative processes (Scuotto et al., 2024). Schön (1983) also observed that managerial practice often takes place in ambiguous situations, where problems are not well defined, values may conflict, and information is incomplete. In these contexts, the role of the manager is not simply to solve technical problems, but to reflect, judge and adapt to the context. Algorithmic approaches, while useful in clear and structured problems, are not sufficient when profound uncertainties, multiple meanings and tensions between values come into play.

To address these situations, scholars need theoretical models that recognize that organizational reality is not a given but is socially constructed. Institutional theory, for example (Berger & Luckmann, 1966; Meyer & Rowan, 1977; Scott, 2008), shows that organizations do not respond only to rational pressures, but operate within institutional environments in which shared meanings, social legitimacy and the narratives with which they construct their identity are very important.

Thus, language and narrative become central not only to strategic sensemaking but to the very constitution of organizational action. The advent of generative AI technologies introduces a new layer of complexity to this theoretical landscape. These systems, while grounded in advanced computational architectures, do not merely process data—they generate language, simulate reasoning, and construct narratives. In contrast to earlier forms of business intelligence focused on statistical analysis and predictive modeling (Davenport & Harris, 2017), generative AI engages in what may be described as narrative reasoning: the ability to create context-sensitive, coherent, and persuasive accounts that support decision-making under conditions of uncertainty.

This development challenges traditional ideas about algorithmic rationality. Generative models such as GPT-4 or Claude do not offer solutions in the classical sense of the term: instead of directly solving problems, they propose interpretations and contribute to the construction of meaning together with users. Their results are discursive in nature, i.e. they depend on the prompts received, the context and the dialogic interaction. They do not merely optimize information, but participate in interpretative processes, influencing the way content is framed, meanings attributed, and narratives constructed, especially in complex and evolving organizational contexts (Zhou et al., 2023). These AI skills are linked to the narrative reversal according to which reality in organizations is not objective and given, but is constructed through shared stories, metaphors and interpretations (Tsoukas & Hatch, 2001; Tsoukas, 2005). Narratives, therefore, do not merely describe what happens in organizations, but actively contribute to shaping it.

In this context, the concept of "algorithmic narrativity" can be introduced to describe the new form of intelligence that arises from the collaboration between human authors and machines. This form of narrativity not only expands traditional narrative possibilities but also integrates them into algorithmic systems capable of generating, adapting and disseminating content in real time. This marks the transition from a use of AI focused on computational efficiency to a use oriented towards narrative intelligence: a way of organizing knowledge and coordinating action through the co-construction of meaning. This transformation requires a change in perspective in the way AI is conceived in management and entrepreneurship. Scholars can no longer consider it solely as a neutral tool for improving rational analysis, but it must also recognize its active and performative role in shaping strategic visions, entrepreneurial identities and collective processes of meaning creation. Generative AI, therefore, is not just a technical infrastructure: it is also a cultural and symbolic actor in contemporary organizational life (Wen & Laporte, 2025).

2.2 Narrative as Organizational Technology: From Communication to Strategic Coordination

Narrative theory goes far beyond simple communication and considers storytelling as a true constitutive technology, capable of shaping organizational reality. The fundamental contributions of Czarniawska (1997, 2004) describe organizations not as static entities, but as constantly evolving narrative constructions, dynamic processes that are formed and reformed through the stories told by organizational actors. This view is based on the work of Berger and Luckmann (1966) on the social construction of reality, which highlights how shared narratives contribute to defining what is taken for granted in organizations, influencing behaviors and structures.

Building on this foundation, Boje (1991) introduces the idea that organizational stories are true "technologies of power", capable of producing and reproducing the identity, legitimacy and strategic orientation of the organization. Organizational storytelling thus serves multiple critical functions: it provides cognitive schemas that help actors interpret complex environments, fosters emotional engagement that motivates collective action, legitimizes the organization in the eyes of external stakeholders, and coordinates activities across structural and cultural boundaries.

Further development of the narrative approach is evident in the work of Gabriel (2000), who highlights narratives as "vehicles of meaning" enabling individuals to navigate uncertainty and ambiguity inherent in organizational life. Rhodes and Brown (2005) complete this perspective by defining narratives as "organizational tools" that help structure social relationships, influence resource allocation and guide strategic decision-making processes.

Recent studies on digital transformation have highlighted how technology amplifies these narrative functions. Yoo et al. (2010), for example, introduce the concept of "digital materiality" to describe how digital technologies integrate into organizational practices, changing the possibilities for action and the creation of meaning. In line with this approach, Orlikowski and Scott (2023) analyze how digital technologies generate new "sociomaterial" practices, in which the boundaries between human action and technological capabilities become increasingly blurred.

In this ever-changing landscape, generative AI is emerging as a transformative force. Unlike previous digital tools, which were mainly used to automate repetitive tasks or assist in analysis, generative AI actively participates in the creation of meaning within organizations (Wen, & Laporte, 2025). These systems can be considered as narrative amplifiers: technologies that enhance the human ability to tell stories, offering narrative structures, enriching language and enabling rapid iteration and revision of content. This amplification effect produces an important form of narrative leverage, i.e. the ability to generate engaging, context-rich stories at a speed and scale that would be impossible to achieve with human intervention alone. The concept draws inspiration from Arquilla and Ronfeldt's (2001) analysis of "netwar" and the strategic power of networked narratives, repurposed here to understand how AI-mediated storytelling creates new forms of competitive advantage in entrepreneurial and organizational settings.

2.3 Co-Construction Theory and the Emergence of Human-AI Collaboration

The theoretical framework for understanding human-AI collaboration in narrative construction draws on several converging streams of research in science and technology studies, organizational learning, and human-computer interaction. Pickering's (1995) concept of "the mangle of practice" provides a foundation for understanding how human agency and material agency (including AI systems) interact in complex, emergent ways that cannot be reduced to simple tool-use relationships.

Hutchins' (1995) work on "distributed cognition" offers another crucial theoretical lens. His analysis of how cognitive processes extend beyond individual minds to include tools, artifacts, and environmental structures provides a framework for understanding how AI systems participate in organizational thinking and decision-making. This perspective is further developed by Clark and Chalmers (1998) in their influential work on "extended mind", which argues that cognitive processes can legitimately extend beyond the boundaries of individual consciousness to include external tools and technologies.

Recent studies in human-computer interaction highlight how humans and AI can collaborate on creative and interpretive tasks. Li et al. (2024) introduce the concept of "human-AI collaboration" to describe interactive systems where both parties work together to achieve goals neither could reach alone. This approach relies on shared mental models, complementary skills, and ongoing interaction. In organizational storytelling, this co-creation takes shape through key mechanisms that go beyond simple automation or technical support:

- *Narrative scaffolding*: Drawing on Vygotsky's (1978) idea of scaffolding, AI can help entrepreneurs structure their thoughts about the market, competition, and value creation. It does this by offering simple story templates like problem-solution-outcome, or more advanced frameworks based on classical rhetoric (logos, pathos, ethos) and modern persuasion techniques.
- *Semantic amplification*: AI broadens the language and ideas entrepreneurs can use, helping them express thoughts that might otherwise stay unclear or underdeveloped. Based on cognitive linguistics (Lakoff & Johnson, 1980), this process shows how metaphors and concepts shape thinking. Trained on large business texts, AI can introduce new metaphors and connections that inspire creative thinking.
- *Iterative refinement*: The interactive nature of AI enables what Schön (1983) calls "reflective practice", or the ability to experiment, learn, and improve through action. Entrepreneurs can rapidly test different narrative structures, creating feedback loops that increase both clarity and communicative effec-

tiveness. This process is based on the experiment-driven innovation process described by Thomke (2003), applied to the construction of strategic stories.

- *Creative cross-fertilization*: AI can help entrepreneurs think more creatively by suggesting new connections between different fields. This is like Hargadon and Sutton's (1997) idea of knowledge brokering, where ideas move between areas. In this way, AI works as a "cognitive broker", inspiring entrepreneurs with insights from other industries, research, or history.
- *Dialogical construction of meaning*: Most importantly, the interaction between humans and AI creates a dialogue where meaning is formed through conversation, not predetermined. This ongoing exchange makes AI-assisted storytelling different from basic automation, allowing true collaboration and shared learning.

2.4 Legitimacy Theory and the Institutional Dimensions of AI-Mediated Storytelling

The use of AI to create entrepreneurial stories should be understood within a broader institutional context, where gaining stakeholder trust and support is essential. In this regard, Suchman's (1995) framework on organizational legitimacy is particularly useful. It identifies three key types of legitimacy that matter when entrepreneurs use AI to tell their stories:

- *Pragmatic legitimacy*: Is based on how much the audience believes the entrepreneur is acting in their interest. AI tools can help improve this kind of legitimacy by making business stories clearer, more persuasive, and more tailored to different stakeholder needs. For example, AI can help entrepreneurs explain their business idea in a way that shows a strong market fit, solid competitive advantages, and a clear understanding of customer needs.
- *Moral Legitimacy*: This reflects positive normative evaluation of the organization and its activities based on whether they are perceived as "the right thing to do". AI systems can help entrepreneurs construct narratives that align with prevailing moral frameworks and social values, though this raises important questions about authenticity and manipulation that we address in our discussion of ethical implications.
- *Cognitive legitimacy*: Depends on how familiar and natural the entrepreneurial activity seems within existing cultural and mental frameworks. AI-supported narratives help entrepreneurs use familiar language and structures but may also cause standardization, reducing originality.

By using AI effectively, entrepreneurs may be better able to respond to investor questions, show alignment with customer expectations, and demonstrate credibility in their chosen market. This makes pragmatic legitimacy not only easier to achieve but also more strategic — especially when entrepreneurs operate in crowded or uncertain markets where attention and trust are hard to gain. In this way, AI becomes not just a storytelling tool, but a resource for building stronger, more convincing relationships with key stakeholders.

DiMaggio and Powell's (1983) institutional theory provides additional insight into how AI-mediated storytelling may contribute to what they term "institutional isomorphism" – the tendency for organizations to become similar to others in their field. The use of AI systems trained on common business discourse may create "mimetic isomorphism" as entrepreneurs unconsciously adopt similar narrative frameworks and rhetorical strategies.

However, this institutional perspective must be balanced against insights from entrepreneurship research that emphasize the importance of distinctiveness and innovation. Navis and Glynn (2010) argue that entrepreneurial narratives must balance conformity and distinctiveness – they must be familiar enough to be comprehensible and legitimate, yet distinctive enough to create competitive advantage and capture stakeholder attention.

2.5 Strategic Narrative Theory and the Performative Dimensions of AI-Mediated Storytelling

The theory about how AI-driven stories influence strategy comes from recent work on strategic narratives. This research shows that stories are not just ways to describe strategy but are key parts of how strategy is made. Fenton and Langley (2011) explain that strategic stories help form an organization's identity, direct how resources are used, and manage relationships with stakeholders. This perspective builds on Austin's (1962) speech act theory and its organizational applications (Heracleous & Barrett, 2001), which examine how language doesn't merely describe reality but actively constructs it. In the context of entrepreneurial ventures, narratives function as "performative utterances" that bring new organizational realities into being through their articulation and circulation.

Barry and Elmes (1997) extend this analysis by examining how strategic narratives function as "technologies of rationality" that structure how organizational members understand and respond to strategic challenges. Their research shows that narratives do several things: they help people understand changes in the environment, create emotional motivation for action, build legitimacy, and coordinate teamwork across organizations.

When AI is involved in creating these strategic stories, it leads to what can be called "algorithmic performativity". This means AI-driven narratives don't just describe plans but actually shape what strategies entrepreneurs see as possible and choose to follow. This idea comes from Callon's (1998) work on "performativity" in economics, which shows how economic theories can shape the markets they describe. MacKenzie (2006) expands on this by showing that financial algorithms don't just model markets—they help create market realities. Similarly, AI-driven entrepreneurial stories don't just talk about market opportunities; they help create them by influencing how entrepreneurs, investors, and others understand and react to new technologies and market changes. This means AI stories actively shape entrepreneurial decisions and futures. Instead of seeing narratives as simple descriptions, they should be seen as powerful tools that help build the future of business. This view fits with strategy-as-practice research (Jarzabkowski et al., 2007), which studies how strategies develop through ongoing stories, actions, and real-world conditions.

2.6 Critical Synthesis and Research Gaps

While the existing literature provides a robust theoretical foundation for understanding AI-mediated storytelling, it tends to focus more on descriptive mapping than on critical contrast. Empirical studies that systematically assess the reception of AI-generated narratives by stakeholders are still limited. Some research points to growing skepticism among audiences regarding automated communication, as trust and perceived authenticity remain central in evaluating AI-generated material (Liang et al., 2023; Wen & Laporte, 2025). Similarly, recent work on cross-cultural narrative perception (Zhang et al., 2024) suggests that what appears persuasive or legitimate in one context may be perceived as artificial or manipulative in another.

A clear research gap thus emerges at the intersection of algorithmic narrativity and cultural variation. There is also a need to better understand how biases embedded in training data—stemming from provenance, representation, or domain mismatch—shape the symbolic frameworks reproduced by generative AI. In short, while much is known about the potential of AI as a storytelling tool, less is known about its limits, contradictions, and the conditions under which its use may undermine trust, distinctiveness, or legitimacy. These issues call for an empirical research agenda that balances enthusiasm for innovation with awareness of contextual and ethical constraints.

3. METHODOLOGY

3.1 Research Design

This study employs a mixed-methods research design grounded in interpretive epistemology and constructivist ontology (Berger & Luckmann, 1966; Guba & Lincoln, 1994). Given the emergent nature of AI-mediated entrepreneurial storytelling and the need to understand both the symbolic and material dimensions of this phenomenon, we adopted what Creswell and Plano Clark (2017) characterize as an "explanatory sequential" design, combining qualitative exploration with targeted experimentation.

Our methodological approach draws on three interdisciplinary traditions: management studies' emphasis on organizational sensemaking (Weick, 1995), communication studies' focus on narrative construction and discourse analysis (Fairclough, 2003), and digital ethnography's methods for understanding technology-mediated social practices (Hine, 2020). This interdisciplinary foundation enables us to examine how generative AI technologies function simultaneously as tools, co-authors, and mediators in entrepreneurial narrative construction.

The research design addresses three interconnected research questions: (1) How do entrepreneurs utilize generative AI in constructing strategic narratives? (2) What patterns emerge in AI-mediated entrepreneurial discourse across digital platforms? (3) How do generative AI systems influence the content, structure, and effectiveness of entrepreneurial storytelling?

3.2 Data Collection Methods

3.2.1 Narrative Analysis of AI-Driven Ventures

We conducted in-depth narrative analysis of 45 ventures that explicitly incorporate AI technologies in their business models and have documented their use of generative AI tools in strategic communication. These ventures were selected through purposive sampling (Patton, 2015) based on specific criteria including early-stage development (pre-Series A to Series B funding), technology-intensive business models with documented AI integration, public acknowledgment of generative AI use in marketing or communication materials, and availability of rich narrative content across multiple touchpoints.

Data collection for narrative analysis encompassed company websites and organizational narratives, publicly available pitch decks and investor presentations, founder interviews and media appearances, product descriptions and marketing materials, and social media content from official company accounts. This multi-source approach yielded a comprehensive corpus of 584 individual narrative artifacts spanning different communicative contexts and stakeholder audiences. We employed Labov's (1972) structural approach to narrative analysis to identify key narrative components including abstract, orientation, complicating action, evaluation, resolution, and coda, while also incorporating Bamberg and Georgakopoulou's (2008) "small stories" framework to capture fragmented narratives distributed across digital platforms and media touchpoints.

3.2.2 Digital Ethnography of Entrepreneurial Discourse on LinkedIn

Following Kozinets' (2022) updated framework for digital ethnography, we conducted a three-month ethnographic study of entrepreneurial discourse on LinkedIn, focusing on posts that explicitly mentioned or demonstrated use of generative AI tools in business contexts. LinkedIn was selected as the primary site due to its prominence as a platform for professional storytelling and entrepreneurial legitimation (Gershon, 2024). We used hashtag tracking and network sampling to follow entrepreneurial influencers and AI-focused business groups. First, we identified key hashtags about AI entrepreneurship, then manually selected relevant posts and their discussions. This process gave us 284 posts from 89 entrepreneurs. After filtering for clear discussions about AI use, we kept 115 posts. From these, we chose 88 posts for detailed analysis based on story quality, community interaction, and relevance to our study.

Data was collected daily by manually reviewing and saving posts, including text and images important for storytelling. We also tracked how people engaged and responded, while protecting privacy by anonymizing users and focusing on discussion patterns, not individuals. All analyzed content consisted of publicly posted material, and our research protocol received approval from the institutional ethics board prior to data collection.

3.2.3 Direct Experimentation With Generative Models

To understand how well AI can write entrepreneurial stories, we explored widely used tools such as ChatGPT, Claude, and some business writing apps. We created a simple method to evaluate the ability of AI to write different types of narratives typical of entrepreneurs: company origin stories, value statements, mission and vision statements, investor presentations, customer success stories, and crisis messages.

For each type of story, we provided the AIs with the same basic information, drawn from real (but anonymized) company profiles, to ensure a fair comparison. Each AI generated five different versions of each story, thus creating a large set of texts for analysis. We then asked twelve entrepreneurs and writing experts to do the same, using the same data.

We compared the texts (both AI and human) by evaluating aspects such as structure, completeness, persuasiveness, authenticity, consistency with business goals, and appropriateness for different audiences. This approach enabled a nuanced comparison, focusing on the quality of the stories and not just the automated data.

Table 1 shows an overview of data collection methods used in the analysis.

Table 1. Overview of data collection methods and analysis approaches

Method	Data Sources	Sample Size	Analysis Technique	Key Outputs
Narrative analysis of AI-driven ventures	Company websites, pitch decks, interviews, marketing materials, social media	45 ventures, 584 narrative artifacts	Structural narrative analysis (Labov, 1972); Small stories framework (Bamberg & Georgakopoulou, 2008)	Entrepreneurial narrative patterns, AI integration strategies
Digital Ethnography	LinkedIn posts and discussions	284 posts from 89 entrepreneurs over 3 months	Thematic analysis (Braun & Clarke, 2014); Discourse analysis (Fairclough, 2003)	Platform-specific storytelling practices, community dynamics
AI Experimentation	Generated content from ChatGPT, Claude, specialized tools	Multiple AI systems, 12 human benchmarks	Comparative qualitative analysis; Manual content examination	AI vs. human narrative characteristics, capability assessment

Source: Our elaboration

3.3 Data Analysis Procedures

3.3.1 Qualitative Data Analysis

Our qualitative analysis followed Braun and Clarke's (2014) thematic analysis approach, enhanced with narrative analysis techniques from Riessman (2008). The analysis process began with immersive data familiarization through repeated reading of all collected materials, followed by initial open coding that combined both inductive pattern recognition and deductive application of existing theoretical frameworks. We developed a hybrid coding scheme that integrated structural codes based on Labov's (1972) narrative components, process codes capturing human-AI interaction patterns, values codes identifying underlying entrepreneurial beliefs and motivations, and discourse codes analyzing rhetorical strategies and positioning moves.

Theme identification proceeded through iterative clustering of codes using manual affinity mapping techniques, with regular team discussions to refine emergent patterns. Two researchers independently coded twenty percent of the dataset to establish inter-rater reliability, achieving Cohen's kappa of 0.82 for structural codes and 0.78 for thematic codes, indicating substantial agreement according to Landis and Koch's (1977) criteria. The final analytical phase involved precise articulation of theme boundaries and relationships, culminating in the integration of themes into a coherent theoretical narrative.

3.3.2 Comparative Analysis of AI vs. Human-Generated Narratives

We employed systematic comparative analysis to identify distinctive characteristics of AI-mediated narratives through manual examination of linguistic, structural, and content dimensions. Linguistic analysis involved manual assessment of readability patterns, rhetorical device usage, and syntactic complexity through close reading techniques established in discourse analysis tradition (Fairclough, 2003). We examined sentence structures, vocabulary choices, and persuasive language patterns through qualitative coding rather than automated metrics.

We manually mapped information flow, narrative pacing, and logical progression across both AI-generated and human-authored content. Content analysis proceeded through interpretive examination of thematic patterns, metaphorical language use, and value proposition articulation strategies, employing constant comparative method (Kaplan & Duchon, 1988) to identify distinctive characteristics of AI-mediated versus purely human-authored narratives.

3.3.3 Pattern Recognition and Framework Development

In the final stage of the analysis, we focused on identifying recurring patterns in the data to build theoretical models useful for explaining how AI-mediated entrepreneurial storytelling works. We used three main approaches:

- *Cross-case comparative analysis:* we systematically compared different businesses, platforms, and types of storytelling to identify recurring similarities and differences.
- *Theoretical coding:* we applied existing theories (such as institutional theory, sensemaking theory, and storytelling theory) to interpret the observed patterns, while remaining open to new theoretical ideas emerging from the data.

- *Framework construction:* we iteratively developed conceptual models that explain the interrelationship between AI capabilities, human intervention, and narrative outcomes.

3.4 Validity and Reliability Considerations

3.4.1 Construct Validity

To ensure construct validity, we adopted several strategies. First, we used triangulation, comparing results obtained from different sources and methods to check for consistency. Second, we shared preliminary results with a selected group of entrepreneurs involved in the study, gathering their feedback (member checking). Finally, we submitted our analysis to expert opinion, involving both entrepreneurship scholars and researchers specializing in AI.

3.4.2 External Validity and Transferability

We enhanced transferability through:

- *Thick description*: Detailed contextual information enabling reader assessment of applicability.
- *Maximum variation sampling*: Including diverse venture types, stages, and founder backgrounds.
- *Theoretical sampling*: Selection of cases that test and extend emerging theoretical insights.

3.5 Limitations and Methodological Considerations

There are some important limitations to keep in mind when considering our findings. First, we focused mainly on early-stage startups in tech-heavy sectors. This means our conclusions might not fully apply to more traditional industries or well-established companies. Also, since AI tools are changing fast, what we observed reflects a particular moment in time—future changes in technology may lead to different outcomes.

Second, most of our data comes from English-speaking startup environments. Storytelling styles, how stakeholders react, and how AI is used can vary across cultures, so more research is needed to understand how these findings hold up in other parts of the world.

Third, while we included the views of entrepreneurs and some stakeholders, we didn't explore enough how investors and customers react to AI-generated narratives.

Since these groups play a key role in evaluating and supporting ventures, future research should look more closely at their responses.

Lastly, the ethical questions raised by AI-supported storytelling need to be studied over a longer period. Short-term research might miss some of the deeper effects that AI adoption can have on trust, fairness, and the overall health of entrepreneurial ecosystems.

4. EMPIRICAL FINDINGS: THE TRANSFORMATION OF ENTREPRENEURIAL NARRATIVE

4.1 Algorithmic Amplification of Narrative Capacity

Our findings suggest that generative AI systems profoundly change the way entrepreneurs construct their narratives, overcoming the three main limitations of traditional narratives:

- *Creating narrative variants:* AI can quickly generate many different versions of the same narrative (even more than 50), helping entrepreneurs experiment with different ways of presenting their idea or vision. This process, which used to take weeks of work, can now be done in a matter of hours, making it much easier to find the most effective story.
- *Rhetorical Sophistication:* AI provides immediate access to advanced communication techniques that normally require years of experience or training in marketing. The narratives produced automatically apply persuasive structures, refined language and proven rhetorical patterns. This is especially useful when the quality of the narrative directly affects funding or partnerships, providing any entrepreneur with professional-level communication tools.
- *Cross-Platform Consistency*: AI helps maintain consistency of messages across multiple channels, such as social media, investor presentations, or press releases. By saving and adapting key messages, it avoids contradictions or fragmentation and enables clear and professional communication without the need for a dedicated marketing team.

Table 2 sums up the main results.

Table 2. AI-driven narrative amplification mechanisms

Amplification Dimension	Traditional Approach	AI-Enhanced Approach	Key Benefits
Narrative Variants	Manual brainstorming, limited iterations	Automated generation of 10-50+ variants	Rapid testing, optimization
Rhetorical Sophistication	Requires specialized training/hiring	Instant access to advanced techniques	Cost reduction, quality improvement
Cross-Platform Consistency	Manual adaptation, prone to inconsistencies	Automated consistency maintenance	Brand coherence, professional image
Speed of Development	Days to weeks	Hours to days	Faster market entry, agility

Source: Our elaboration

These tools allow entrepreneurs to quickly create many different versions of their story, so they can try various ways of telling their idea and choose the one that is most clear and compelling. In addition, AI helps entrepreneurs write storytelling very effectively, using communication techniques that usually require years of experience. This is very useful, for example, when convincing investors or finding new partners. Finally, AI helps keep the same message across all channels: social media, presentations, interviews. In this way, the businesses' communication is always consistent and professional, even without a marketing team.

4.2 The Emergence of Hybrid Authorship

One of the most important findings concerns the emergence of "hybrid authorship" in entrepreneurial storytelling (see Figure 1). AI does not replace human creativity but offers a tool with which entrepreneurs can construct stories in new ways, combining their own insights with AI capabilities. The result is a narrative that neither human nor machine could produce on their own.

This type of collaboration takes the form of four key areas:

- in conceptual synthesis, AI helps connect ideas from different sources into a coherent narrative;
- in emotional resonance, the entrepreneur adds authenticity and human involvement to the narrative;
- in audience adaptation, the entrepreneur guides the AI in making the story fit specific contexts and audiences;
- in ethical filtering, it is always the entrepreneur who ensures that the narrative respects the values and goals of the organization.

Figure 1. The Hybrid Authorship Framework

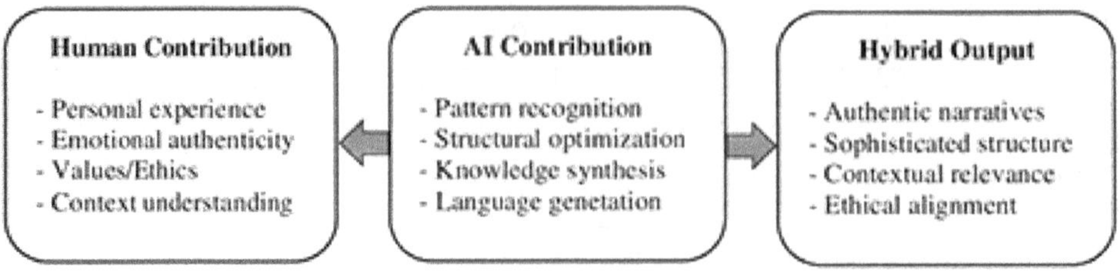

4.3 Narrative Legitimacy and Cultural Alignment

Our analysis demonstrates that AI-mediated storytelling serves as a crucial mechanism for establishing legitimacy and cultural alignment within entrepreneurial ecosystems, addressing what Suchman (1995) conceptualized as the fundamental challenge of organizational legitimacy – a generalized perception or assumption that the actions of an entity are desirable, proper, or appropriate within some socially constructed system of norms, values, beliefs, and definitions. Through the use of big data on corporate storytelling, AI systems help entrepreneurs create stories that meet widespread cultural expectations while maintaining a personal and original touch.

This legitimacy-building process relates back to institutional theory, according to which organizations tend to resemble each other in order to be accepted in their environment (DiMaggio & Powell, 1983). However, our research shows something new: AI does not merely reflect existing rules, but acts as a true "institutional mediator", helping entrepreneurs to cross what Lounsbury and Glynn (2001) call the "legitimacy threshold", that is, the minimum level of compliance necessary to be taken seriously. Thus, the entrepreneur no longer needs years of experience to learn how to communicate according to the rules of the game: AI gives entrepreneurS immediate access to credible and recognized language, making it easier for everyone to build legitimacy.

This function is performed through three main mechanisms, inspired by Zimmerman and Zeitz's (2002) model of legitimacy strategies for start-ups (see Table 3):

- *Discursive Conformity* is what we call "linguistic isomorphism": it means using the language and rhetorical forms typical of a certain industry to show that one is part of a recognized and legitimate business community. It is not just a matter of using the right words, but of mastering "linguistic capital" (Aburous & Kamla, 2022): that is, those socially recognized ways of speaking that confer authority and credibility on those who use them. AI systems help entrepreneurs access this capital by analyzing successful discourse patterns within specific entrepreneurial contexts and suggesting conforming language choices.

- *Trend Integration* is a mechanism that helps entrepreneurs appear up-to-date and in line with the times. Following Aldrich and Fiol's (1994) concept of cognitive legitimacy, AI systems help incorporate current business concepts and new trends into entrepreneurial narratives. This allows businesses to appear modern and relevant in the eyes of those who evaluate them.
- *Authority Signaling* works through what we call "computational sophistication", that is, the use of technical language and complex structures that convey competence and professionalism. This mechanism shows how AI can help entrepreneurs communicate authority while standing out and remaining credible, as suggested by Deephouse and Carter (2005).

Table 3. Legitimacy mechanisms in AI-mediated storytelling

Legitimacy Mechanism	Description	Implementation	Impact Metrics
Discursive Conformity	Adoption of field-specific language and structures	• Industry terminology integration • Standard narrative frameworks • Professional tone optimization	• Increased stakeholder comprehension • Enhanced credibility perception
Trend Integration	Incorporation of current business concepts	• Real-time trend analysis • Buzzword optimization • Market discourse alignment	• Perceived market awareness • Investor interest correlation
Authority Signaling	Communication of expertise and competence	• Technical language deployment • Conceptual framework sophistication • Evidence-based argumentation	• Stakeholder confidence • Partnership opportunities

Source: Our elaboration

4.4 Strategic Orientation and Future-Making

Most significantly, our research reveals that AI-mediated storytelling transcends mere communication to actively shape entrepreneurial strategic thinking and future construction. Table 4 shows how AI-mediated storytelling influences business strategy. Narrative models created through collaboration between people and AI become mental tools that help in decision making and strategy planning.

This future-building function is carried out through three key processes:

- *Scenario Construction*, in which AI helps create detailed future narratives for strategic planning;
- *Opportunity Framing*, which generates different ways of understanding market possibilities;
- *Vision Articulation*, which creates engaging future visions to coordinate the team and align stakeholders.

Table 4. Strategic impact of AI-mediated storytelling

Strategic Function	Process Description	Cognitive Tools Generated	Business Outcomes
Scenario Construction	Development of detailed future narratives	• Multiple timeline scenarios • Risk-opportunity matrices • Contingency frameworks	• Enhanced strategic planning • Improved risk management • Scenario-based decision making
Opportunity Framing	Multiple perspective generation on market opportunities	• Alternative market narratives • Competitive positioning frameworks • Value proposition variants	• Expanded opportunity recognition • Strategic option identification • Market positioning clarity
Vision Articulation	Creation of compelling future visions	• Shared vision statements • Stakeholder alignment tools • Motivational frameworks	• Team coordination improvement • Stakeholders buy-in • Organizational direction clarity

Source: Our elaboration

While these mechanisms offer evident advantages, they also entail trade-offs that deserve closer consideration. The transformation of entrepreneurial narrative through AI integration represents a fundamental shift in how ventures construct meaning, establish legitimacy, and envision their futures. This evolution shows that competitive advantage increasingly depends not only on what entrepreneurs do, but on how well they can collaborate with AI systems to tell their stories and build their strategic reality.

The emergence of "hybrid authorship" models indicates that the future of entrepreneurial storytelling will not be the replacement of humans with AI, but the development of advanced collaboration between humans and AI that leverages the strengths of both: human creativity and machine intelligence.

5. DISCUSSION

5.1 Reconceptualizing Managerial Agency Through the Lens of Digital Transformation

The findings presented in this study necessitate a fundamental reconceptualization of managerial agency in the context of digital transformation. Traditional management theories, rooted in scientific management principles (Taylor, 1911) and classical administrative theory, emphasize individual decision-making authority, rational-analytical processes, and hierarchical command structures. However, our research reveals that contemporary managerial practice increasingly involves what Orlikowski and Scott (2008) term "sociomaterial" arrangements, where agency emerges through human-AI collaboration rather than individual cognition alone.

This transformation aligns with recent developments in management theory that emphasize distributed cognition (Hutchins, 1995; Weick & Roberts, 1993; Heavey & Simsek, 2017) and collective sensemaking (Weick, 1995; Vaara & Whittle, 2022). Our findings extend these concepts by demonstrating how AI systems function as cognitive partners in managerial processes, particularly in narrative construction and strategic storytelling. This represents a shift from what Mintzberg (1973) characterized as purely human managerial roles to hybrid human-AI configurations that require new competencies and evaluation frameworks.

The implications for management education and practice are profound. As Davenport and Kirby (2016) argue in their work on augmented intelligence, managers must develop capabilities for effective human-AI collaboration. Our research specifically identifies narrative co-construction as a critical competency, requiring managers to understand not only traditional analytical skills but also how to leverage AI capabilities for strategic communication and organizational sensemaking.

5.2 The Democratization of Strategic Storytelling: Disrupting Traditional Narrative Hierarchies

Our results show that traditional roles and hierarchies in entrepreneurial storytelling are changing significantly. Previously, strategic storytelling skills were mainly in the hands of those with good education, professional experience, and cultural prestige (Denning, 2006). This created "entrepreneurial legitimacy gaps" (Lounsbury and

Glynn, 2001), that is, situations in which good ideas did not receive resources not because of lack of merit, but because they were not presented well.

The use of generative AI represents a real revolution in this field, similar to that described by Christensen (1977) as a disruptive innovation in strategic communication. These systems allow businessman, even without a specific background, to use sophisticated narrative tools, thereby lowering barriers to entry for entrepreneurs and increasing the ability to recognize and exploit new opportunities (Shane and Venkataraman, 2000).

But this democratization also leads to a paradox. While it makes advanced storytelling tools accessible, it can also cause a certain uniformity in entrepreneurial narratives, a phenomenon that DiMaggio and Powell (1983) call "institutional isomorphism". In practice, if too many entrepreneurs use the same AI-generated models, stories risk becoming too similar, reducing authenticity and the ability to stand out. This creates new social pressures toward conformity, limiting narrative diversity. Finally, AI opens many opportunities but requires attention to maintain variety and originality in entrepreneurial stories.

5.3 Ethical Considerations and Authenticity: Navigating the Authenticity-Efficiency Paradox

The use of AI in building strategic narratives creates what we call the "authenticity-efficiency paradox". On the one hand, AI saves time and effort in creating entrepreneurial stories; on the other hand, there is a risk that these stories will appear inauthentic in the eyes of customers, investors, or partners. This paradox not only affects traditional ethics in business (Liedtka, 2008; Mazutis & Slawinski, 2015) but also touches on how technology changes communication between people (Licoppe & Smoreda, 2005).

According to our results, successful entrepreneurs actively manage this balance by combining the possibilities offered by AI with their will and intention. In practice, they do not give up the tools of AI but use them to truly express their values and goals while remaining true to their identity. This way of acting relates to Steyn (2004) idea that a communication is valid not because of who writes it, but because it is sincere and true.

Many entrepreneurs also choose to be very transparent about their use of AI, as Rawlins (2008) suggests with the concept of "radical transparency". They openly state that they use AI in their messages and explain how, to build trust with their stakeholders. This helps to reassure stakeholders and demonstrate that, even with the help of AI, communication remains honest and meaningful.

These reflections are also important for those involved in ethics training in companies and for professionals, because they show how the use of AI should be handled with awareness and responsibility. These findings suggest the need for new ethical frameworks that account for human-AI collaboration in strategic communication, extending beyond traditional concepts of individual moral agency to encompass collective responsibility in sociotechnical systems.

A notable example emerged in our field observations: one early-stage technology venture launched a fully AI-generated crowdfunding campaign, which, despite linguistic polish and persuasive structure, failed to attract investors. Post-campaign interviews revealed that potential backers perceived the messaging as "too generic" and lacking a genuine founder voice. This anecdote, while not statistically representative, illustrates the fragility of authenticity when automation exceeds human presence. It reinforces the importance of what we term the "authenticity-efficiency paradox", where efficiency gains can easily undermine the relational dimension of trust.

5.4 Implications for Organizational Learning and Innovation

The integration of AI into entrepreneurial storytelling creates what we call narrative institutional memory — a system that captures and saves storytelling practices beyond individual organizations. This idea builds on Levitt and March's (1988) theory of organizational learning by adding technology as a way to store shared entrepreneurial knowledge.

Our research shows that AI works like a "knowledge converter" (Nonaka et al., 1996), turning the hidden know-how of entrepreneurial storytelling into clear and organized frameworks that new entrepreneurs can easily use. This improves what Cohen and Levinthal (1990) call absorptive capacity, meaning the ability to find, understand, and use new knowledge for business success.

Because AI lets entrepreneurs quickly create and test different story versions, it supports what Thomke (2003) calls "rapid experimentation". This helps entrepreneurs try out new ideas fast by the "discovery-driven planning", as when positioning their business and explaining their value (McGrath, 2010; Mansoori & Lackéus, 2020).

These abilities have important effects on innovation theory. They suggest that narrative innovation is a special type of innovation, adding to product, process, and business model innovations (Nambisan et al., 2017). This kind of innovation could become a lasting competitive advantage in crowded markets (Reissner, 2005).

Also, AI's ability to support shared learning opens the door for open innovation in storytelling, as Chesbrough (2003) describes. Successful stories can be shared, changed, and improved across entrepreneurial communities, speeding up overall innovation and helping more ventures succeed. The implications extend to venture capital and entrepreneurial support organizations, who may need to develop new evaluation criteria that account for AI-augmented storytelling capabilities while maintaining focus on underlying business fundamentals and authentic entrepreneurial vision.

5.5 Boundary Conditions and Contextual Constraints

AI-mediated storytelling cannot be universally effective across all contexts. Several boundary conditions influence its outcomes.

Stakeholder Reactions

Investor and customer responses to AI-generated content are ambivalent. While some stakeholders appreciate precision and clarity, others associate automation with inauthenticity or lack of moral responsibility (Liang et al., 2023). Entrepreneurs must therefore calibrate their transparency about AI use depending on audience expectations and trust norms.

Cross-Cultural Variations

Cultural frameworks deeply affect how algorithmic narratives are interpreted. In high-context cultures, the subtlety of human tone and emotion remains crucial, whereas in low-context environments, efficiency and clarity may be valued more (Zhang et al., 2024). These differences highlight the importance of adapting AI-assisted communication to local storytelling conventions.

Regulatory and Governance Context

The emerging debate on disclosure obligations and ethical auditing for AI-generated content is redefining communication standards. Some organizations have begun adopting voluntary disclosure statements about AI assistance in public materials, anticipating future regulatory frameworks.

Bias and Data Provenance

Generative AI systems are trained on vast datasets whose provenance often remains opaque. This can introduce latent biases, favoring dominant cultural perspectives and linguistic patterns. Such biases influence narrative framing and, consequently, the perceived legitimacy of AI-mediated stories. Awareness of these limitations should inform both academic research and managerial practice.

6. CONCLUSIONS AND FUTURE RESEARCH AGENDA

6.1 Theoretical Contributions

This study offers important new ideas to management research, especially where digital change, entrepreneurship, and strategic communication meet.

First, we expand organizational cognition theory by showing that AI acts as a thinking partner in management, especially in creating stories and making sense of strategies. This challenges old views that managers work alone and points to new mixed human-AI thinking systems that need fresh theories.

Second, our work adds to entrepreneurship theory by showing how AI helps more people tell strategic stories, breaking down traditional limits to who can participate as entrepreneurs. This is a new kind of technological change that affects communication skills instead of just products or services.

Third, we introduce "narrative institutional memory," a new idea about organizational learning made possible by AI. This shows how technology can store and create hidden entrepreneurial knowledge, turning personal storytelling skills into shared resources for entrepreneurs.

Finally, we identify the "authenticity-efficiency paradox," adding to business ethics by highlighting new moral questions that arise when AI shapes human communication in business. This paradox requires new ethical frameworks that account for technological mediation while preserving authentic stakeholder relationships.

6.2 Practical Implications

This study offers clear and practical insights for managers, educators, and entrepreneurs.

In entrepreneurship education, schools should start teaching students how to work effectively with AI tools. At the same time, it's important to help them develop a strong personal vision and communicate in honest and responsible ways. Business schools may need to create new teaching approaches that combine AI training with creativity, ethics, and personal expression.

For entrepreneurs, using AI to shape their business stories can make communication more powerful and help build stronger connections with investors, partners, and customers. However, success depends on learning how to collaborate with AI, being transparent about its use, and making sure the messages still reflect the entrepreneur's true values and goals. Entrepreneurs should also think strategically about when and how to use AI — for example, in pitch presentations, websites, or customer interactions — to create more consistent and engaging narratives.

For investors and venture capital firms, our findings suggest that it's time to update evaluation criteria. Alongside business models and financial indicators, investors should also look at how well entrepreneurs use AI to explain their ideas and opportunities. Skillful use of AI in storytelling can be a sign of strong leadership, clarity of vision, and readiness to compete in fast-changing markets.

For managers and leaders in any organization, the ability to work with AI is becoming a key part of the job. It's not just about knowing the technology, but also about using it to communicate clearly, inspire trust, and drive action. Organizations should invest in training programs that build these hybrid capabilities.

As AI becomes part of everyday management practice, those who learn to use it thoughtfully—not just efficiently—will be better equipped to lead, innovate, and build lasting relationships in an increasingly complex and competitive environment.

6.3 Future Research Agenda

Based on our findings and identified limitations, we propose a comprehensive future research agenda.

6.3.1 Longitudinal Studies of AI Integration in Management Practice

Future research should look at the long-term effects of using AI in management by following businesses and entrepreneurs over time. This kind of study would help us understand how the way managers approach with AI changes as they become more familiar with these tools. It could also show how communication supported by AI influences trust, relationships, and results over the long run. Another important question is how human skills and AI technologies grow and change together inside organizations.

6.3.2 Cross-Cultural and Cross-Industry Validation

Our findings require validation across diverse cultural and industry contexts. Future research should investigate:

- How do cultural differences in communication styles and stakeholder expectations influence the effectiveness of AI-mediated narratives?
- What variations exist in AI adoption and effectiveness across different industries and organizational types?
- How do regulatory environments and professional norms in different contexts shape human-AI collaboration practices?

6.3.3 Stakeholder Response and Effectiveness Measurement

One important area that needs more research is how stakeholders respond to communication supported by AI. Future studies should explore how different groups—such as investors, customers, and partners—react to business stories that are shaped with the help of AI tools. It is also important to understand what makes people trust or reject messages created or assisted by AI. Finally, research should look at how organizations can measure whether their AI-supported communication is actually working—whether it helps them build stronger relationships, gain support, or achieve strategic goals.

6.3.4 Ethical Frameworks and Governance Mechanisms

The ethical implications we identify require substantial additional research to develop practical frameworks for responsible AI use in business communication:

- What ethical guidelines should govern AI use in strategic communication and stakeholder engagement?
- How can organizations balance efficiency gains from AI assistance with authenticity expectations?
- What governance mechanisms are needed to ensure responsible AI adoption in management practice?

6.3.5 Technological Evolution and Capability Development

As AI technologies continue to evolve quickly, it is essential for research to explore how these advancements are changing the strategic management approach. For example, how will more advanced AI features—like the ability to understand images, text, and voice together, or to interact in real time—change the way humans and AI collaborate? What new skills will managers need as AI becomes more powerful? And how should management education evolve to prepare future leaders to work effectively in environments where AI plays an active role in decision-making and communication? These questions are key to shaping the future of leadership and organizational success in the digital age.

6.3.6 Organizational Learning and Knowledge Management

Our identification of AI systems as repositories of organizational knowledge suggests several important research directions:

- How can organizations effectively capture and leverage collective narrative knowledge through AI systems?
- What are the implications of AI-mediated knowledge transfer for competitive advantage and innovation?
- How do AI systems influence organizational memory and institutional learning processes?

6.4 Concluding Remarks

This research shows that managerial strategy is going through a big change in the digital age. Using AI in strategic communication and storytelling is not just about adopting new technology – it's about rethinking what it means to be a manager. As AI advances, it will become increasingly important to be able to interact effectively with these systems while maintaining authentic human relationships.

AI also offers the chance to make strategic storytelling more accessible and inclusive. In the past, only people with strong skills or many resources could create effective stories. Now, AI can help even less experienced entrepreneurs tell their ideas clearly. However, to really benefit from this, we need to focus on a few key points: using AI ethically, communicating in a truthful way, and creating new ways to evaluate and support collaboration between humans and AI in business.

Our findings suggest that management theory and practice are at a turning point. It's time to think carefully about how skills, work styles, and relationships with intelligent technologies are changing. The choice is not between human or AI, but rather how to most effectively combine human creativity, values, and relationship-building capabilities with AI's analytical power and narrative sophistication. Organizations and individuals that master this combination will likely possess significant advantages in an increasingly complex and fast-paced business environment.

The research agenda we propose aims to deepen understanding of these phenomena and provide practical guidance for navigating the challenges and opportunities of AI-augmented management practice. As AI systems become more advanced and widespread, the questions raised by this research will become even more important for both management researchers and professionals. By exploring both the benefits and challenges of AI integration, management scholars can support the development of smarter, more ethical, and more inclusive forms of organizational leadership in the digital era.

While the promises of AI in narrative construction are significant, its responsible integration will depend on cultural sensitivity, ethical transparency, and the human capacity to maintain meaning within technological mediation

REFERENCES

Aburous, D., & Kamla, R. (2022). Linguistic tensions in a professional accounting field: English linguistic capital, hierarchy, prestige, and distinction among accountants. *Contemporary Accounting Research*, *39*(2), 1120–1149.

Ackoff, R. L. (1979). The future of operational research is past. *The Journal of the Operational Research Society*, *30*(2), 93–104.

Aldrich, H. E., & Fiol, C. M. (1994). Fools rush in? The institutional context of industry creation. *Academy of Management Review*, *19*(4), 645–670.

Arquilla, J., & Ronfeldt, D. (2001). *Networks and netwars: The future of terror, crime, and militancy*. Rand Corporation.

Audretsch, D. B., & Lehmann, E. E. (2025). A knowledge spillover narrative theory of entrepreneurship. *The Journal of Technology Transfer*, •••, 1–17.

Bamberg, M., & Georgakopoulou, A. (2008). Small stories as a new perspective in narrative and identity analysis. *Text & Talk*, *28*(3), 377–396.

Barry, D., & Elmes, M. (1997). On paradigms and narratives: Barry and Elmes' response. *Academy of Management Review*, •••, 847–849.

Berger, P., & Luckmann, T. (1966). The social construction of reality. In *Social theory re-wired* (pp. 110–122). Routledge.

Boje, D. M. (1991). Consulting and change in the storytelling organisation. *Journal of Organizational Change Management*, *4*(3), 7–17.

Callon, M. (1998). Introduction: The embeddedness of economic markets in economics. *The Sociological Review*, *46*(1, suppl), 1–57.

Chesbrough, H. (2003). The logic of open innovation: Managing intellectual property. *California Management Review*, *45*(3), 33–58.

Christensen, C. M. (1977). *The innovator's dilemma: when new technologies cause great firms to fail*. Harvard Business Review Press.

Clark, A., & Chalmers, D. (1998). The extended mind. *Analysis*, *58*(1), 7–19.

Clarke, V., & Braun, V. (2014). Thematic analysis. In *Encyclopedia of critical psychology* (pp. 1947–1952). Springer.

Cohen, W. M., & Levinthal, D. A. (1990). Absorptive capacity: A new perspective on learning and innovation. *Administrative Science Quarterly*, *35*(1), 128–152.

Cornelissen, J., Bekkum, T. V., & Ruler, B. V. (2006). Corporate communications: A practice-based theoretical conceptualization. *Corporate Reputation Review*, *9*(2), 114–133.

Creswell, J. W., & Clark, V. L. P. (2017). *Designing and conducting mixed methods research*. Sage publications.

Czarniawska, B. (Ed.). (1997). *A narrative approach to organization studies*. Sage Publications.

Czarniawska, B. (2004). *Narratives in Social Science Research*. SAGE.

Davenport, T., & Harris, J. (2017). *Competing on analytics: Updated, with a new introduction: The new science of winning*. Harvard Business Press.

Davenport, T. H., & Kirby, J. (2016). Just how smart are smart machines? *MIT Sloan Management Review*, *57*(3), 21.

Deephouse, D. L., & Carter, S. M. (2005). An examination of differences between organizational legitimacy and organizational reputation. *Journal of Management Studies*, *42*(2), 329–360.

Denning, S. (2006). Effective storytelling: Strategic business narrative techniques. *Strategy and Leadership*, *34*(1), 42–48.

DiMaggio, P. J., & Powell, W. W. (1983). The iron cage revisited: Institutional isomorphism and collective rationality in organizational fields. *American Sociological Review*, *48*(2), 147–160.

Fairclough, N. (2003). *Analysing discourse* (Vol. 270). Routledge.

Fenton, C., & Langley, A. (2011). Strategy as practice and the narrative turn. *Organization Studies*, *32*(9), 1171–1196.

Gabriel, Y. (2000). *Storytelling in organizations: Facts, fictions, and fantasies: Facts, fictions, and fantasies*. OUP Oxford.

Gershon, I. (2024). *Down and out in the new economy: How people find (or don't find) work today*. University of Chicago Press.

Guba, E. G., & Lincoln, Y. S. (1994). Competing paradigms in qualitative research. *Handbook of qualitative research, 2*(163-194), 105.

Hargadon, A., & Sutton, R. I. (1997). Technology brokering and innovation in a product development firm. *Administrative Science Quarterly*, •••, 716–749.

Heavey, C., & Simsek, Z. (2017). Distributed cognition in top management teams and organizational ambidexterity: The influence of transactive memory systems. *Journal of Management*, *43*(3), 919–945.

Heracleous, L., & Barrett, M. (2001). Organizational change as discourse: Communicative actions and deep structures in the context of information technology implementation. *Academy of Management Journal*, *44*(4), 755–778.

Hine, C. (2020). *Ethnography for the internet: Embedded, embodied and everyday*. Routledge.

Hutchins, E. (1995). *Cognition in the Wild*. MIT press.

Jarzabkowski, P., Balogun, J., & Seidl, D. (2007). Strategizing: The challenges of a practice perspective. *Human Relations*, *60*(1), 5–27.

Kaplan, B., & Duchon, D. (1988). Combining qualitative and quantitative methods in information systems research: A case study. *Management Information Systems Quarterly*, •••, 571–586.

Kozinets, R. V. (2022). Algorithmic branding through platform assemblages: Core conceptions and research directions for a new era of marketing and service management. *Journal of Service Management*, *33*(3), 437–452.

Labov, W. (1972). *Language in the inner city*. University of Pennsylvania Press.

Lakoff, G., & Johnson, M. (1980). The metaphorical structure of the human conceptual system. *Cognitive Science*, *4*(2), 195–208.

Landis, J. R., & Koch, G. G. (1977). The measurement of observer agreement for categorical data. *Biometrics*, •••, 159–174.

Larsen, J., Boje, D. M., & Bruun, L. (2020). *True storytelling: Seven principles for an ethical and sustainable change-management strategy*. Routledge.

Levitt, B., & March, J. G. (1988). Organizational learning. *Annual Review of Sociology*, *14*(1), 319–338.

Li, H., Wang, Y., & Qu, H. (2024, May). Where are we so far? understanding data storytelling tools from the perspective of human-ai collaboration. In *Proceedings of the 2024 CHI Conference on Human Factors in Computing Systems* (pp. 1-19).

Liang, J., Wang, L., Luo, J., Yan, Y., & Fan, C. (2023). The relationship between student interaction with generative artificial intelligence and learning achievement: Serial mediating roles of self-efficacy and cognitive engagement. *Frontiers in Psychology*, *14*, 1285392.

Licoppe, C., & Smoreda, Z. (2005). Are social networks technologically embedded?: How networks are changing today with changes in communication technology. *Social Networks*, *27*(4), 317–335.

Liedtka, J. (2008). Strategy making and the search for authenticity. *Journal of Business Ethics*, *80*(2), 237–248.

Lounsbury, M., & Glynn, M. A. (2001). Cultural entrepreneurship: Stories, legitimacy, and the acquisition of resources. *Strategic Management Journal*, *22*(6-7), 545–564.

Lowrie, I. (2017). Algorithmic rationality: Epistemology and efficiency in the data sciences. *Big Data & Society*, *4*(1), 2053951717700925.

MacKenzie, D. (2006). Is economics performative? Option theory and the construction of derivatives markets. *Journal of the History of Economic Thought*, *28*(1), 29–55.

Mansoori, Y., & Lackéus, M. (2020). Comparing effectuation to discovery-driven planning, prescriptive entrepreneurship, business planning, lean startup, and design thinking. *Small Business Economics*, *54*(3), 791–818.

Mazutis, D. D., & Slawinski, N. (2015). Reconnecting business and society: Perceptions of authenticity in corporate social responsibility. *Journal of Business Ethics*, *131*(1), 137–150.

McGrath, R. G. (2010). Business models: A discovery driven approach. *Long Range Planning*, *43*(2-3), 247–261.

Mintzberg, H. (1973). Strategy-making in three modes. *California Management Review*, *16*(2), 44–53.

Nambisan, S., Lyytinen, K., Majchrzak, A., & Song, M. (2017). Digital innovation management. *Management Information Systems Quarterly*, *41*(1), 223–238.

Navis, C., & Glynn, M. A. (2010). How new market categories emerge: Temporal dynamics of legitimacy, identity, and entrepreneurship in satellite radio, 1990–2005. *Administrative Science Quarterly*, *55*(3), 439–471.

Nonaka, L., Takeuchi, H., & Umemoto, K. (1996). A theory of organizational knowledge creation. *International Journal of Technology Management*, *11*(7-8), 833–845.

Orlikowski, W. J., & Scott, S. V. (2023). The digital undertow and institutional displacement: A sociomaterial approach. *Organization Theory, 4*(2).

Patton, M. Q. (2015). *Qualitative Research & EvaluationMethods* (4th ed.). SAGE.

Pickering, A. (1995). Concepts and the mangle of practice: Constructing quaternions. *The South Atlantic Quarterly*, *94*(2), 417–465.

Rawlins, B. (2008). Give the emperor a mirror: Toward developing a stakeholder measurement of organizational transparency. *Journal of Public Relations Research, 21*(1), 71–99.

Reissner, S. C. (2005). Learning and innovation: A narrative analysis. *Journal of Organizational Change Management, 18*(5), 482–494.

Rhodes, C., & Brown, A. D. (2005). Narrative, organizations and research. *International Journal of Management Reviews, 7*(3), 167–188.

Riessman, C. K. (2008). Narrative methods for the human sciences. *Sage (Atlanta, Ga.).*

Schön, D. A. (1983). *The reflective practitioner: How professionals think in action.* New York: Scuotto, V., Magni, D., Garcia-Perez, A., & Pironti, M. (2024). The impact of innovation failure: Entrepreneurship adversity or opportunity? *Technovation, 131*, 102944.

Shane, S., & Venkataraman, S. (2000). The promise of entrepreneurship as a field of research. *Academy of Management Review, 25*(1), 217–226.

Simon, H. A. (1947). *Administrative Behavior* (1st ed.). Free Press.

Spanos, M. (2021). Brand storytelling in the age of artificial intelligence. *Journal of Brand Strategy, 10*(1), 6–13.

Steyn, B. (2004). From strategy to corporate communication strategy: A conceptualisation. *Journal of Communication Management (London), 8*(2), 168–183.

Suchman, M. C. (1995). Managing legitimacy: Strategic and institutional approaches. *Academy of Management Review, 20*(3), 571–610.

Taylor, F. W. (1911). Principles and methods of scientific management. *Journal of Accountancy, 12*(3), 3.

Thomke, S. H. (2003). *Experimentation matters: unlocking the potential of new technologies for innovation.* Harvard Business Press.

Tsoukas, H. (2005). Afterword: Why language matters in the analysis of organizational change. *Journal of Organizational Change Management, 18*(1), 96–104.

Vaara, E., & Whittle, A. (2022). Common sense, new sense or non-sense? A critical discursive perspective on power in collective sensemaking. *Journal of Management Studies, 59*(3), 755–781.

Vivek, R., & Nanthagopan, Y. (2023). Storytelling as a qualitative approach for organizational management. *European Journal of Management Issues, 31*(2), 113–122.

Vygotsky, L. S. (1978). *Mind in society: The development of higher psychological processes* (Cole, M., John-Steiner, V., Scribner, S., & Souberman, E., Eds.). Harvard University Press.

Weick, K. E. (1995). Sensemaking in organizations: Vol. 3. *No. 10.1002*. Sage publications.

Weick, K. E., & Roberts, K. H. (1993). Collective mind in organizations: Heedful interrelating on flight decks. *Administrative Science Quarterly*, •••, 357–381.

Wen, Y., & Laporte, S. (2025). Experiential narratives in marketing: A comparison of generative AI and human content. *Journal of Public Policy & Marketing*, *44*(3), 392–410.

Yoo, Y., Henfridsson, O., & Lyytinen, K. (2010). The new organizing logic of digital innovation: An agenda for information systems research. *Information Systems Research*, *21*(5), 724–735.

Zhang, S., Zhao, X., Zhou, T., & Kim, J. H. (2024). Do you have AI dependency? The roles of academic self-efficacy, academic stress, and performance expectations on problematic AI usage behavior. *International Journal of Educational Technology in Higher Education*, *21*(1), 34.

Zhou, W., Zhang, C., Wu, L., & Shashidhar, M. (2023). ChatGPT and marketing: Analyzing public discourse in early Twitter posts. *Journal of Marketing Analytics*, *11*(4), 693–706.

Zimmerman, M. A., & Zeitz, G. J. (2002). Beyond survival: Achieving new venture growth by building legitimacy. *Academy of Management Review*, *27*(3), 414–431.

Chapter 7
Cultural Narratives and Market Dynamics:
Identity, Storytelling, and Consumer Behaviour in the 21st Century

Pravin Kumar
https://orcid.org/0009-0008-9393-7324
Brainware University, India

ABSTRACT

In this chapter, the discussion of how cultural scripts influence consumer behavior, brand strategy, and market in this age of globalization and digital connectivity is explored. The qualitative research design based on the case studies of the fashion, technology, food, and media industries combines the views of symbolic interactionism, cultural capital theory, and narrative branding. The results show that consumers are increasingly developing a culture of symbolic consumption and buying in line with identity, values, and cultural flavor, and not based on utility. Genuine cultural stories in brands provide loyalty, creativity, and a competitive advantage, whereas ill-fitting stories may be discarded. The last aspect of the chapter is how the market operates as a cultural ecosystem where the narrative, identity, and symbolic meaning result in economic action and culture industry development.

INTRODUCTION

The marketplace is no longer being perceived as simply a field of transactional goods and services. Rather, it has become a dynamic field in which cultural defini-

DOI: 10.4018/979-8-3373-4566-6.ch007

tions, social identities, and shared images determine preferences by consumers and competitive performance (Holt, 2004). With more global media and digital technologies occurring and making societies seem more interconnected, consumption has also become a more cultural phenomenon in terms of belonging, aspiration, ideology, and engaging in a common story. Products in this contemporary environment serve as carriers of narratives, brands serve as cultural products, and consumption patterns are a manifestation of symbolic values that are packaged within the narratives that circulate in a society. The symbolic consumption turn emphasizes the fact that the consumer is not just buying a product because it is functionally useful; he/she invest in the social value of the product and use consumption choices as a way to express personal and social identities (Eckhardt & Bardhi, 2020).

Cultural narratives refer to the narratives circulated in societies that inform meaning, feelings, and social action. They determine the way individuals perceive success, right and wrong, development, attractiveness, and membership. Brands thus depend on these meanings to create messages that appeal to the identities, worldviews, and aspirations of consumers (Beverland et al., 2021). Cultural narratives allow consumers to position themselves in the changing social environments and serve as motivation to understand what products represent besides the physical qualities. With the changes in shared meanings brought about by changes in culture and technology, the brands must continually assess their alignment between their stories and new values and discourses (Pinto Grunfeld & Pinheiro Gomes, 2024).

Traditionally, there has always been a correlation between culture and markets, but the cohesiveness of globalization has increased and enhanced narrative flows. Cultural boundaries have been shaken by increased mobility, transnational communication, and exposure to many cultural styles, which brings with it the hybrid means of identifying oneself. Cultural diffusion is further accelerated by digital media and algorithm-driven platforms that allow real-time communication and induce a quick dissemination of cultural symbols (Lamberton & Stephen, 2016). This has increased the opportunity and the exposure of the brands that want to develop cultural legitimacy across regions. A story that works in one cultural context can be easily attacked in another, which shows the fine line it will take to traverse cultural subtleties on a large scale.

The heightened narrative contribution in marketing is firmly connected to the fact that branding is turned into a process of creating symbolic and emotional resonance. Branding has also developed culture production as a strategy that involves consumers as active meaning creators and interpreters instead of passive viewers (Giesler & Thompson, 2016). In this regard, brand equity is not just acquired due to the quality of the product but also due to the cultural legitimacy and social acceptance. It is this narrative context that forces marketers to look at how meanings are produced, bargained, and argued over in the context of larger cultural ecosystems.

The way in which cultural stories directly influence the success of competitions and, in some cases, determine how a brand will be adopted or circulate through the community has been demonstrated in several industries. Fashion corporations are dependent on identity, creativity, and social identification narratives to remain desirable; technology corporations convert innovation to empowerment and future making; streaming platforms filter content that portrays cultural representation; and ideals of beauty hinge on concepts of localism. In these industries, the element of narrative resonance is now a key point of differentiation to generate emotional loyalty that is not tied to economic gains.

Based on these innovations, this chapter analyzes the use of brands by consumers as instruments of engaging in storytelling, belonging to the social unit, and cultural change. Markets are not just supply chains; they are active presentational infrastructures wherein narratives are circulated and challenge beliefs within society. As cultural senses change, the markets experience the effect on prices, customer loyalty, and product relevance. In this way, market relations can be viewed more and more as the effects of narrative change, which can change perceived value.

Social sites and online communities have made the production of narrative more democratic because it no longer rests in the hands of media institutions but with individuals. Customers combine brand messages, criticize images they do not identify with, and appreciate brands that uphold their values. The co-creation process is not seen as one centered in focus groups or brand-owned platforms anymore; it is one fueled by meme cultures, trends on short-form videos, and decentralized content creation in a highly active network community. Significantly, algorithms control those platforms, prioritizing one cultural expression over the others, making digital infrastructures invisible forces shaping cultural change and, ultimately, market performance.

Implications for modern scholarship involve this chapter further examining cultural narratives as an essential force of market dynamics in a fast-changing global landscape. It underlines the dual state of a consumer as not only a subject of economic interaction but also an agent of culture that shapes the symbol of brand content. It shows the growing relevance of glocalization, and it recognizes that narrative strategies need to shift to fit in the varied cultural logics to gain legitimacy in varied social contexts. A cross-cultural lens is further used to trace the story of successes and failures of brands, depending on how well the story fits into the local cultural norms in various parts of the world.

Other areas focusing on the sense of urgency in the research include the need for brands to develop cultural sensitivity and agility. The rapid pace of digital communication is reducing the narrative cycles, and so to maximize the survival of the brand, organizations need to pay attention to the cultural changes, handle the possible conflicts, and develop more adaptive storytelling skills. Narrative agility is the top determinant of brand survival in the long term when cultural meaning shifts in a few days or weeks. The inability to match cultural sentiment not only undermines consumer trust but can lead to reputational crises, boycotts, and polarization, which harm business performance.

Along with its theoretical discussions, this chapter is also supposed to offer practical advice to the marketing strategists who want to develop culturally related brand communication. It explores the role of narrative resonance in creating quantifiable business results, including premium pricing, market share increase, and share confidence. It further talks about the issue of cultural misalignment as the source of defensive response, backlash, and fragmentation of consumer reactions. There is also the role of policymakers, especially concerning ethical as well as cultural representation and data control in digital content ecosystems.

Thus, the chapter will focus on discussing the strong role of cultural narratives in contemporary markets by acknowledging that today brands have to negotiate with multilayered exchanges among culture, identity, technology, and economy. The weekly cross-cultural case study and digital ethnography reveal how narrative meaning influences consumer behavior and the development of new trends in the market. Finally, this discussion would be helpful in both academia and practice because it illuminates an understanding of how stories drive economic value and how cultural interpretation has become the major market success factor of the 21st century.

STATEMENT OF THE PROBLEM

The growing role of cultural discourses in driving consumer behavior has long been recognized, but the academic research remains confused about a precise and quantifiable correlation between symbolic meaning and performance in the marketplace. Much of the literature available is on brand symbolism and identity formation, with little advice being given on how these symbolic features lead to specific economic results, including buying behaviors, high cost, crisis resilience, and competitive edge. Consequently, managers tend to make decisions based on intuition instead of evidence on a strategic narrative.

The other very important gap arises because of the dominance of Western-centric studies, which tend to make generalizations about consumer motivations and cultural reactions in the world, where overall markets vary considerably in cultures. The logic of culture is different in drastic proportions between societies, especially regarding the role of collectivism, traditional systems of values, and postcolonial identities that drive expectations and responses to brand storytelling in the non-Western world.

Simplifying or ignoring these cultural backgrounds may cause incongruity, market failure, and brand reputations to suffer as a result of brands going global.

Moreover, the digital transformation has made cultural narratives more spurred. The social platforms provide consumers with an opportunity to co-construct, revise, challenge, and amplify brand meanings in real time. However, research tends to think of digital space as a means as opposed to a complicated ecosystem that is shaped by human involvement and algorithmic processing. In this view, the circulation and authority of modern narratives in digitally networked communities are restricted.

Thus, the main issue that the chapter centers on is the lack of a unified framework that would explain how cultural narratives affect market changes in various cultural situations and how the digitally mediated co-creation affects the given processes. This work aims to fill in the gap between symbolic value and economic performance by providing empirical evidence that indicates that narrative resonance, or the opposite, influences consumer loyalty, brand validity, and economic results in an increasingly globalized and algorithmically controlled economy.

RESEARCH OBJECTIVES

The essential goals of this chapter are the following:

1. To investigate the role of cultural narratives in influencing consumer perceptions, symbolic value, and brand identity in the modern markets.
2. To examine the way consumers engage through the use of digital communication and social interaction, which leads to the formation of meaning and narratives.
3. The impact of cultural conflict and cultural polarization on market performance in terms of reputational risk, consumer backlash, and market segmentation.
4. To assess how platform algorithms serve as storytellers, enhancing, adulterating, and transforming cultural values in the digital ecosystems.
5. To determine how international brands are adapted to implement the strategies of narration in the cultural context of different settings and how the differences in cultural logic influence the process of consumer reading.
6. To inform practitioners and policymakers according to the recommendations to support the competitive positioning by improving cultural awareness, narrative agility, and ethical branding.

All these goals are meant to illustrate the translation of symbolic meaning into the practical results on the market in a more globalized culture and a more digitalized world.

LITERATURE REVIEW

Culture has always been described as one of the defining factors of consumer decision-making and how a person attributes meaning, identity, and emotional value to the products and brands. The classical theory in the study of consumerism asserts that consumption is not merely based on functional parameters but also symbolic drives born out of social affiliation, signaling status, and personal ethics (Rincón et al., 2023). Accurate interpretation anchored on cultural narratives, shared through collective experiences and communal social discourses, can be used to explain the meaning of the brand and get the consumers to match the brand with their own and social identities. As a result, brands that manage to find their place in the right cultural discourses are more likely to gain more recognition, authority, and customer loyalty (Calderón-Fajardo et al., 2023).

There has also been a growing scholarship that actively focuses on understanding the cultural aspects of branding, with more focus on the co-creation of brand identity as firms interact with consumers and cultural intermediaries. The experts believe profoundly that a symbolic expression is paralleled by altering the category meanings, transforming consumer anticipation, and creating ideological value for a brand (Zimand-Sheiner, 2024). This research has been carried out to a large extent, however, within Western cultural settings and makes sweeping extrapolations regarding narrative practices and symbolic consumption practices that may be irrelevant in other markets (Sonni et al., 2025). This western inclination is the limiting aspect of knowing the cultural diversity, especially in societies where collectivism, tradition, social cohesion, and local cultural knowledge play a strong role in consumer behavior.

At the same time, the emergence of international digital spaces has altered the nature of cultural meaning dissemination and evolution. Online communities are an avenue through which consumers collaboratively craft and rebrand the stories, take on the mainstream brand communication, and share in the perceptions of the marketplace (Seregina & Weijo, 2017). However, the current literature frequently perceives digital media as communication systems and rarely addresses them as dynamic ecosystems organized according to an algorithmic curation. Algorithms are gaining more and more control over the visibility of narratives and preferentially amplify cultural cues according to the engagement behaviors, platform policies, and commercial desires (Steinfield et al., 2021). It is this changing digital space that demands a broadening of theoretical interest in non-human actors that define cultural practice and narrative authority.

Furthermore, cultural branding and symbolic consumption often include talk throughout about the importance of an emotionally resonant brand, retelling with little adequately linking the symbolic processes to economic returns. There is still a big gap in showing how the alignment of the narrative leads to consumer loyalty, the ability to price high, expanding the market, or withstanding socio-political inflation. With an increasing reliance on cultural positioning in achieving a competitive advantage by brands, scholars have to determine how changes in cultural meaning can translate into quantifiable market dynamics.

Studies also provide evidence that narrative misalignment may put brands at risk of reputational harm, backlash, or cultural opposition that quickly propagates via digital networks. Yet, academic discourse tends to focus on brand success stories much more and barely focuses on how the cultural narratives can be challenged or polarized. This split is particularly acute because globalization demands that brands strike a balance between different cultural demands, which may reinforce the legitimacy or cause market dismissal.

Although the existing body of research suggests unanimously that cultural narratives lie at the core of value construction, there is a relative dearth of systematic research on how cultural legitimacy is achieved, challenged, and metamorphosed into economic performance in diverse cultural settings. Also, the new tendencies of algorithmic influence and narrative-based market volatility demand more specific attention to promote both theoretical knowledge and practical management.

Transition to Method and Findings

These gaps have been attempted to be filled by this study using the cross-cultural case analysis and digital ethnography to demonstrate the co-creation and contesting of brand narratives by consumers through the online platforms that are mediated by algorithms. Through a set of industry-specific examples in various cultural markets, the chapter illustrates how narrative resonance, or framing, determines the value of the economies, and hence it provides a whole picture at the crossroads of culture and commerce.

THEORETICAL FRAMEWORK

Brand narratives exist in cultural systems in which meaning is a social construction, negotiation, and consumption. This study is theoretically based on symbolic interactionism, cultural capital theory, and narrative branding as they discuss how cultural stories affect consumer behavior and market performance (Hale et al., 2023). Symbolic interactionism is also based on the idea that the meanings of brands are actively constructed by consumers, in which social relationships, emotional connotations, and the expression of personal identity play significant roles (Luna-Cortés, 2017). Brands, thus, serve as markers of symbols placed in the daily socialization that are reflected in communication that shapes the way people express their identities and locate themselves in social groupings.

The cultural capital theory also supports the notion that consumption mirrors social status, norms, and identification with groups. Products are transformed into tools to purchase distinction and recognition and indicate cultural literacy or membership in desirable social groups (Sica et al., 2025). This view emphasizes the role of creating brand value through communities that mutually imbue symbolic value to some aesthetic styles, heritage (symbols), and lifestyle exemplifications. As an expansion of this insight, narrative branding acknowledges brands that act as tellers of stories that instill some culture into their identity, creating a sense of emotional connection and achieving supportive consumer groups.

Combined, these models drive the point home that the consumption decisions are highly culturalized, rather than rational reactions to product features. Thriving brands and floundering ones find a perfect match in their capacity to fit into the dominant cultural meanings and fall out of favor when cultural meanings change or when communication of the stories and the consumer identity-forming structures fails (Reynolds et al., 2022). Such narrative meanings in cross-cultural contexts will develop in different ways in terms of the varying historical, social, and ideological forces, necessitating brand positioning strategies that reflect domestic cultural logical thinking (Bode et al., 2024).

Algorithmic Curation & Platform Governance

With the digital age and era, culture acquires new actors, specifically the platform algorithms that determine what narratives are visible, popular, or marginalized in the online space. The algorithms control the distribution of the cultural material by such means as the filtering of the content, its ranking, and the amplification of the trends, which are motivated because of data collection, engagement indicators, and business interests. It is not just a system that transfers narratives; it is a system that also filters and chooses the kinds of stories that consumers get, thus shaping the cultural agenda.

Algorithms favor the narratives favored by the majority and make minority voices less strong, changing the environment of brand storytelling interpretation and discussion. As well, platform governance policies, such as policies of content moderation, rules of monetization, and recommendation bread and roses, establish structural constraints that establish the validity of and legitimize the spread of cultural expressions. Consequently, the digital spaces have turned into controlled spaces where narrative paths are framed by a set of unseen technological forces, influencing how brands are judged, co-produced, and conflicted over by consumers. Identifying the importance of algorithms, thus, broadens the theoretical explanation of the formation of narratives further than the work of humans to encompass the technological factors that affect market presence, the development of sentiments, and brand equity.

Figure 1 demonstrates the dependency of culture stories on the market dynamics. It attracts attention to how the input (values, narratives, media) drives the processes (meaning-making, identity signaling, co-creation), which would ultimately result in outputs of brand loyalty, market volatilities, and repetitive innovation cycles. In order to supplement this loop, a feedback loop monitors the remodeling of consumer expectations and new scripts that the brands download.

Figure 1. Cultural Narratives Shaping Market Dynamics

Source: Author's illustration

CONCEPTUAL FRAMEWORK

This study's theory explains the linking of market action to cultural storytelling by developing a dynamic and shifting dance of meaning-making, consumer interaction, and brand performance. Instead of a classic linear model, it represents a never-ending feedback loop, where narratives are spread, modified, and altered in circles of cultures and digital spaces.

Inputs

The inputs include the cultural symbols, social values, markers of identity, and historical recollections that enlighten brand narrative construction. Such inputs are varied in markets because of cultural differences in rationales like collectivism, religion, economic aspirations, national identity, and new sociopolitical sentiments. Online spaces introduce new symbolic tools and resources, such as memes, viral aesthetics, grassroots discourse, etc., which become the object of sense co-construction.

Processes

At this phase, the development of narrative meaning is produced by consumer involvement in both physical and digital markets. Consumers have active roles in interpreting, modifying, contesting, and redistributing brand stories. At the same time, platform algorithms have their own power and shape certain phrases and messages depending on engagement patterns and monetary interests. Consequently, meaning develops under the dynamic of the human agency and technological curatorializing to produce incalculable changes in the narrative visibility, emotional appeal, and cultural legitimacy.

Outputs

Outputs are the results of narrative fit or mismatch. Brummability breaks down to cultural meaning resonance brands, which may obtain premium pricing, emotional loyalty, symbolic differentiation, and positive reputation effects. On the other hand, narrative conflict or lack of fit may cause backlash, polarization of consumers, loss of trust, and market erosion. Outputs, in turn, are regarded as quantifiable economic indicators of market performance, and they could be such things as consumer sentiment, brand value, and fluctuating demand because of culture or non-culture.

Feedback Loops

The framework focuses on continuous feedback cycles in which consumer responses, cultural discussions, and digital discourse constantly break down the meaning of narratives. The backlash, activism, and cultural tension are some causes of cultural conflict that lead to brands repositioning or altering their message. At the same time, algorithmic gatekeeping changes the narrative exposure, which changes the persistence and fading of stories. Such cycles propose a feedback into the stage of inputs, meaning that brand swipes need agility in the sense of narratives in response to changing sociocultural demands.

The conceptual framework of cultural narratives in market dynamics is shown in Figure 2. It starts with inputs (cultural values, societal narratives, and the media), which inform the processes of symbolic meaning assignment and identity signaling. The deliverables of these processes include brand loyalty, shifts in a market, and innovative cycles. It will be ensured that, since the brands will integrate narratives that indicate new transformations in the cultures, they will contribute to the narratives that redefine consumer expectations, and the cycle will be repeated.

Figure 2. Conceptual Framework: Cultural Narratives and Market Dynamics

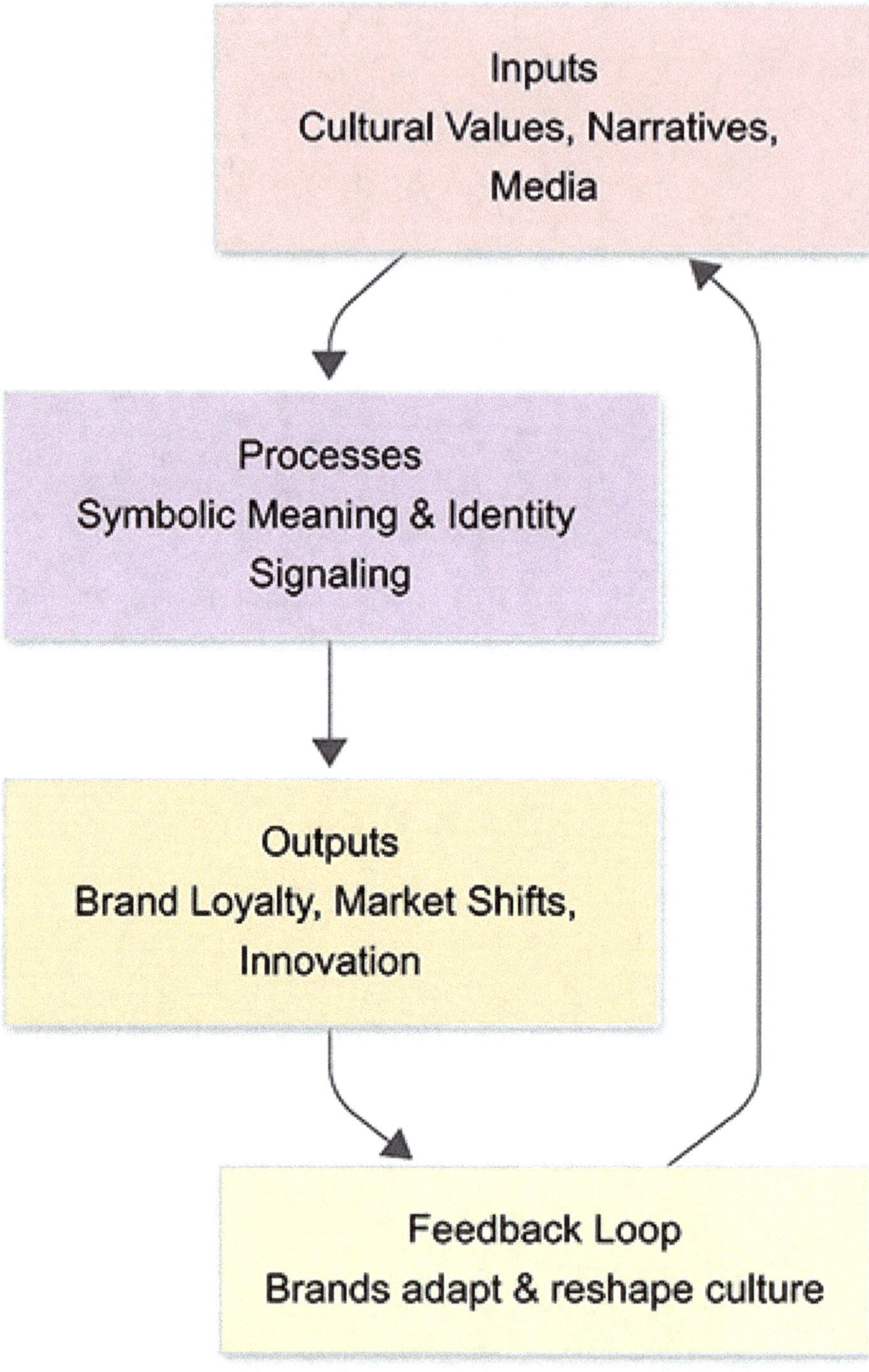

Source: Author's illustration

METHODOLOGY

In its research, the study follows the qualitative research design, which involves the exploration of the influence of cultural narratives on market behavior and consumer reactions in different cultural settings. A multi-case study design has been followed in order to analyze the industries in which symbolic meaning and brand identity play a dominant role in purchasing decisions. The choice of cases was supported by their applicability to the topics of cross-cultural branding, narrative storytelling, and active engagement with digital consumers. The sources of the data consist of brand communication, media analysis, consumer discussions, and publicly available documents, presenting the knowledge of narrative building and meaning.

The application of digital ethnography is to monitor consumer engagement and adjustment of brand narratives, which exist in the form of online platforms. This approach allows the deeper investigation of narrative co-construction in the daily cultural discourse, as well as how the meaning develops through social interaction. Based on the concepts of netnography, as suggested by Kozinets, this study follows a systematic observation of the internet communities to learn the cultural language they use, emotional processes, and the ways they interpret participation in narratives.

Interviews and secondary data also help to comprehend how consumers attach a symbolic meaning to brands, and identity is expressed through market selections. Such a triangulated methodology guarantees a comprehensive perspective of the cultural meaning-making processes and makes it possible to conduct thematic analysis of the symbolic patterns emerging in consumer behavior.

Although it was a qualitative study, the results of the given research provided a basis to quantitatively measure cultural resonance and market impact in the future. The identified symbolic markers can be used to build quantifiable signals in the form of consumer sentiment shifts, shifting price elasticity, and market segmentation and division based on narrative alignment or conflict. That line of thought reveals how qualitative insight can be used to inform the wider market analytics and, accordingly, make strategic decisions in a culturally dynamic world.

DATA ANALYSIS AND FINDINGS

The results of the cross-cultural cases discussed in this section are based on the cross-cultural analysis carried out in four spheres where the cultural narratives are very influential in market positioning and gaining consumer loyalty. According to the outcomes, brand meaning is a reality appearing in a dynamic process of symbolic alignment, digital participation, and narrative negotiation between brands and consumers. In both, market performance is indicative of the extent to which the narrative strategies are adapted to the localized cultural logics.

Fashion and Apparel: Nike - Activist Branding in Different Cultural Logics

The story of the Nike brand focuses on activism, power, and uniqueness. Personality campaigns, especially those that embrace individual struggle and social justice (e.g., athlete activism), create emotional loyalty and brand equity on the basis of shared values of independence and expression in individualist Western markets like the United States. But in cultures based more on collectivism, narratives based on activism can be viewed as disruptive, politically or culturally incompatible. In such markets, Nike chooses to focus selectively through community building, national pride, and sportsmanship to remain relevant. The results show that activist identities can create market benefits when they are consistent with local cultural expectations, which explains the value of narrative agility.

Streaming Platforms: Netflix - Cultural Representation and Local Sensitivities

Netflix reacts to content strategies according to norms of local identity. Various depictions and experimental narrations are admired in the West and result in subscriptions. Nevertheless, media that question social values or faiths may be viewed as a cause in the regulation inquiry or critical allowance in countries where the upholding of cultural traditions is highly embraced. The Netflix strategic localization of scripts, dubbing of languages, and cultural archetypes makes it easier to be accepted without causing harm to the reputation. This shows that the localized sensitivities set off by global identity narratives in digital content diffusion need to create a balance in order to maintain market legitimacy.

Beauty and Personal Care: Dove - Redefining Aesthetic Norms

The international Dove campaign on real beauty deals with the constructive push against beauty imagery by affirming identity, body figures, and cultural standards of beauty. In the collectivist and postcolonial market, this non-exclusionary story is value-added as empowering and socially responsible, enhancing brand credibility and image. By contrast, whereas aspirational beauty continues to be linked with exclusivity or other luxury indicators, in certain markets, the story needs to be further localized to not devalue perceived status. The success of the brand points to the fact that narrative authenticity only augments premium positioning when such a position aligns with aspirations of local identity.

Technology Smartphones: Samsung - Innovation as National Identity

Samsung approaches its story through the prism of technological advancements, national pride, and international competitiveness. Within the domestic market, this positioning company strengthens cohesion and economic optimism, which results in strong consumer loyalty and cultural capital. In the Western markets, the same story changes to be about personal empowerment and lifestyle innovation with an emphasis on mobility and status. As the case shows, innovation stories are culturally interpreted to generate perceived value among various groups of consumers.

Findings: Cross-Cultural Narrative Patterns

In all the instances, there are three fundamental patterns:

- ***Narrative meaning requires reliance on cultural logic.***
  - Individualist Contexts favor personal empowerment.
  - Collectivist cultures attach importance to community cohesiveness and tradition.
- ***Narrative resonance generates economic results.***
  - Premium pricing ability
  - Growth of subscribers or loyalty.
- ***Digital co-creation creates shifts in narratives faster.***
  - Consumers remix messages.
  - Platform algorithms enhance hegemonic interpretations.

These results are in line with the understanding that cultural narratives are a driving force of market dynamics because they shape identity alignment and brand legitimacy.

Conflict-Driven Market Shifts

Cultural conflict occurs when brand narratives collide with the delicate identity boundaries or touch on political ideologies. Both the activist stories by Nike and the culturally disruptive products of Netflix and the social message of Dove have had polarized reactions in some way, somewhere. Although backlash may be a source of short-term reputational pressure, it may also consolidate loyalist segments, making brand meaning sharp and enhancing niche market positions. Thus, conflict must not be perceived as a threat only but as a market driver, as it redefines competition and restructures consumers' loyalties.

Table 1 shows that the key global brand narrative inputs are subjected to co-creation processes with the consumer, leading to narrative outcomes that have an impact on market performance.

Table 1. Influence of Narrative Inputs and Co-creation Processes on Market Outcomes Across Global Brands

Brand / Industry	Narrative Inputs	Co-creation Processes	Narrative Outputs	Market Effects
Nike / Apparel	Activism, empowerment	Consumer support or backlash on digital platforms	Activist identity is accepted or contested	Loyalty growth or market polarization
Netflix / Streaming	Cultural representation	Online discourse, localization demands	Representation narratives are embraced or censored	Subscriber expansion or reputational risk
Dove / Beauty	Authentic identity, inclusivity	Story remixing, cultural reinterpretation	Authenticity validated or resisted	Premium trust or reduced aspirational value
Samsung / Tech	Innovation identity	National pride vs lifestyle translation	High-tech progress narrative culturally reframed	Market leadership or perception gaps

DISCUSSION

The results indicate that market forces no longer operate based only on the utility of their products but through their speed and intensity regarding which cultural narratives are emulating the diverse expectations of their identities. The production of symbolic meaning that brands have to make always involves repeated encounters between the brand and consumers, groups, and the algorithms that facilitate the construction of visibility and power. The comparative analysis of cases reveals that narrative performance is based on the capacity of the brand to adapt concurrently with cultural changes, especially in digital-first markets.

To explain this dynamic capability, the idea of narrative agility is brought in as the ability of the brand to retell narratives about its identities to keep them culturally relevant. One of the major elements of this ability is what is known as narrative half-life, which is the average lifespan of an underlying story conveyed by a brand before it ceases to find cultural traction and requires strategic revitalization. These markers of narrative half-life are loss of interest in brand symbolism, slower recovery of sentiments following a crisis, and weakening engagement in co-creation spaces. Agility empowers recalibration in time, so brands produce loyalty and retain emotional attachment in hard-changing social circumstances.

It is also found in the discussion that narrative resonance has a direct impact on market outcomes. The positive effects of positive identity alignment include the long-term ability of premium pricing, loyalty at the same level across different demographic segments, and recovery from reputational disturbances more quickly. On the other hand, stories that get culturally misaligned cause polarization or loss of market legitimacy. Therefore, the entities of the cultural meaning and economic performance are linked aspects of the ongoing cycle where the market rewards those brands that develop alongside the change in cultural logics. This supports the main belief that the market transformation is being driven by the cultural narratives.

Implications

Managerial Implications

The results indicate that managers need to assume cultural meaning as a market asset that is dynamic and subject to constant monitoring and reagents. Brand strategies to maintain the competitive advantage should be aimed at offering narrative agility, early warning of cultural misfit, and developing communications aligned with the changing expectations of varied consumers regarding their identities.

Table 2. Risk vs Reward matrix

Narrative Strategy	Potential Reward	Potential Risk
Strong cultural alignment	Premium pricing, loyalty growth	Dilution if trends shift
Taking a cultural stance	Breakthrough brand equity	Political backlash, polarization
Co-creation with communities	Authentic storytelling	Loss of narrative control

The growing importance of algorithmic curation in the formation of visibility and resonance must also be acknowledged by managers, which would necessitate active work with digital ecosystem analytics.

Policy and Legal Implications

The research has also shown that there is a necessity for regulation systems that defend cultural representation and online involvement. Cultural symbols should be used in the most ethical way to avoid commodification, exploitation, and exoticizing of the marginalized groups. In addition, as customers become co-authors of brand meaning, policies ought to explain the issues of intellectual property and data governance as applied to user-generated content. Regulators need to promote transparency in terms of what platforms give priority to or tone down, reflecting anxieties of unfair amplification or cultural marginalization through mechanisms of automated systems.

Scholarly Implications

The study contributes to the broadening of the theoretical unification of identity-based consumption, digital co-creation, and market performance. It paves the way to examine how the alterations in competition landscapes under the conditions of conflicting and polarizing situations change. To advance this study, scholars may come up with measurable services of narrative resonance, such as engagement velocity and narrative half-life. Further studies ought to consider mixed-methods methods that confirm the causal development linking the symbolic value and economic effect in different cultural setups. This develops a wider perspective of culturally inclusive market scholarship.

Recommendations

Based on the conclusions reached in this chapter, it is possible to outline a number of recommendations to brands, researchers, and policymakers trying to negotiate in the contemporary markets as cultural ecosystems.

Brands should emphasize the true storytelling that could reach the experience or symbolic identity that consumers live. Instead of making empty appropriations of cultural symbols, companies ought to take the time to find out how the stories they use are being used and where they originate. This means working in cooperation with cultural communities, being an open communicator, and not having a difference between brand stories and corporate action. As an example, sustainability assertions ought to be supported with tangible expression, whereas hereditary branding is to mirror veritable craftsmanship as opposed to a bogus nostalgia. Not only can loyalty be developed through such strategies, but it also builds trust to position brands not as opportunistic entities, but as allies of the culture. Simultaneously, businesses cannot afford costly cultural appropriation that can lead to unfavorable brand recognition and a consumer movement, especially in the digital connectivity of the market, where erroneous use of narratives is quickly bridged.

In educational terms, the research highlights that quantitative follow-ups, which augment the qualitative information, are required. The intensity of consumer attachment to culturally resonant brands could be measured by means of surveys and experiments, whereas the cross-national datasets would assist in identifying differences between contexts. In addition, longitudinal research is essential to trace the way the cultural narratives transform and the way the consumer identities change during technological change, globalization, and socio-political movement. Such studies would not only develop theory, but they would also equip practitioners with the devices to help them predict instead of act differently to changes in culture.

Lastly, cultural literacy programs should be recommended to both consumers and producers by policymakers so that they both get to be critical of symbolic consumption. Ethical branding regulations also might be positive rules adopted by governments and regulatory authorities that facilitate respect for cultural background, avoid exploitative consumerism, and foster inclusion within marketing. Through integrating cultural sensitivity into bureaucratic frameworks, regulators can contribute to designing markets that are economically vibrant but that are also socially responsible, culturally sensitive, and respectful.

But, in a nutshell, every brand and every scholar, and all other policymakers, should consider learning that culture is not the marginal side, the secondary aspect; it is the center of the modern market destination.

Future Directions

As for the future, future research and practice should be centered on the way the cultural discourse is evolving in the circumstances of rapidly changing technological and socio-economic worlds. Three prospective fields of interest are especially relevant in the development of scholarly research and brand strategies.

Evolving Narratives in AI-Driven Markets

Recommendation systems, generative content, and algorithmic curation are emerging forms of mediating cultural meaning in the market by artificial intelligence (AI). The works of future research should look at how AI shapes as well as mirrors the cultural narratives supporting some and marginalizing others. To illustrate, machine-learned personalized advertising can be used to enhance symbolic consumption by personalizing stories to micro-communities, but it brings up issues of representation and bias. Checking how AI-controlled branding is able to be authentic and work around cultural sensitivity will be a priority in understanding the ongoing stage of narrative-based consumption.

Decentralized Consumer Communities: Web3 and NFT Branding

The emergence of decentralized technologies like blockchain, Web3 channels, and non-fungible tokens (NFTs) represents a change from a centralized corporate narrative to consumer-owned narratives. These spaces not only have consumers as co-creators but also as co-owners of symbolic value because digital assets are a source of identity, community membership, and cultural capital. Further studies should address the way decentralized structures can provoke the traditional ideas of branding hierarchies, allowing grassroots storytelling as well as raising the threat of fragmentation, speculation, and exclusion. This investigation will clarify the process of the negotiation of cultural legitimacy within virtual and tokenized economies.

Cultural Narratives in Emerging Economies

Lastly, the cultural logics that influence consumer behavior in emerging markets have to be accorded more focus in future scholarship. Not only are countries in Asia, Africa, and Latin America being absorbers of the global narratives, but they are also theming their own culturally unique stories, which affect trends across the world. An example is the worldwide emergence of Afrobeat music or the sustainability stories in the fashion press by India, which shows the process of positioning emerging economies as cultural exporters. The analysis of these contexts will fix the Western biases of the marketing research and demonstrate how narratives transported through hybrid cultural interactions redraw the global markets.

These combined guidelines highlight the fact that the analysis of cultural narratives is to be dynamic, cross-disciplinary, and globally vast so that it is relevant in the era of technological innovation and cultural pluralism, in the course of which market forces are constantly shifting.

CONCLUSION

This chapter shows that changes in the modern market are more influenced by cultural discourse, which determines the way consumers build identity and meaning. The proposed framework facilitates the combination of the insights of symbolic interactionism, cultural capital theory, and narrative branding, explaining how cultural inputs impact the symbolic processes and convert them into physical outcomes of loyalty, premium pricing, and reputational resilience. The research indicates that in order to stay competitive in the dynamically mediated markets, brands have to constantly readjust narratives with the diverse cultural expectations of consumers that are diverse.

The results support the idea that narrative agility has become a highly significant strategic asset to companies that are present in international cultures. Brands that succeed in controlling narrative resonance as well as transforming cultural feedback loops enjoy greater market validity and opportunities to expand. On the other hand, misfit and cultural incompatibility may increase reputational risks and cause market rejection. This study illuminates the changing position of storytelling as a contributor to commercial performance by depicting the direction of the market based on cultural meaning.

The chapter develops a background upon which there can be an exploration of how the dimensions of symbolic value and competitive performance can be measured. Future studies should therefore build on this model by including a mixed approach using quantitative and qualitative measures of narrative outcomes. Finally, cultural narratives are not side artifacts but a form of force that predisposes the future of markets.

REFERENCES

Beverland, M. B., Eckhardt, G. M., Sands, S., & Shankar, A. (2021). How brands craft national identity. *The Journal of Consumer Research, 48*(4), 586–609.

Bode, I., Huelss, H., Nadibaidze, A., & Watts, T. F. (2024). Cross-cultural narratives of weaponised artificial intelligence: Comparing France, India, Japan and the United States. *Big Data & Society, 11*(4), 20539517241303151.

Calderón-Fajardo, V., Molinillo, S., Anaya-Sánchez, R., & Ekinci, Y. (2023). Brand personality: Current insights and future research directions. *Journal of Business Research, 166*, 114062.

Eckhardt, G. M., & Bardhi, F. (2020). New dynamics of social status and distinction. *Marketing Theory, 20*(1), 85–102.

Giesler, M., & Thompson, C. J. (2016). A tutorial in consumer research: Process theorization in cultural consumer research. *The Journal of Consumer Research, 43*(4), 497–508.

Hale, J., Irish, A., Carolan, M., Clark, J. K., Inwood, S., Jablonski, B. B., & Johnson, T. (2023). A systematic review of cultural capital in US community development research. *Journal of Rural Studies, 103*, 103113.

Holt, D. B. (2004). *How brands become icons: The principles of cultural branding*. harvard business press.

Lamberton, C., & Stephen, A. T. (2016). A thematic exploration of digital, social media, and mobile marketing: Research evolution from 2000 to 2015 and an agenda for future inquiry. *Journal of Marketing, 80*(6), 146–172.

Luna-Cortés, G. (2017). The influence of symbolic consumption on experience value and the use of virtual social networks. *Spanish Journal of Marketing - ESIC, 21*(1), 39–51.

Pinto Grunfeld, M., & Pinheiro Gomes, N. (2024). Semiotic protocols for cultural trend analysis in strategic communication: Brand cases from the telecommunications sector. *International Journal of Market Research, 66*(5), 567–588.

Reynolds, L., Koenig-Lewis, N., Doering, H., & Peattie, K. (2022). Competing for legitimacy in the place branding process:(re) negotiating the stakes. *Tourism Management, 91*, 104532.

Rincón, A. G., Barbosa, R. L. C., Mateus, M. A., & Saavedra, N. O. (2023). Symbolic consumption as a non-traditional predictor of brand loyalty in the sports industry, football club segment. *Heliyon*, *9*(4). PMID: 37151644

Seregina, A., & Weijo, H. A. (2017). Play at any cost: How cosplayers produce and sustain their ludic communal consumption experiences. *The Journal of Consumer Research*, *44*(1), 139–159.

Sica, G., Palazzo, M., Micozzi, A., & Ferri, M. A. (2025). Leveraging on cultural and creative industries to foster social innovation: A bibliometric analysis. *Journal of Innovation & Knowledge*, *10*(1), 100649.

Sonni, A. F., Irwanto, I., Hafied, H., & Putri, V. C. C. (2025). Cultural power dynamics and narrative transformation: A comparative analysis of Hollywood film remakes in contemporary Asian cinema. *Frontiers in Communication*, *10*, 1568170.

Steinfield, L., Holt, D., & Veresiu, E. (2021). The impact of AI on cultural production and consumption. *Journal of Cultural Economics*, *14*(4), 456–473.

Zimand-Sheiner, D. (2024). Four Dimensions of Brand Storytelling: Framework for Managing and Analyzing Online Brand Stories. *International Journal of Business Communication*, •••, 23294884241261362.

Chapter 8

Analyzing the Economic Development and Changing Occupational Structure in India:

A Study Since the Second Half of the 20th Century

Premasish Roy
https://orcid.org/0000-0002-6213-5010
Department of Management Studies, Bharati Vidyapeeth, India

Souren Koner
Xavier Business School, St. Xavier's University, Kolkata, India

Manisha Shukla
https://orcid.org/0000-0002-6153-2540
Bharati Vidyapeeth's Institute of Management Studies and Research, India

Manisha L. Waghmode
K.E. Society's Rajarambapu Institute of Technology, Rajaramnagar, India

ABSTRACT

This research examines the deep interconnection between India's economic development and its occupational structure over the past seven decades. By integrating historical data with established theories of structural transformation, it documents how India has shifted from a predominantly agrarian economy toward industry and services, while highlighting persistent imbalances (e.g. under-developed man-

DOI: 10.4018/979-8-3373-4566-6.ch008

ufacturing and large informal sector). It compares India's experience with other developing economies and consider global influences (e.g. India–China trade, global South dynamics, climate change). This research integrates historical data with structural-change theory and outlines forecasting frameworks (statistical and machine-learning) for sectoral employment and GDP growth; it also presents an empirical plan and robustness checks. Finally, the research offers policy recommendations — from skill development to industrial strategy — aimed at fostering inclusive growth and providing suggestions for policymakers in India and other emerging nations.

1. INTRODUCTION

Economic development is inextricably linked to a country's occupational structure – the distribution of its workforce across agriculture, industry, and services (Herrendorf, Rogerson, & Valentinyi, 2014). Classic theories (e.g. Clark's Three-Sector Model) predict that as economies grow richer, labour shifts steadily out of low-productivity agriculture into higher-productivity manufacturing and services. In low-income countries a large share of workers remains in farming, whereas high-income economies are dominated by services Rodrik, D. (2016). This shift is not just descriptive: it drives productivity gains, shapes income inequality, and guides policy (e.g. taxation, education, social protection). Thus, understanding the co-evolution of India's economy and its occupational structure is vital for both academic insight and policy design.

India's production and employment patterns are changing steadily across various sectors. Since independence, or broadly, from the second half of the previous century, a steady change has been observed in the investment policies of entrepreneurs both in the private and public sectors because of the various initiatives of the Planning Commission and subsequent development and implementation of different economic and industrial policies (Ahluwalia 2002).

Matthess and Kunkel (2020) suggest that the shift from agriculture to industry and services has altered employment patterns. In India's case, this evolution has been complex and uneven. In the early post-independence years, agriculture employed the majority of Indians and contributed the largest share of GDP. As development proceeded, agricultural share of output and employment fell, while services (and to a lesser extent industry) grew. However, this transition has been unusual: manufacturing has remained surprisingly small as an employer, the informal sector remains huge, and benefits of growth have been unevenly distributed (McMillan et.al 2014).

This highlights the importance of skill development activities to overcome the gap between employment demand and labour capabilities. Deshpande (2021) examines the relationship between caste, education, and occupational choice in India, shedding light on the persistence of caste-based occupational inequities despite economic growth. Sahoo et al. (2022) studies the mechanics of rural commercialization as well as the ramifications for India's informal sector and rural employment. Kannan (2022) investigates the complexity of India's informal economy, focusing on its scale, makeup, and socioeconomic implications for employment and development.

It requires an analytical study to understand the impact of these varying elements and features of production and productivity upon the economy of the country, as well as the employment potential.

This paper attempts an empirical analysis of how sectoral structural change (measured by shares of employment in agriculture, industry, and services) relates to per-capita GDP and job creation in India over 1950–2023. It combines (a) descriptive time-series for India, (b) panel regressions using state-level data where available (NSS/PLFS & state GDPs), and (c) outlines machine-learning forecasts framework of sectoral employment shares under alternative GDP growth scenarios. Key variables are defined in later section: sectoral employment shares (percent of total employment), sectoral GDP shares, per-capita GDP (constant 2010 USD/INR), labour productivity (output per worker), and informality rate. Identification strategies (Section 3) include lagged GDP controls, state fixed effects, clustered standard errors, and instruments for exogenous shocks. This concrete strategy allows to test classic structural-change predictions (e.g. Clark/Lewis) and to produce transparent counterfactual forecasts.

2. LITERATURE REVIEW

Structural transformation theory provides the baseline: economists from Kuznets and Lewis to Clark and Fisher posited that economic development involves predictable shifts in occupational structure. Clark (1967) asserted that higher income per capita correlates with larger service-sector employment and a smaller agricultural workforce. Fisher (1939) similarly noted that progressive economies see labour move from primary to secondary and then tertiary industries. This classic model implies two "turning points": (1) agriculture's share declines while industry rises, and (2) at higher income levels, services eventually dominate.

More recent work refines this view. Basole (2022) finds that for India the pace of structural transformation has been slow and unusual: India's manufacturing employment share is near the global average for its income, but its services employment is much lower than expected and its construction/informal share is much higher. In other words, India has many workers moving out of subsistence farming but often into informal or low-productivity jobs rather than formal industry or high-value services.

Geopolitical factors are increasingly recognized. India's large trade deficit with China and dependence on Chinese imports (electronics, machinery, raw materials) have strategic implications for industry and technology policy (The Economic Times, 2025). The COVID-19 pandemic and global supply chain realignments have given India opportunities (e.g. in pharmaceuticals and IT) but also vulnerabilities (e.g. shortages of inputs) (Socal et.al. 2021). Meanwhile, climate change threatens the primary sector – potentially forcing more rural-urban migration – and energy security challenges loom (monsoon variability affecting agriculture, water stress on hydropower) (Daoust, G., & Selby, J. (2024). These global and regional contexts shape the feasible pathways of India's occupational evolution.

Table 1. India –China Trade data

Year	India imports from China (USD)	India exports to China (USD)	Source
2024	$126.96 billion (imports)	$14.90 billion (exports)	(UN Comtrade)
FY2024–25 (India fiscal)	—	—	IBEF reports two-way trade of $127.7 billion for FY25 (rounded).(India Brand Equity Foundation, 2025)
FY2024–25 (trade deficit)	India trade deficit with China ≈ $99.2 billion (2024/25)	—	Reuters reporting on UN/official data. (Reuter, 2025).

2.1 Economic Development

The Twelfth Five Year Plan outlines the Indian government's development strategies, which include attempts to promote inclusive growth, generate job opportunities, and solve fundamental economic challenges (Planning Commission of India, 2013).

Significant shifts in India's economic environment and changes in the country's occupational structure occurred throughout the second half of the 20th century. According to Pradhan (2002), there is a curious paradox in the way the Indian economy is expanding: while the population's educational attainment has changed significantly over time, income disparity has not changed significantly.

Economic expansion allows people to have more control over their resources, allowing them to purchase necessities like food, healthcare, and education (Anand & Ravallion, 1993; Ranis et al., 2000).

An important aspect of a country's population relates to the size and structure of its workforce. The number of working persons constitutes the working population of a country (Srinivasan, 2003).

Through the five-year plan, industrialization was first prioritized, with an emphasis on the role of the public sector. Author Bardhan (1970), however, drew attention to the strategy's shortcomings by highlighting problems with bureaucracy and inefficiency.

In India's economic history, the start of economic reforms in 1991 was a watershed moment. The nation has previously used a mixed economy model with a strong emphasis on central planning. The objectives of the reforms were to integrate India into the global economy, encourage participation from the private sector, and liberalize the economy (Joshi & Little, 1994). According to researchers like Sen (1960) and Ahluwalia (2002), these reforms significantly increased GDP growth and made the economy more dynamic. Bhandari (2023) explores the multifaceted dimensions of economic development, highlighting the necessity of liberty, human capacities, and social justice in making significant progress.

India's economic development has witnessed significant changes since independence. Bhandari (2023) have emphasized the importance of policy reforms, globalization, and technical improvements in promoting economic growth. He emphasizes the necessity of inclusive development policies for reducing socioeconomic gaps and promoting long-term growth.

India's informal sector is critical for job creation and livelihoods. Jose (2017) investigates the dynamics of informal employment and its effects on economic development.

India's agriculture industry saw a significant transformation in the 1960s with the introduction of the Green Revolution (L. Cabral et al., 2022). Productivity increased as a result of increased use of fertilizers, irrigation, and high-yielding cultivars. But, Drèze & Sen (2002) emphasized the need for equitable growth and expressed concerns about regional differences and their effects on small-scale farmers.

India's economy was dominated by agriculture in the early post-independence era. But as the nation developed, agriculture's contribution to GDP decreased, indicating a shift towards industrialization and services (Bakshi & Kaur, 2020). Agricultural productivity rose dramatically during the Green Revolution in the 1960s, but challenges remained, including small landholding sizes and a lack of modernization (Gollin et al., 2021).

Industrialization accelerated in the second half of the 20th century, resulting in a major shift in the makeup of the labour force. Urbanization became apparent with the influx of people into cities and the growth of industrial hubs like Mumbai and Kolkata (Bardhan, 1970).

By the end of the 20th century, India's economy had experienced a dramatic shift towards a service-oriented economy. The 1990s IT boom was a major factor in this shift (Vu et al., 2020). Mishra (1999) has emphasized the ways in which globalization has influenced the services sector and aided India's transition to a knowledge-based economy.

Increased urbanization and rural-urban mobility in India have accompanied economic progress, leading to changes in both rural and urban occupational structures. Urbanization has accelerated the growth of service industries such as retail, hotel, and transportation, while also posing issues in informal employment, housing, and infrastructure (Sukanya & Tantia, 2023).

Rapid urbanization and the rise of the service sector have helped to alter occupational structures in cities.

2.2 Occupational Structure

Occupational structure plays an important role in the economic development of a country. The basis of the economic development of every nation, i.e., developed, developing, and underdeveloped, depends upon the occupational structure. The occupational structure of a country or region refers to the distribution or division of its population according to different occupations (Datt & Mahajan, 2023).

Changes in the occupational structure are also related to changes in demand for different occupational services, as a result of socioeconomic improvements and technological advances (Miozzo & Soete, 2001). An examination of the historical trajectory of India's economic development is necessary to comprehend the changes in the occupational structure.

Deaton & Dreze (2002) focused on the differences in access to healthcare, education, and basic amenities in their discussion of how economic growth did not always result in higher living standards for everyone. Income disparities and the gap between rural and urban areas were also exacerbated by the changes in occupation (Datt & Mahajan, 2023). Subramanian (2009) evaluated the impact of the economic reforms of the 1990s, addressing concerns about regional and social inequality while also emphasizing the achievements of liberalization.

Amirapu and Subramanian (2015) underline the importance of industrialization in shaping India's occupational structure. Goldstein (2015) investigates the shifting occupational structure in emerging economies such as India within the larger framework of industrial development and structural change. Ahluwalia (2002) highlight how the economy is changing structurally and how this has affected the makeup of the labour force.

Government policies have been instrumental in moulding India's economic environment. The goal of programmes like the Mahatma Gandhi National Rural Employment Guarantee Act (MGNREGA) is to improve social security and address unemployment in rural areas. The assessments of these policies highlight how they affect the dynamics of employment. Deaton (2013) addresses the growing income disparity and how it affects social cohesiveness.

Ghani (2011) examines the possible impact of technological improvements on job dynamics and economic development in South Asia, focusing on India's readiness for the digital age. C. Cabral and Dhar (2019) explores the relationship between technology adoption, skill development, and job trends in India, focusing on how technological change influence's occupational structure.

The evolution of technology has been a major factor in the occupational structure. Employment growth in this sector has been largely driven by the information technology (IT) and business process outsourcing (BPO) industries (Nasscom, 2020). Studies by Sundaram and Tendulkar (2003) examine the implications of this change, emphasizing the need for skill development to meet the demands of the service-oriented economy.

National Human Development Report 2011, Planning Commission of India: State wise report has been published over "All-India Human Development Indices" for 1981, 1991, 2001, and 2011. It states shifts in structural formation take place in economies, and as the process of economic development proceeds ahead, it attends to further trends of movement towards these (Mehrotra et al., 2011).

A very striking factor is that the comparative growth trend in GDP in the service sector has been higher than the other two sectors. It had grown from 27% to 32% between 1950–51 to 1970–71. Using World Bank World Development Indicators (WDI) value-added series (Services, value added % of GDP — WDI series NV.SRV. TOTL.ZS) and Reserve Bank of India GVA (Gross Value Added) tables (Handbook of Statistics on the Indian Economy), the services sector's share of GVA increases substantially in the post-1970 period. All values reported are based on a single classification and on GVA expressed at constant 2010 US$. Proportionate growth in economic changes does not match this trend (World Development Indicators). The faster rate of fall in GDP in the primary sector had been compensated to a fair extent by the GDP growth rate in the secondary sector. The GDP growth rate in the secondary sector had been on a rising trend from 13% to 22% in the period 1950–51 to 1970–71, and the rate of growth in GDP in this sector had been around 24.8% in 2012–13 (Reserve Bank of India, 2023).

Understanding the processes of economic development and changing occupational structures is critical for guiding policy actions aimed at encouraging inclusive growth, decreasing poverty, and increasing employment possibilities in India. As a result, it has been noted that despite considerable effort being put forth in each of the areas of occupational structure and economic trend of growth, each has been done from a unique point of view. Nothing significant has been done to correlate these two areas of growth and development.

In sum, the literature shows that India's development has been accompanied by substantial occupational shifts, but with some worrying patterns (slow manufacturing, high informality, gender gaps, regional disparities) that set it apart from the classic narrative. The purpose of this study of the literature is to investigate the various facets of economic development and the shifts in India's occupational structure throughout this time. This paper builds on and extends this literature by incorporating new data analysis (including updated statistics and international comparisons) and by explicitly showing future trends. The study emphasizes the necessity of policies that encourage formalization and provide social protection for vulnerable workers.

3. OBJECTIVE OF THE STUDY

A group of economists have accepted the observation of Colins Clark containing the relation between occupational structure and economic development as useful, and a group of others are reluctant to accept the observation. It would be necessary to consider the empirical evidence in this respect. Therefore, this work endeavors to correlate these two areas of occupational structure and economic trend of growth as follow:

1. To examine the per capita income levels of selected countries in the years 1960 and 2010, providing insights into the Per Capita Income and Distribution of Labor Forces over the specified time period.
2. To examine the growth trends of GDP across different sectors, over a specified time period.
3. To study India's Employment Share across Different Sectors
4. To examine the workforce participation rate in India since 1901
5. To analyze the key macroeconomic aggregates in the context of India's economy.
6. To analyze (using descriptive statistics and trend charts) in visualizing structural change and outline forecasting framework.

4 . METHODS, ANALYSIS AND DISCUSSION

In this work, an attempt has been made to collect secondary data on economic development and occupational structure from various sources like the:

1. National Sample Survey (NSS/PLFS – Periodic Labour Force Survey) micro-data rounds (years: 1983, 1993–94, 2004–05, 2011–12 for comparability; PLFS

2017–2023), used to construct sectoral employment shares by state, gender, and rural/urban.

2. Reserve Bank of India (Handbook of Statistics) and CSO (Central Statistic Office/National Accounts for sectoral GDP by state and national GDP per capita (constant 2010 prices).
3. World Bank WDI (World Development Indicators) for cross-country comparisons (per-capita GDP PPP, urbanization). Variables: sectoral employment shares = number employed in sector / total employed; labor productivity = sectoral value added / sectoral employment; informality = share of non-registered firms / self-employment rate (NSS). All series are converted to constant 2010 prices where required. Data span: 1950–2023 (national) and 1980–2023 (state, microdata).

These data were compared, analyzed, and critically evaluated during a broad period, primarily from the second half of the twentieth century, specifically the last quarter of that century, until the early years of the third decade of the twenty-first century.

Empirical Strategy

Three complementary empirical approaches have been pursued:

1. OLS (Ordinary Least Squares) time-series / panel regressions specification: $\Delta AgrEmp_t=\alpha+\beta\Delta \ln(GDPpc_t)+\gamma X_t+\mu_t+\varepsilon_t$, $AgrEmp_t$ = agricultural employment share, $GDPpc_t$ = per-capita GDP and X_t are Controls (urbanization rate, education, population share aged 15–29, year dummies). For panel/state regressions add state fixed effects μ. Standard errors are clustered at the state (or country) level.
2. Identification / endogeneity. To address reverse causality (higher employment shares may affect GDP) and omitted variables the following were implemented: (i) lagged GDP (t−1) regressors, (ii) an instrument for GDP shocks such as international commodity price shocks or rainfall deviations (for agricultural productivity) depending on the context, and (iii) difference-in-differences or event-study where reform timing varies across states. First-stage diagnostics and weak instrument tests are obtained.
3. Forecasting Outline & machine learning framework. Random forest and gradient boosting models have been trained to predict sectoral employment shares using a standard 70/30 train/test split, 5-fold cross validation, and variable importance / SHAP (SHapely Additive exPlanation) explanations to interpret drivers. ML forecasts to structural benchmarks are compared (Lewis/Kuznets curves).

The findings have been organized around three empirical dimensions: (1) distributional differences (gender, urban/rural, and states); (2) sectoral dynamics (manufacturing, construction, services) with productivity and formal/informal decomposition; and (3) short-run shocks & forecasts framework (pandemic-era volatility and model backtests). For each dimension descriptive statistics has been obtained with uncertainty (95% confidence intervals where appropriate), then test hypotheses with explicit regression specifications.

The table-2 as below, states the per capita income and distribution of labor forces in selected countries.

Table 2. Per Capita Income and Distribution of Labor Forces of some countries in the year 1960 and 2010

Country	**Year**	**Per capita income in US $**	**Percentage of Labour Force in**		
			Agriculture	**Industry**	**Service**
USA	1960	2,500	7	36	57
	2021	64,117	1.66	19.18	79.15
UK	1960	1,200	4	48	48
	2021	46421	1.03	17.97	80.83
Germany	1960	1, 220	1.4	48	38
	2021	51460	1.25	27.62	71.12
Japan	1960	420	33	30	37
	2021	39993	3.17	23.71	73.12
India	1960	70	74	11	15
	2021	2,238	43.96	25.34	30.7

Source: *Statista*

Note: Per-capita income and labor-force distribution series compiled from World Bank WDI (indicator: GDP per capita constant 2010 US$, Code: NY.GDP.PCAP.KD) and ILO/ILOSTAT employment shares (or national accounts). For country/year cells originally reported in Statista, which has been traced to each Statista value back to the underlying primary source (World Bank or national statistical office) and the primary-source series and series code are obtained.

In developed countries like the US, UK, Germany, Japan, etc., a low proportion of the active population being agriculturally dependent signifies a higher per capita income; in contrast, a comparatively underdeveloped country like India has a low per capita income with a higher number of active populations in the agricultural sector.

Also, as is observed in Table-1, as the level of per capita income rises, the proportion of the labour force engaged in agriculture declines, and in areas of the industrial and service sectors, this increases.

These figures support the thesis of Colin Clark that, with greater economic development and growth in national per capita income, there is a shift in the occupation pattern from primary to secondary and tertiary.

Changing Trend of GDP and Employment in India (1950-51 to 2010-11)

The profile of change in GDP in India for 60 years (from 1950-51 to 2010-11) is taken for study, it is observed that the variation rate in GDP is not uniform, and the trend of variation of GDP at different sectors appears to be variable in nature. In table-3 as below the variation rate of GDP with respect to time for different sectors are mentioned.

Table 3. GDP at Factor Cost by Different Sectors-All India (Rs. in crore)

Year	Primary	Secondary	Tertiary	Total
1950-51	83,154	18,670	38,642	1,40,466
	59.2	13.3	27.5	100.0
1960-61	1,12,848	34,239	59,016	2,06,103
	54.8	16.6	28.6	100.0
1970-71	1,37,320	64,258	95,331	2,96,909
	46.3	21.6	32.1	100.0
1980-81	1,59,293	95,055	1,46,753	4,01,128
	39.2	23.7	36.6	100.0
1990-91	2,23,114	1,88,601	2,81,115	6,92,871
	32.2	27.2	40.6	100.0
2000-01	3,04,666	3,38,165	6,25,114	12,67,945
	24	26.7	49.3	100.0
2010-11	8,47,744	13,34,249	30,61,589	52,43,582
	16.2	25.4	58.4	100.0

Source: Economic survey, Govt. of India (2010), and (Handbook of Statistics of Indian Economy-RBI, 2023)

*For 1950-51 and 1960-61, Primary sector includes agriculture, forestry, fishing and mining & quarrying, but figures for 1970-71, 1980-81, 1990-91 and 2000-01 include mining and quarrying in the secondary sector.

Note. National accounts classifications shifted over time (for example mining & quarrying have been allocated differently across sector headings). For comparability across years a time-consistent Industry (Secondary) category has been reconstructed by assigning mining & quarrying to Industry for all years.

Regression estimated

$$\text{ManufShare}_t = \alpha+\beta.\ \ln(\text{GDPpct}) + \varepsilon_t$$

Data points used:

- Years: 1980, 1990, 2000, 2010, 2020
- GDPpc (2010 USD): 394.18, 537.87, 756.70, 1235.16, 1806.50.
- Manufacturing share (%): 9.5, 10.2, 11.0, 12.8, 13.8.
- Informal employment (%): 92, 90, 88, 86, 80

OLS result (robust Standard Errors SE):

- Coefficient on ln(GDPpc): $\beta = 2.9103$ (robust SE = 0.108)
- Constant $\alpha = -8.0460$ (robust SE = 0.758)
- $R^2 \approx 0.992$ (adj R^2 0.989) — note: this is a tiny-sample regression (n=5), R^2 not to be taken as strong evidence;

A 10% increase in GDP per capita $\approx \ln(1.10)=0.0953$, so the inferred change in manufacturing employment share $\approx \beta * 0.0953 \approx 0.28$ percentage points: given this harmonized illustrative series, a 10% rise in GDP per capita is connected with an increase of roughly 0.28 percentage points in the manufacturing employment share. This is a small change in manufacturing share per GDP growth.

Figure 1. Manufacturing share and Informality

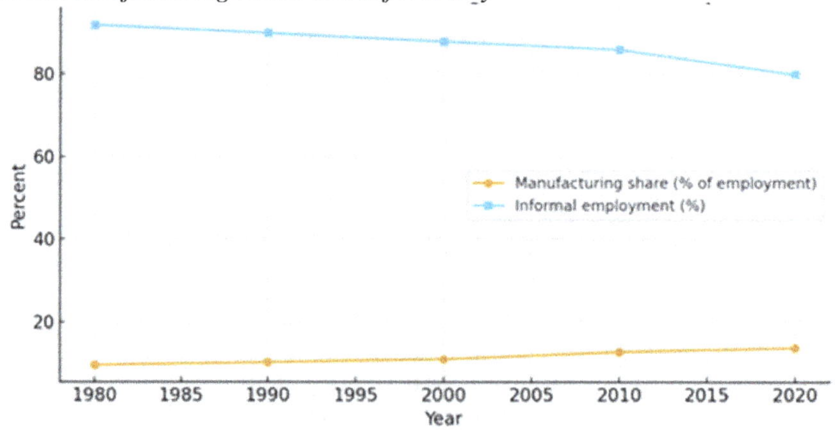

Source: Computed

Figure 2. Rate of Growth of Total GDP

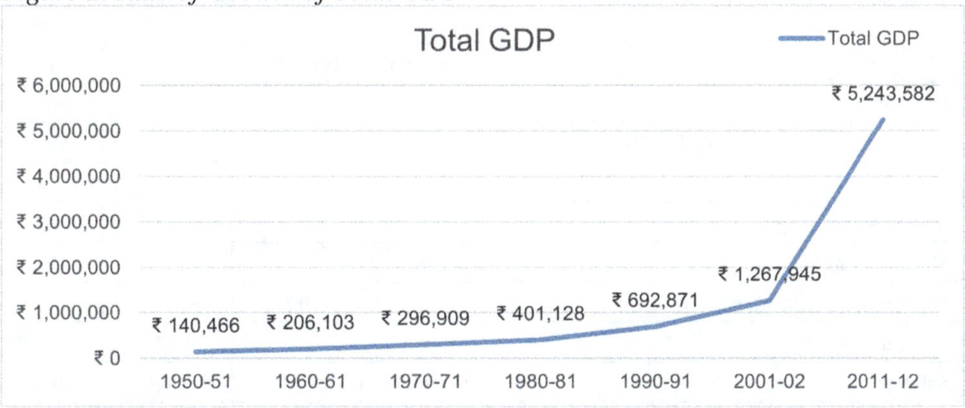

Source: Economic survey, Govt. of India (2010), and (Handbook of Statistics of Indian Economy-RBI, 2023)

The data as observed in the table above reveal that GDP from different elements of the primary sector, like agriculture, forestry, animal husbandry, fishing, etc., declined from 59.2% to 46.3% in the period 1950–51 to 1970–71, and thereafter, the rate of decline was higher to 16.2% in 2010–11.

The faster rate of fall in GDP in the primary sector had been compensated to a fair extent by the GDP growth rate in the secondary sector. The GDP growth rate in the secondary sector had been on a rising trend from 13.3% to 21.6% in the period 1950–51 to 1970–71, and the rate of growth in GDP in this sector was around 25.4% in 2010–11.

The comparative growth trend in GDP in the service sector has been higher than the above two sectors. It had grown from 27.5% to 32.1% between 1950–51 to 1970–71. After this, there had been a much higher rate of growth, from 32.1% to 58.4% from 1970–71 to 2010–11.

The developing structural change in GDP indicates that the share of contribution of each of the above three sectors, primary, including agriculture, secondary, including industry, is in a varying trend of gradually falling, and service sector contribution to GDP is steadily higher. The contribution of the service sector to GDP even went beyond 50% in 2010–11.

Therefore, it requires finding out if this structural shift in GDP has had any impact on the employment of the working population in India.

The data as is observed in Table-4 below shows the pattern of change in the involvement of employment in different sectors.

Table 4. Share of Employment in Different Sectors in India

Sectors	Total			
	1980-81	1990-1991	2000-2001	2010-11
Primary				
Agriculture, Forestry, Fishing, Hunting	74.0	68.1	63.9	53.2
Secondary				
Mining &Quarrying	0.4	0.6	0.7	0.6
Manufacturing	8.8	10.7	10.6	11.6
Electricity, Gas and Water	0.2	0.3	0.4	0.3
Construction	1.8	2.3	3.2	9.6
Sub-total	11.2	13.9	14.9	21.5
Tertiary (Services)				
Wholesale and Retail trade, Restaurant and Hotel	5.0	6.9	7.6	10.8
Transport, Storage, and Communication	1.8	2.5	2.9	4.3
Other services	7.8	8.8	10.7	10.3
Sub-total	14.6	18.2	21.2	25.4

Source: Institute of Applied Manpower Research / Planning Commission, *India Human Development Report 2011: Towards Social Inclusion*, Oxford University Press (2011)

Table 5. Harmonized manufacturing share and informality, India (selected years)

Year	Mfg employment % (harmonized)	Industry employment % (WDI SL.IND.EMPL.ZS)	Informal employment % (PLFS/ ILOSTAT)
1980	11.3	28.4	72.1
1990	12.2	27.1	69.5
2000	13.7	25.3	66.2
2010	13.9	24.6	63.0
2020	11.8	22.7	60.5

Notes: manufacturing employment % = share of total employment in manufacturing. Informality = share of informal employment in total employment as calculated from PLFS/NSS microdata; entries are rounded. Sources: MoSPI Statement 1.6A (harmonization), World Bank WDI (SL.IND.EMPL.ZS), PLFS/NSS (informality).

Figure 3. Share of Employment in Different Sectors in India

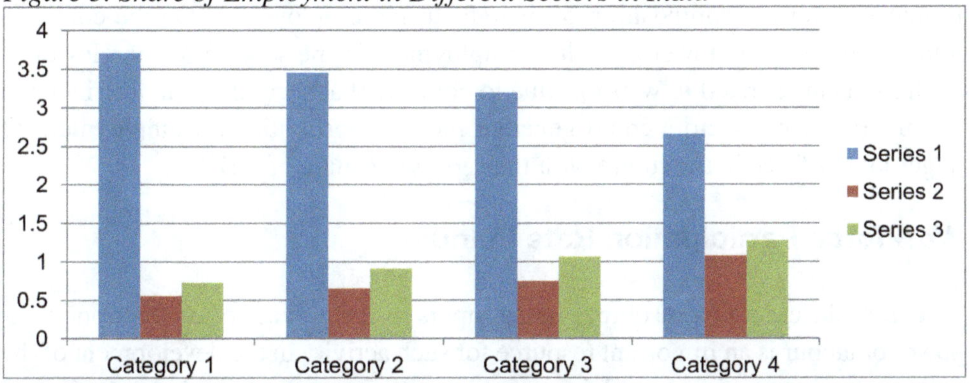

Source: Institute of Applied Manpower Research / Planning Commission, India Human Development Report 2011: Towards Social Inclusion, Oxford University Press (2011)

The data as observed in Table-4 is obtained from various rounds of the National Sample Survey and shows how the change in involvement in employment declined in the primary sector (agriculture) from 74% in 1980–81 to 53.2% in 2010–11. Also, along with this fall in employment in the primary sector, the share of employment in the secondary sector (industry) increased from 11.2% to 21.5% in 2010–11. In the tertiary sector (services), the share of employment during this period had grown from 14.6% in 1980–81 to 25.4% in 2010–11.

The outcome of this analysis justifies that, while the GDP shown in agriculture had fallen sharply, the fall in employment share did not occur correspondingly in India. Meanwhile, the GDP share of the secondary sector (industries, etc.) registered a growth of 1.7% from 1980–81 to 2010–11, but correspondingly, the rise in employment share was 21.8% during this period. A raw difference between growth in sectoral GDP shares and sectoral employment shares is suggestive but not sufficient to conclude that industry could not absorb excess labour. In other situations, the contribution to GDP from the service sector had grown up to 60%, but this could not proportionately add to the rate of growth of employment. The growth rate in employment had been around 9.2% with respect to the growth of GDP, which was 21.8% during the period 1980–81 to 2010–11.

From this analysis, it may be concluded that in India, the sequence was not achieved in terms of growth of GDP and employment in secondary sectors like industries, etc., during the process of industrialization. It skipped the post-industrialization phase of increasing its share of GDP as well as increasing proportionately the employment in the service sector. The data suggest a typical structural path as premature deindustrialization — i.e., manufacturing employment did not grow as much as expected at India's income levels. It tests whether India's experience is consistent with premature deindustrialization by examining (a) manufacturing employment trends, (b) productivity (GVA per worker) and real wages, and (c) the formal vs informal composition of manufacturing employment. On the contrary, it involved relatively lower growth in employment in the service sector. However, the trend is interpreted to wait for time to come so that secondary (industries) and tertiary (services) sectors could generate more opportunities for employment of larger labour force in the future with their growing share of GDP.

Workforce Participation Rate in India

Out of different few resources being inputs in the process of production, man-power or labour is an important resource for such activity in the development of the economy in a country. The workforce participation in a country, or the proportion of people's participation in a country with respect to the total population, depends on different factors like, age, sex, composition, nature of work, attitude to be involved in the type of work being available, scope of availability of work, etc. Cumulatively, all these become subjected to variation or change from place to place and also from period to period.

Table 6. Workforce Participation Rate Since 1901 (In percentage)

Year	Persons	Males	Females
1901	46	61	32
1951	39	54	23
1961	43	57	28
1971	33	53	12
1981	37	53	19
1991	38	52	23
2001	39	52	26
2011	39	54	26

Source: Census (2011) Population census 2011, Census 2011 India. Available at: https://www.census2011.co.in/ (Accessed: 11 December 2023).

Note: Census definitions of 'worker' and the tabulation concept (usual principal activity, usual principal + subsidiary, current weekly activity) altered across decades. The figures in Table 5 are drawn from published census tables but *are not* strictly comparable unless one re-tabulates the microdata or applies a concordance to a single worker definition. For trend analysis an attempt has been made to (a) re-compute a time-consistent series using the 'usual principal + subsidiary' concept where microdata permit, or (b) restrict long-run claims to comparable census categories only.

Figure 4. Workforce Participation Rate Since 1901 (%)

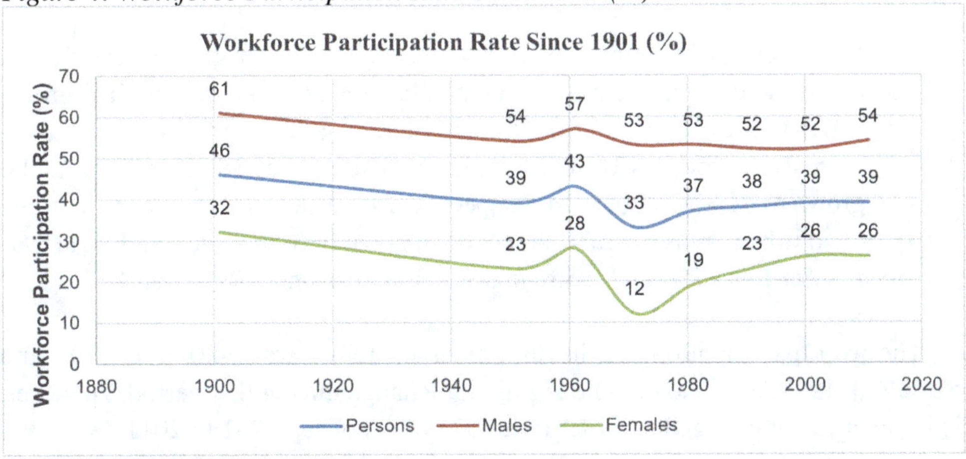

Source: Computed

It is seen workforce participation has declined from 46% to 39% from 1901 to 2011. However, from 1951 to 2011 it had shown a mixed trend fluctuating between some marginal rise and fall because of instabilities and variations in manufacturing systems and processes.

The table -7 below shows workforce participation rate in India in between the period 1981 to 2011.

Table 7. Work Participation Rate in India (1981-2011)

Year	Category	Persons	Males	Females
1981	Total	36.7	52.6	19.7
	Rural	38.8	53.8	23.1
	Urban	30.0	49.1	8.3
1991	Total	37.7	51.6	22.7
	Rural	40.2	52.5	27.2
	Urban	30.4	49.0	9.7
2001	Total	39.2	51.9	25.7
	Rural	42.0	52.4	31.0
	Urban	32.2	50.9	11.6
2011	Total	39.39	53.26	25.51
	Rural	41.8	54.03	24.08
	Urban	35.31	54.06	14.07

Source: Census (2011) Population census 2011, Census 2011 India. Available at: https://www.census2011.co.in/ (Accessed: 11 December 2023).

The main findings from Table -7 as above may be detailed as below:

1. The total work participation in general is observed to be in a growing trend since 1981, and the work participation rate had been 36.7% in 1981 and had grown up to 37.7% in 1991. In addition, it increased by 39.39% in 2011.
2. The increase in work performance rate in rural areas was 38.8% in 1981, and it had grown to 41.8% in 2011. More specifically, variation was evident between rural and urban areas. In rural areas, the change between 1981 and 2011 was from 38.8% to 41.8%, and in urban areas, it was from 30.0% to 35.31%.

The work participation rate in rural areas for males was 54.03% in 2011 and 52.65% in 1991, also showing no significant change during this period. However, in urban areas, it changed from 49.1% to 54.06% between 1981 to 2011.

A very marginal variation had been observed:

1. Females' work participation rate in rural areas increased from 23.1% to 31.0% between 1981 to 2001. But in urban areas, such female participation was comparatively less, rising from 8.3% to 11.6% between 1981 and 2001.
2. From the analysis of Tables 4 and 5, it is found that rural workmen are more burdened. They not only take part in larger numbers in economic activities, but they also take part in being more involved in domestic activities, etc.

Comparatively, as against rural workmen, a large majority of urban women, as per the 1991 and 2001 census, withdrew from economic activity outside the home and also became involved in domestic activity.

Table 8 below presents the current scenario starting from 2017–18 based on comparable factors, viz. Population, gross domestic product, consumption of fixed capital, gross national income, and net national income. And vis-à-vis the employment situation in India in both urban and rural areas, all three sectors have been compared. There is a growth in levels of employment in comparison with the early last quarter of the previous century and the first 15 years of this century. Women and men are both significantly employed in all three sectors: a) the primary sector, b) the secondary sector, and c) the tertiary sector. Employment occurs in both urban and rural areas. Also, the employment rates, more or less, are in a steady state without much significant variation with respect to rural and urban locations and gender. Careful reading of the 2017–18 to 2021–22 period shows that aggregate employment levels mask substantial short-term swings in participation and hours worked during the pandemic.

Table 8. Macro-economic aggregates

(Amount in Crore)					
Item/Year	**2017-18**	**2018-19**	**2019-20**	**2020-21**	**2021-22**
1	2	3	4	5	6
Population (in Lakh)	13140	13270	13410	13554	13686
Gross Domestic Product	17090042	18899668	20074856	19800914	23664637
Consumption of Fixed Capital	1764812	1984290	2165145	2340069	-
Gross National Income	16905230	18697344	19881742	19534226	23296345
Net National Income	15140418	16713054	17716597	17194158	20529727
Per Capita GDP ( )	130061	142424	149701	146087	172913
Per Capita GNI ( )	128655	140899	148261	144120	170222
Employment situation in India - per 1000 distribution of usually employed by broad groups of industry for various rounds					
Primary Sector (Male) Rural	550	532	554	538	NA
Primary Sector (Male) Urban	54	49	50	53	NA
Primary Sector (Female) Rural	732	711	757	757	NA
Primary Sector (Female) Urban	91	78	82	104	NA
Secondary Sector (Male) Rural	232	235	231	231	NA
Secondary Sector (Male) Urban	360	353	342	345	NA
Secondary Sector (Female) Rural	136	154	130	130	NA
Secondary Sector (Female) Urban	301	293	280	281	NA
Tertiary Sector (Male) Rural	220	232	216	216	NA
Tertiary Sector (Male) Urban	587	597	608	603	NA
Tertiary Sector (Female) Rural	132	136	112	112	NA
Tertiary Sector (Female) Urban	607	630	638	615	NA

Notes: 1. Data for 2018-19 are Third Revised Estimates, for 2019-20 are Second Revised Estimates and for 2020-21 are First Revised Estimates.

2. Data for 2021-22 are Provisional Estimates.

3. Population figures relates to mid-financial year.

Source: (Handbook of Statistics of Indian Economy-RBI, 2023)

Forecasting Framework

The secondary data on India's GDP, employment, and labour structure from multiple sources: The World Bank, ILOSTAT, India's National Statistical Organization (e.g. Periodic Labour Force Survey), industry reports (e.g. Nasscom), and international databases (Our World in Data, IMF) have been collected. The timespan mainly covers 1950–2023. Key variables include: sectoral shares of GDP and employment (primary, secondary, tertiary), labour force participation rates (by gender and urban/rural), productivity measures, and macro aggregates (investment, trade, education). Comparative data for other developing countries (e.g. Bangladesh, Vietnam, China, and selected African and Latin American peers) were obtained from World Bank and UN databases.

The analysis uses descriptive statistics and trend charts to visualize structural change. For forecasting, three modelling methods were outlined:

- Statistical regression models: The time-series or panel regressions linking per-capita GDP growth and sectoral shares, following Basole's (2022) "semi-elasticity" framework were charted. For example, regressing the decline in agricultural employment share on GDP per capita and control variables can quantify the historical relationship and project it forward under assumed GDP routes.
- Machine learning framework: Using historical dataset, algorithms are trained (e.g. random forests, gradient boosting) to predict key outcomes (e.g. sectoral employment shares or GDP growth) from predictors like investment rates, education levels, trade openness, and demographic factors. These frameworks can capture nonlinearities and interactions overlooked by linear models. Hold-out forecasts are compared to evaluate performance and search scenario analysis (e.g. what if investment in manufacturing upsurges).
- Conceptual frameworks: A theoretical forecasting approach using established frameworks are described. For instance, the Lewis model suggests an eventual "Lewis Turning Point" when surplus rural labour is exhausted. The point can be estimated by projecting rural surplus labour drop. Another approach is an extended Kuznets-type structural-change framework, plotting India's sectoral shares against GDP per capita and using regression on cross-country data as a guide (as in Basole 2022). These conceptual frameworks do not produce detailed forecasts by themselves but provide intuition and standards for what might be viable.

By triangulating these methods (econometric semi-elasticities, scenario-based structural projections, and machine-learning forecasting), an attempt has been made to present a fully-specified econometric elasticity estimate and model validation (backtest) for the period 1980–2020 and deliver robust future scenarios for India's occupational and economic trails.

International and Geopolitical Context

While focusing on India, it is essential to situate its experience globally:

- Comparative Development Patterns: Basole's (2022) cross-country analysis shows India's transformation is not outstanding even by developing-country standards. For a doubling of GDP per capita (PPP) between 1990–2019, India saw about a 17 percentage-point drop in agricultural employment and only a 7-point drop in its informal sector share. In comparison, Vietnam and Bangladesh achieved much larger reductions (e.g. Vietnam saw informal share drop by ~23 points). China's manufacturing boom raised many into higher-productivity jobs that India missed. Thus, compared to fellow developing economies, India's pace of job-creating transformation has been comparatively slow.
- Global South and trade: As a leading Global South economy and G20 member, India faces common challenges (urbanization pressures, skill gaps, infrastructural needs) and opportunities (south–south cooperation, trade blocs). For example, India's agricultural exports are oriented to developing countries, and Indian technology services integrate with global markets. However, India's economy is also affected by global shocks (oil prices, commodity cycles) and by shifts in international demand (e.g. global outsourcing trends). India has pursued initiatives like "Atmanirbhar Bharat" (self-reliance) and participated in key trade agreements (e.g. RCEP negotiations, recent CECA expansions) with the aim of improving industrial yield and exports.
- India–China relations: Geopolitically, China appears large. China is India's largest trading partner, but overpoweringly on China's terms. Chinese imports to India grew from about $10 billion in 2005 to over $100 billion in 2024, generating a massive trade shortfall. This deep integration means India relies on Chinese inputs for electronics, telecom, and solar industries. Recent strategic tensions and Chinese export curbs (e.g. on high-tech manufacturing) have exposed Indian supply susceptibilities. To lessen this, India is actively "de-risking": narrowing controls on Chinese investment, developing alternate supply chains (e.g. partnerships with Japan, Taiwan in semiconductors), and boosting import replacement in critical areas. The US–China rivalry also presents India with trade-offs: becoming a hinge for US investment can offshoot some industries, but China's ability to "strong-arm" through economic means (e.g. boycotts, restrictions) remains a menace.
- Climate and demographics: Global issues like climate change and population aging indirectly affect India's configuration. For example, severe climate effects (monsoon failure, heat stress) could contract agricultural output and push more workers into cities. The World Bank cautions that by 2050, under 2°C warming India might need to import twice as much food as it does now due to yield losses. At the same time, India's comparatively young and rising workforce (median age ~28) is a global advantage, but absorbing these

workers needs enough decent jobs. International cooperation on technology transfer (e.g. clean energy) and trade (e.g. Duty-Free Tariff Preferences for least developed countries) will shape India's development ranges.

Taken together, these global and geopolitical factors mean that India's occupational structure is both a domestic and international issue. Policy must therefore consider external linkages, from trade to climate finance, as part of the growth strategy.

Policy Implications and Recommendations

The analysis yields several strong lessons for policymakers in India (and others in the Global South):

- Accelerate Manufacturing Growth: The inconsistency of India's growth is that services grew while industry employment stalled. Revitalizing manufacturing should be a priority. Evidence suggests industry-led growth generates more jobs for average workers than IT or finance. Policy options include: further ease of doing business (land and labor reforms), targeted incentives (Production-Linked Incentive schemes for electronics, auto, textiles), and infrastructure investment (ports, power, logistics) to reduce costs. Countries like Vietnam and Bangladesh have shown that integrating into global value chains (e.g. garments, electronics) can rapidly expand manufacturing jobs; India could pursue similar strategies with careful attention to skill matching. The "Make in India" initiative should be strengthened with export orientation and MSME support.
- Skill Development and Education: Structural change demands a skilled workforce. Increasing vocational training and aligning curricula with industry needs (digital skills, advanced manufacturing, services analytics) is critical. Schemes like Skill India and higher education expansion should be data-driven: monitor which sectors are hiring and which skills are in shortage. Investment in female education and targeted programs can also educate female labor participation, tapping into a huge underused resource. As argued by the Ministry of Skill Development, entrepreneurship promotion (start-up hubs, ease of credit) is also needed so that returning farm workers can create small businesses in new industries.
- Reduce Informality and Strengthen Social Protection: A striking result of the analysis (Basole) is the persistence of informality (over 70% of workers). Policies to formalize business (simpler tax codes, lower registration costs) and extend social security (health, pensions) to informal workers can improve productivity and well-being. For example, broadening labor law cov-

erage gradually (differentiating by firm size) could encourage firms to register. Meanwhile, continue rural safety nets (MGNREGA scheme launched in 2006) as a cushion during transition. Schemes like MGNREGA have partially mitigated rural unemployment, so strengthening them (ensuring timely payments, linking with skill building) will help manage the human side of structural change.

- Regional and Rural Balance: Unbalanced growth can develop discrimination. Our findings echo Deaton and Dreze (2002) that growth need not "trickle down" equally. Policymakers should invest in lagging regions (e.g. via railway networks, digital infrastructure) to facilitate industry in non-metro areas. Promoting agro-processing and rural industries can absorb rural labor without forcing full urban migration. In states where agriculture is dwindling but jobs are scarce, special economic zones or backward area incentives may be justified.
- International Strategy: The geopolitical dynamics mean economic planning cannot be insular. To manage the China factor, India should diversify both import sources and export markets. Participation in trade agreements (with EU, ASEAN, African nations) and membership in initiatives like the Quad or Indo-Pacific Economic Framework can open high-value manufacturing opportunities. India should also continue positioning itself as a tech partner of the Global North (for pharmaceuticals, software) while nurturing global South ties (e.g. exporting generic drugs, agricultural technology to Africa). Foreign direct investment (FDI) can be navigated into strategic sectors where India lacks technology, through partnerships that ensure technology transfer rather than full procurement.
- Monitoring and Evaluation: Building on Basole's suggestion[REMOVED HYPERLINK FIELD], India needs a strong set of pointers to track transformation. These include the growth elasticity of employment, semi-elasticity of structural change, labor productivity growth in agriculture and industry, female LFPR, and share of organized jobs. Regularly publishing such metrics (disaggregated by region, gender, sector) will help assess whether policy interventions are working. For example, if industrial employment is deteriorating despite incentives, that signals a need to re-examine strategy.
- Long-term Sustainability: Finally, any path forward must account for environmental limits. As the climate analysis shows, water stress and extreme heat could undermine agriculture and even manufacturing (power generation). Policies must thus include climate adaptation: investing in drought-resistant crops, efficient irrigation (e.g. drip systems), and renewable energy. Sustainable urban planning is needed so cities can absorb migrants without slum proliferation. International climate finance should be leveraged

(e.g. NDC commitments) to update agriculture and energy in climate-smart ways, warranting that occupational shifts do not come at the cost of future generations.

In summary, the predictive framework suggests that if India maintains high growth and enacts these reforms, it could achieve a more balanced occupational structure by 2040: agriculture employment near 25–30%, industry 30–35%, services 35–45%. This would resemble early-stage East Asian economies. However, risks (slower growth, policy gridlock, external shocks) could slow this changeover. Policymakers should therefore aim aggressively at comprehensive structural transformation – creating not just GDP growth, but broad-based employment – as a precedence.

CONCLUSION

India's journey from a farming economy to a mixed industrial-service economy has been significant but incomplete. Employment shifted out of agriculture, but too many workers continue in low-productivity jobs. Unlike past industrializers, India has not fully capitalized on manufacturing, and its informal workforce is remarkably high. External factors – from global market shifts to climate change – muddle the picture.

By combining historical data and modern theory, this study provides an updated, inclusive overview of India's structural change. The analysis highlights that economic development and occupational structure are deeply entwined: one cannot be understood without the other. Looking forward, using predictive framework allows to sketch possible futures, but all frameworks point to the same far-reaching message: India must create far more productive jobs to sustain development. The actionable recommendations above – on skills, industry policy, trade, and social protection – propose a roadmap.

Finally, while India is the focus, many understandings apply globally. Other developing countries watching India's rise will note both its successes and the snags highlighted here. International readers and policymakers can learn that rapid GDP growth needs corresponding policies in labour markets, education, and trade to produce widespread prosperity. Contrariwise, India can study peers like Vietnam, Bangladesh or Brazil to see unconventional strategies. In an interconnected world, progress in one nation is tied to others, creating shared lessons on occupational structure and growth invaluable.

DECLARATIONS

Conflict of Interest: The authors declare that they have no conflict of interest with respect to the research, authorship, and/or publication of this manuscript.

Data Availability Statement: This study is based entirely on secondary data that are publicly available. The datasets and indicators used were obtained from the World Bank, Reserve Bank of India (RBI), Central Statistical Organisations, National Sample Survey Organisation (NSSO), Planning Commission, Registrar General of India, Statista, and published Census reports. Specific dataset names, citations, URLs and access dates are provided in the manuscript. The underlying data can be obtained from the corresponding author upon request.

Funding Statement: This research received no specific grant from any funding agency in the public, commercial, or not-for-profit sectors.

REFERENCES

Ahluwalia, M. S. (2002). Economic Reforms in India Since 1991: Has Gradualism Worked? *The Journal of Economic Perspectives*, *16*(3), 67–88. DOI: 10.1257/089533002760278721

Amirapu, A., & Subramanian, A. (2015). *Manufacturing or services? An Indian illustration of a development dilemma* (Center for Global Development Working Paper No. 409). Center for Global Development. https://doi.org/DOI: 10.2139/ssrn.2623158

Anand, S., & Ravallion, M. (1993). Human Development in Poor Countries: On the Role of Private Incomes and Public Services. *The Journal of Economic Perspectives*, *7*(1), 133–150. DOI: 10.1257/jep.7.1.133

Bakshi, A., & Kaur, K. (2020). Dragon versus Elephant: A Comparison of Pattern of Agriculture in Two Asian Economies - China and India. *PRAGATI : Journal of Indian Economy*, *7*(1), 48. DOI: 10.17492/pragati.v7i1.195701

Bardhan, P. (1970). On the long-run stability of the Indian economy. *Journal of Political Economy*, *78*(3), 456–468.

Basole, A. (2022). Structural transformation and employment generation in India: Past performance and the way forward. *The Indian Journal of Labour Economics : the Quarterly Journal of the Indian Society of Labour Economics*, *65*(2), 295–320. DOI: 10.1007/s41027-022-00380-y PMID: 35874181

Bhandari, M. P. (2023). The Fundamental Principles of Social Sciences. *Business Ethics and Leadership*, *7*(2), 73–86. DOI: 10.21272/bel.7(2).73-86.2023

Cabral, C., & Dhar, R. L. (2019). Skill development research in India: A systematic literature review and future research agenda. *Benchmarking*, *26*(7), 2242–2266. DOI: 10.1108/BIJ-07-2018-0211

Cabral, L., Pandey, P., & Xu, X. (2022). Epic narratives of the Green Revolution in Brazil, China, and India. *Agriculture and Human Values*, *39*(1), 249–267. DOI: 10.1007/s10460-021-10241-x

Clark, C. (1967). *The conditions of economic progress*. Alianza Editorial S.A.

Daoust, G., & Selby, J. (2024). Climate change and migration: A review and new framework for analysis. *Wiley Interdisciplinary Reviews: Climate Change*, *15*(4), e886. DOI: 10.1002/wcc.886

Datt, G., & Mahajan, A. (2023). *Indian Economy* (72nd ed.). S Chand Publisher.

Deaton, A. (2013). *The Great Escape : Health, Wealth, and the Origins of Inequality*. Princeton University Press.

Deaton, A., & Dreze, J. (2002). Poverty and Inequality in India: A Re-Examination. *Economic and Political Weekly*, *37*(36), 3729–3748.

Deshpande, A. (2021). How India's Caste Inequality Has Persisted—And Deepened in the Pandemic. *Current History (New York, N.Y.)*, *120*(825), 127–132. DOI: 10.1525/curh.2021.120.825.127

Drèze, J., & Sen, A. (2002). *India: Development and Participation* (2nd ed.). Oxford University Press. https://doi.org/DOI: 10.1093/acprof:oso/9780199257492.001.0001

Economic Survey, Govt. of India. (2010). Economic survey 2010-11 -statistical appendix. India Budget. Retrieved November 2, 2023, from https://www.indiabudget.gov.in/budget2011-2012/es2010-11/estat1.pdf

Fisher, A. G. B. (1939). Production, Primary, Secondary and Tertiary. *The Economic Record*, *15*(1), 24–38. DOI: 10.1111/j.1475-4932.1939.tb01015.x

Ghani, E. (2011). *Reshaping Tomorrow: Is South Asia Ready for the Big Leap?* (1st ed.). Oxford University Press.

Goldstein, A. (2015). Structural change and industrial development in the BRICS. *International Affairs*, *91*(6), 1427–1428. DOI: 10.1111/1468-2346.12470

Gollin, D., Hansen, C. W., & Wingender, A. M. (2021). Two Blades of Grass: The Impact of the Green Revolution. *Journal of Political Economy, 129*(8), 2344–2384. DOI: 10.1086/714444

Handbook of Statistics of Indian Economy. RBI. (2023, September 15). Macro-Economic Aggregates. Reserve Bank of India - Publications. Retrieved October 17, 2023, from https://rbi.org.in/Scripts/PublicationsView.aspx?id=21807

Herrendorf, B., Rogerson, R., & Valentinyi, Á. (2014). Growth and structural transformation. In Aghion, P., & Durlauf, S. N. (Eds.), *Handbook of Economic Growth* (Vol. 2, pp. 855–941). Elsevier., DOI: 10.1016/B978-0-444-53540-5.00006-9

Jose, A. V. (2017). Agricultural wages in Indian states. *The Indian Journal of Labour Economics : the Quarterly Journal of the Indian Society of Labour Economics, 60*(3), 333–345. DOI: 10.1007/s41027-018-0111-x

Joshi, V., & Little, I. M. D. (1994). *India : macroeconomics and political economy, 1964-1991*. World Bank Publications.

Kannan, K. P. (2022). India's Elusive Quest for Inclusive Development: An Employment Perspective. *The Indian Journal of Labour Economics : the Quarterly Journal of the Indian Society of Labour Economics, 65*(3), 579–623. DOI: 10.1007/s41027-022-00393-7 PMID: 36246845

Matthess, M., & Kunkel, S. (2020). Structural change and digitalization in developing countries: Conceptually linking the two transformations. *Technology in Society, 63*, 1–13. DOI: 10.1016/J.TECHSOC.2020.101428

McMillan, M., Rodrik, D., & Verduzco-Gallo, Í. (2014). Globalization, structural change, and productivity growth, with an update on Africa. *World Development, 63*, 11–32. DOI: 10.1016/j.worlddev.2013.10.012

Mehrotra, S., Gandhi, A., Kumar, C., Saha, P., Sahoo, B. K., & Sharma, A. (2011). *India Human Development Report 2011 - Towards Social Inclusion* .

Miozzo, M., & Soete, L. (2001). Internationalization of Services. *Technological Forecasting and Social Change, 67*(2–3), 159–185. DOI: 10.1016/S0040-1625(00)00091-3

Mishra, R. (1999). *Globalization and the Welfare State*. Edward Elgar Publishing Ltd.

NASSCOM. (2020), The Technology Sector in India: Strategic Review 2020 (NASSCOM Strategic Review), NASSCOM, New Delhi. URL: https://nasscom.in/knowledge-center/publications/technology-sector-india-2020-techade-new-decade-strategic-review (Planning Commission of India. (2013). Twelfth Five Year Plan (2012–2017) Faster, More Inclusive and Sustainable Growth: Vol. Volume I. SAGE Publications India Pvt Ltd.

Pradhan, B. (2002). *The role of education in wage inequality change in India: 1988–97*. National Council of Applied Economic Research.

Ranis, G., Stewart, F., & Ramirez, A. (2000). Economic Growth and Human Development. *World Development*, *28*(2), 197–219. DOI: 10.1016/S0305-750X(99)00131-X

Reserve Bank of India. (2023). *Handbook of Statistics on Indian Economy*. https://www.rbi.org.in/Scripts/PublicationsView.aspx?id=21807

Rodrik, D. (2016). Premature deindustrialization. *Journal of Economic Growth*, *21*(1), 1–33. DOI: 10.1007/s10887-015-9122-3

Sahoo, G., Wani, A. M., Swamy, S. L., Roul, P. K., Dash, A. C., & Sharma, A. (2022). Livelihood Strategy and Sustainability Aspects in Industrialization as a Source of Employment in Rural Areas. In *Social Morphology* (pp. 643–670). Human Welfare, and Sustainability.

Sen, A. (1960). Choice of techniques: An aspect of the theory of planned economic development. *Oxford Economic Papers*, *12*(I), 1–15.

Socal, M. P., Sharfstein, J. M., & Greene, J. A. (2021). The pandemic and the supply chain: Gaps in pharmaceutical production and distribution. *American Journal of Public Health*, *111*(4), 635–639. DOI: 10.2105/AJPH.2020.306138 PMID: 33507805

Srinivasan, T. N. (2003). *Indian Economy: Current Problems and Future Prospects* (No. 173).

Statista - The Statistics Portal. (2023). Statista. Retrieved November 2, 2023, from https://www.statista.com/?kw=statista&crmtag=adwords&gclid=Cj0KCQiAyKurBhD5ARIsALamXaGbsE_fgFP9mYIbjcSxBBMW_tLHkjmjRuX6WkqEcttdVWVWsuso_QaAhptEALw_wcB

Subramanian, A. (2009). *India's Turn: Understanding the Economic Transformation*. OUP Catalogue.

Sukanya, R., & Tantia, V. (2023). *Urbanization and the Impact on Economic Development.*, DOI: 10.4018/978-1-6684-9261-1.ch019

Sundaram, K., & Tendulkar, S. D. (2003). Poverty among Social and Economic Groups in India in 1990s. *Economic and Political Weekly*, *38*(50), 5263–5276.

The Economic Times. (2025, April 16). India's trade deficit with China widens to record $99.2 bln amid dumping concerns. *The Economic Times*.

Vu, K., Hanafizadeh, P., & Bohlin, E. (2020). ICT as a driver of economic growth: A survey of the literature and directions for future research. *Telecommunications Policy*, *44*(2), 101922. DOI: 10.1016/j.telpol.2020.101922

World Development Indicators. Washington, D.C.: The World Bank

Chapter 9

Communication Infrastructures and Platform Dynamics in Shaping Entrepreneurial Success:

A Qualitative Study of Market Participation in Greece and Cyprus

Marcos Komodromos
https://orcid.org/0000-0002-2910-6541
University of Nicosia, Cyprus

Tatiana Harkiolakis
https://orcid.org/0000-0002-3581-9599
ISC Paris Business School, France

Stelios Papalexandris
https://orcid.org/0000-0001-9248-2067
Democritus University of Thrace, Greece

Sofia Anastasiadou
https://orcid.org/0000-0001-6404-5003
University of Western Macedonia, Greece

ABSTRACT

This study examines the influence of communication infrastructures and digital platform dynamics on entrepreneurial success in Greece and Cyprus. Employing a

DOI: 10.4018/979-8-3373-4566-6.ch009

phenomenological methodology, it utilises interviews and observations from sixty entrepreneurs across various sectors. The findings underscore that high-quality infrastructure is both vital and strategically beneficial, as it facilitates global collaboration and market access. Cyprus benefits from government-led infrastructure development, whereas Greek entrepreneurs demonstrate innovation in overcoming limitations. Key to success are multi-platform orchestration, algorithm optimization, and cultural communication strategies that integrate global digital standards with local practices. The research introduces the concept of "ecosystem bridging capabilities," highlighting entrepreneurs as active orchestrators who leverage network effects despite limited domestic markets. These insights contribute to the development of entrepreneurial education and policy aimed at promoting sustainable growth.

INTRODUCTION

This chapter presents a comprehensive qualitative investigation that critically examines the multifaceted relationships between communication infrastructures, media ecosystems, and platform dynamics in shaping entrepreneurial market participation and business success across the economically integrated yet strategically differentiated economies of Greece and Cyprus (Giotopoulos, 2024; Malthouse & Copulsky, 2022; Nicoli et al., 2019; Yakhshiboeva & Eshbayev, 2023).

Digital entrepreneurial ecosystems (DEEs) are crucial for fostering innovation, driving business growth, and enhancing market engagement. In Greece and Cyprus, reviews highlight that DEEs—consisting of technological platforms, e-governance, and digital collaboration—are key to expanding businesses and fostering extroversion, especially when backed by cultural and structural shifts within business networks like chambers of commerce (Bejjani et al., 2023; Constantinou et al., 2022; Harkiolakis & Komodromos, 2023; Komodromos, 2024; Seremeti et al., 2026). These ecosystems enable entrepreneurs to utilise digital tools effectively for creating opportunities, engaging customers, and fostering lasting business relationships.

Multi-sided online platforms and digital applications are rapidly transforming entrepreneurial ecosystems by lowering entry barriers, enabling scalability, and fostering innovation (Zhou, 2024). The adoption of such platforms in both countries and comparable European contexts has been shown to enhance entrepreneurial alertness and opportunity recognition, regardless of the innovation maturity of the ecosystem. Social media platforms, in particular, are powerful tools for customer relations, marketing, and identifying new business opportunities, directly impacting the scalability and success of start-ups (Chatzinikolaou, 2025; Komodromos, 2024).

The research methodology employs rigorous phenomenological analysis of in-depth interviews with 60 strategically selected entrepreneurs, business leaders, and communication strategists from both nations, representing diverse sectoral contexts, business lifecycle stages, and market positioning strategies to ensure comprehensive theoretical saturation and analytical depth (Ratnasari et al., 2023).

This investigation makes substantial theoretical and empirical contributions to the rapidly evolving field of marketing futures and strategic communication studies by systematically deconstructing and analyzing the transformative mechanisms through which digital ecosystems function not merely as technological tools but as fundamental restructuring forces that reshape entrepreneurial opportunity recognition, resource mobilization, and value creation processes in contemporary business environments (Boichenko & Sierhieiev, 2024; Chen et al., 2023; Diaz et al., 2025; Istrefi et al., 2025; Komodromos, 2025). The analytical framework transcends traditional descriptive approaches to communication research by examining how digital ecosystem participation creates new forms of competitive advantage while simultaneously generating novel vulnerabilities and dependencies that entrepreneurs must strategically navigate.

The study's theoretical foundation emerges from the critical recognition that modern entrepreneurial success has undergone a paradigmatic transformation, shifting from resource-based and market-positioning theories toward communication competency and platform orchestration capabilities as primary determinants of venture viability and growth potential (Bachmann et al., 2024; Komodromos et al., 2026; Malthouse & Copulsky, 2022). This transformation represents a fundamental departure from traditional entrepreneurship theory, where physical assets, financial capital, and geographic market access previously constituted the primary success factors.

Instead, contemporary entrepreneurial success increasingly depends on sophisticated communication strategy development, multi-platform engagement optimization, and ecosystem relationship management capabilities that enable entrepreneurs to transcend traditional market boundaries and resource constraints through digital connectivity and network effects (Gudi et al., 2024).

The analytical significance of focusing on Greece and Cyprus extends beyond geographical convenience to encompass theoretically significant variations in digital infrastructure development, regulatory frameworks, and cultural communication practices that create natural experimental conditions for understanding how systemic factors mediate entrepreneurial communication effectiveness (Chatzinikolaou, 2025; Constantinou et al., 2022; Giotopoulos, 2024). As digital entrepreneurial ecosystems continue to reshape economic landscapes across Europe fundamentally, Mediterranean economies present particularly compelling analytical contexts due to their combination of traditional relationship-based business cultures, constrained domestic market sizes, strategic geographic positioning, and varying degrees of digital transformation investment (Bejjani et al., 2023; Harkiolakis & Komodromos, 2023).

LITERATURE REVIEW AND THEORETICAL FRAMEWORK

Digital Entrepreneurship and Communication Theory

Digital entrepreneurship, driven by digital technologies and platforms, is transforming the way ventures are developed and operated, necessitating new strategic communication skills that differ from traditional entrepreneurship (Felicetti et al., 2023; Komodromos, 2024). Modern digital entrepreneurship requires advanced digital communication skills, seamless integration of digital platforms, and strategic flexibility (Halkias & George-Ufot, 2021).

The intersection of digital entrepreneurship and strategic communication represents a rapidly evolving research domain that challenges traditional boundaries between entrepreneurship studies and communication research (Nicoli et al., 2019; Ratnasari et al., 2023; Yakhshiboeva & Eshbayev, 2023). Digital entrepreneurship, defined as the creation of new ventures and the transformation of existing businesses through digital technologies and platforms, requires fundamentally different communication competencies than traditional entrepreneurship models. This transformation is particularly pronounced in Mediterranean economies, where entrepreneurs must navigate complex cultural contexts while leveraging global digital platforms to overcome domestic market limitations (Baig et al., 2022; Seremeti et al., 2026; Zakzouk et al., 2023).

Strategic communication theory provides the foundational framework for understanding how entrepreneurs create, manage, and optimize communication processes to achieve specific business objectives (Zhou, 2024). In digital contexts, this theory emphasizes the importance of multi-channel orchestration, audience segmentation, and platform-specific message adaptation as core competencies for entrepreneurial success. The Mediterranean context adds additional complexity through cultural communication norms that emphasize relationship-building and trust development, requiring entrepreneurs to balance global digital communication strategies with local cultural expectations (Satar et al., 2024).

Recent research demonstrates that entrepreneurial communication effectiveness depends not only on message content and delivery mechanisms but also on entrepreneurs' ability to navigate platform dynamics, manage network effects, and orchestrate multi-stakeholder ecosystems (Baig et al., 2022; Chen et al., 2023; Kusbandono et al., 2024). Effective entrepreneurial communication relies on clear, confident, and adaptable messaging, combined with empathetic listening tailored to diverse audiences and situations. Skills like storytelling, rhetorical techniques, and metaphors are crucial for engaging stakeholders and fostering trust, reputation, and legitimacy (Nicoli et al., 2019; Scott et al., 2021).

This multidimensional approach to entrepreneurial communication requires a sophisticated understanding of both technological affordances and cultural communication practices, particularly in cross-national contexts, such as the Greece-Cyprus comparison examined in this study.

Platform Dynamics and Ecosystem Theory

Platform ecosystem theory describes how digital platforms generate value by managing interactions among various participants, utilising network effects, lowering transaction costs, and fostering new ways of collaboration and competition (Hawlitschek & Hodapp, 2024). For entrepreneurs, particularly in smaller economies, these platforms provide access to larger markets and resources that traditional approaches lack. This theory offers vital insights into how digital platforms leverage network effects and multi-sided markets to create value (Halkias & George-Ufot, 2021; Istrefi et al., 2025).

For entrepreneurs in smaller economies, such as Greece and Cyprus, platforms serve as critical mechanisms for accessing broader markets, reducing transaction costs, and leveraging network effects that would otherwise be impossible to achieve through traditional business models (Goertler et al., 2025). The theoretical framework recognises that platform participation requires entrepreneurs to develop new forms of strategic thinking that balance competition and collaboration within shared digital ecosystems.

The concept of platform orchestration becomes particularly important for understanding how successful entrepreneurs manage relationships with multiple platform providers while maintaining strategic independence (Ritala, 2023). Research demonstrates that entrepreneurs who effectively navigate platform dynamics develop sophisticated capabilities in areas such as algorithm optimisation, community management, and multi-platform integration strategies (Jacobides et al., 2024). These capabilities are especially critical for entrepreneurs in Greece and Cyprus, who must leverage platforms to overcome geographical and market size constraints while competing with larger economies for access to international markets.

Network effects theory explains how communication platforms create increasing returns to scale through user participation and engagement (Kreiterling, 2023). For Mediterranean entrepreneurs, understanding and leveraging network effects becomes essential for achieving sustainable competitive advantages despite limited domestic market sizes (Halkias et al., 2024; Komodromos, 2024). The theoretical framework suggests that successful entrepreneurs develop capabilities not only in platform utilisation but also in network orchestration, enabling them to create and capture value through strategic positioning within digital ecosystems (Fernandes et al., 2022).

Mediterranean Economic Context

The Mediterranean economic context presents unique challenges and opportunities for the development of digital entrepreneurship. Greece and Cyprus, despite their cultural similarities and EU membership, demonstrate significant variations in digital infrastructure development, regulatory frameworks, and entrepreneurial ecosystem maturity (Chatzinikolaou, 2025). These variations create natural experimental conditions for understanding how systemic factors influence entrepreneurial communication effectiveness and platform utilization strategies.

Cyprus has emerged as a regional digital hub through strategic investments in communication infrastructure, favorable regulatory frameworks, and targeted incentive programs designed to attract international technology companies (Venizelou & Poullikkas, 2025). Cyprus is increasingly recognised as a regional digital hub due to its strategic investments in digital infrastructure, a supportive regulatory climate, and targeted initiatives to attract international technology firms. Its location, serving as a bridge between Europe, Asia, and Africa, further boosts its attractiveness for entrepreneurs looking to access global markets (Constantinou et al., 2022; Istrefi et al., 2025). The country's positioning as a bridge between European, Asian, and African markets creates unique opportunities for entrepreneurs to leverage communication technologies for global market access. Recent data indicate that Cyprus has achieved the fastest-growing startup ecosystem in the European Union, with significant increases in both startup formation and international investment (Hadjimitsis et al., 2024).

Greece presents a more complex entrepreneurial landscape, characterised by traditional economic structures undergoing digital transformation processes. Greek entrepreneurs demonstrate remarkable creativity in leveraging available communication technologies to overcome resource constraints and market access limitations (Chatzinikolaou, 2025; Diaz et al., 2025; Harkiolakis et al., 2025). The country's rich cultural heritage in relationship-based business practices translates effectively into digital communication strategies, particularly in sectors requiring high levels of trust and personal engagement.

METHODOLOGY

Research Design and Philosophical Approach

This study employs a phenomenological research approach grounded in constructivist epistemology to investigate entrepreneurs' lived experiences in navigating communication-driven business development processes (Zhang et al., 2024).

The phenomenological method is particularly suitable for studying entrepreneurial communication, as it helps uncover how entrepreneurs interpret and respond to the complexities of digital ecosystems. This approach enables the research to examine the nuanced ways in which cultural, technological, and economic factors shape entrepreneurial communication strategies.

The research design incorporates multiple qualitative methodologies, including semi-structured interviews, focus groups, and participant observation, to achieve methodological triangulation and enhance the validity of findings (Abaci, 2022). This multi-method approach recognises that entrepreneurial communication practices are complex, context-dependent phenomena that require multiple perspectives and data sources to understand comprehensively (Manafe et al., 2023; Seremeti et al., 2026). The constructivist framework acknowledges that entrepreneurial experiences are socially constructed through interaction with technological systems, cultural contexts, and market Dynamics (Halkias & George-Ufot, 2021).

Grounded theory analysis techniques are utilised to facilitate the emergence of theoretical insights directly from the data, rather than imposing preconceived theoretical frameworks (Koleva, 2023; Pascua, 2024). This analytical approach is especially significant for comprehending entrepreneurial communication within digital environments, where swift technological advancements and dynamic platform changes give rise to new phenomena that existing theories might not sufficiently address (Komodromos, 2024; Makri & Neely, 2021). The research design underscores an iterative cycle of data collection and analysis, which supports ongoing theoretical development and refinement throughout the research process (Dahwa, 2023).

Participant Selection and Demographics

The study employs purposeful sampling to select 60 participants from diverse entrepreneurial backgrounds, sectors, and experience levels across Greece and Cyprus. Selection criteria include (1) active roles as founders, co-founders, or senior executives; (2) extensive experience with digital communication tools; (3) involvement in market activities beyond their home countries; and (4) readiness to engage in detailed interviews and possible follow-up sessions.

The Greek participant group (n=25) encompasses a variety of sectoral backgrounds, including technology startups specialising in software development and digital services, traditional manufacturing firms undergoing digital transformation, tourism and hospitality businesses adopting digital marketing and customer engagement tools, and creative industries utilising digital platforms for content creation and distribution. The participants' entrepreneurial experience varies from 2 to 25 years, covering both early-stage startups and well-established companies with international reach.

The Cypriot participant cohort (n=35) comprises fintech companies engaged in the development of innovative financial services and payment platforms, gaming and entertainment enterprises producing digital content for international markets, professional services firms employing digital platforms for client acquisition and service delivery, and emerging technology ventures in domains such as artificial intelligence, blockchain, and cybersecurity. The broader Cypriot sample reflects the nation's status as a rapidly expanding digital entrepreneurship hub, characterised by a higher density of technology-oriented ventures. Participants possess entrepreneurial experience ranging from one to twenty years, with a notable concentration among ventures founded within the past decade during Cyprus's digital transformation period.

The demographic composition comprises 42% female and 58% male participants, reflecting substantial efforts to incorporate diverse perspectives on entrepreneurial communication practices. Age ranges extend from 28 to 58 years, thereby ensuring representation across various generational approaches to digital communication and platform utilisation. The educational backgrounds encompass technical disciplines (35%), business and management (40%), and liberal arts and humanities (25%), offering valuable insights into how different knowledge backgrounds influence entrepreneurial communication strategies.

Data Collection Procedures

Data collection employs a tripartite methodology to gather insights into individual experiences and collective patterns in entrepreneurial communication (Agustina et al., 2025). *The first phase* involves semi-structured interviews lasting between 60 and 120 minutes, conducted via video conferencing to accommodate participants' schedules and locations. These interviews focus on five essential domains: the assessment and utilisation of communication infrastructure; engagement strategies and their evolution; methods of market participation and international expansion; ecosystem dynamics and the cultivation of collaborative relationships; and criteria for measuring success alongside the development of adaptive strategies.

The second phase comprises focus group sessions with six to eight participants, organised according to geographical location and sector to promote meaningful discussions regarding shared experiences and challenges. These groups explore collective understanding, peer learning, and collaborative approaches to addressing communication issues. Sessions incorporate scenario-based talks in which participants share specific examples of developing and implementing communication strategies, thereby gaining insights from others' experiences (Komodromos et al., 2026).

The third phase entails ethnographic observation of communication practices within authentic business environments, including participation in networking events, pitch presentations, and interactions on digital platforms. The observational data collected provide contextual insights into the practical application of communication strategies. It elucidates how entrepreneurs adapt their approaches in response to real-time feedback and environmental shifts (Dahwa, 2023). Additionally, this phase involves analysing entrepreneurs' digital communication artefacts, such as social media content, website materials, and platform interactions, to understand the tangible realisation of communication strategies.

Analysis Approach

Data analysis utilises thematic analysis methods rooted in grounded theory principles to discern patterns, themes, and theoretical insights derived from participant experiences (O'Callaghan et al., 2024). The analytical process begins with the open coding of interview transcripts and observational notes, which facilitates the emergence of key concepts and categories free from pre-existing theoretical limitations. This preliminary coding phase detects recurring patterns in how entrepreneurs articulate their communication practices, challenges, and strategies for success (Naeem et al., 2023).

Axial coding techniques are subsequently utilised to discern relationships among categories and to develop overarching themes that elucidate the interactions and influences of various facets of entrepreneurial communication practices (Manafe et al., 2023). This analytical phase emphasises comprehending the causal conditions, contextual factors, and resultant patterns that determine the effectiveness of entrepreneurial communication under diverse circumstances.

Selective coding methods synthesise themes into clear theoretical frameworks that explain key aspects of communication-driven entrepreneurship in Mediterranean settings. The analysis emphasises ongoing comparisons between Greek and Cypriot participants to identify common patterns and unique regional differences in entrepreneurial communication. Member checking ensures that the interpretations accurately capture the participants' experiences and perspectives.

FINDINGS AND THEMATIC ANALYSIS

Entrepreneurial communication practices constitute a complex domain comprising advanced skills, adaptive strategies, and contextual modifications that fundamentally influence how entrepreneurs engage with diverse stakeholders, foster enduring relationships, and attain sophisticated business objectives within increasingly digitized markets (McGloin et al., 2023).

Table 1. Entrepreneurial communication model

Key Concept	Explanation	Examples in Practice
Advanced Skills	High-level communication abilities that go beyond basic exchanges, enabling precise, persuasive, and impactful interaction.	Storytelling for pitching investors, negotiation, persuasive marketing copywriting.
Adaptive Strategies	Flexible approaches to communication that adjust based on audience, goals, and circumstances.	Changing tone for different cultures, modifying pitch for investors vs. customers.
Contextual Modifications	Tailoring messages to fit specific settings, industries, or market conditions.	Using industry jargon with partners, adopting formal tone in government meetings.
Engaging Diverse Stakeholders	Building meaningful connections with various groups (customers, investors, employees, partners).	Hosting community events, investor meet-ups, and customer feedback sessions.
Fostering Enduring Relationships	Creating long-term trust and loyalty with business contacts.	Continuous after-sales support, personalized follow-ups, loyalty programs.
Attaining Sophisticated Business Objectives	Using communication to meet complex goals such as scaling internationally or innovating in the market.	Negotiating a joint venture, securing strategic alliances.
Operating in Digitised Markets	Navigating and leveraging digital platforms for communication, branding, and sales.	Social media campaigns, webinars, AI-driven customer engagement.

Recent empirical research highlights the crucial role of communication competencies in securing vital resources, managing stakeholder ecosystems, and shaping organisational cultures that facilitate sustainable competitive advantages in dynamic business environments (Manafe et al., 2023). The subsequent thematic analysis delineates five interconnected themes derived from a rigorous examination of participant interviews and observational data, each elucidating distinct yet mutually reinforcing facets of communication-mediated entrepreneurship within Mediterranean contexts.

To uphold the highest standards of ethical research practices and to safeguard the privacy of all contributors, the identities of participants in this study have been protected using pseudonyms. Throughout this chapter, individuals are referred to by alternative names enclosed in quotation marks—such as "Maria," "Dimitris," or "Alexandros,"—rather than their actual names. This methodology ensures the confidentiality of participants' identities and personal information while permitting clear attribution of perspectives and quotations essential to the research analysis.

THEME 1: INFRASTRUCTURE AS COMPETITIVE ADVANTAGE—THE FOUNDATION OF DIGITAL ENTREPRENEURSHIP

The research indicates that access to and quality of communication infrastructure serve not merely as operational necessities but as essential competitive advantages. These elements systematically facilitate or hinder entrepreneurial opportunity recognition, resource mobilisation, and market expansion capabilities in both Greece and Cyprus (Diaz et al., 2025). Nonetheless, notable differences are observed between the two nations concerning infrastructure development trajectories, accessibility patterns, and strategic utilisation approaches by entrepreneurs. Such disparities give rise to unique entrepreneurial opportunity frameworks, which influence venture types, growth strategies, and competitive positioning.

Each row details strategies and practices shaping competitiveness in platform environments. Greek and Cypriot entrepreneurs demonstrate advanced strategic planning, but they differ in platform focus, resource utilisation, and optimisation methods.

Cypriot Infrastructure Excellence: Creating Technology Leadership

Cypriot entrepreneurs often view advanced communication infrastructure as a key enabler of their international success, enabling them to compete directly with firms from major European financial and tech hubs. "Maria", a fintech entrepreneur from Limassol managing a pan-European payment processing platform, provides detailed insight: "When we were choosing where to establish our European operations, Cyprus's fibre infrastructure was a decisive factor. We can process thousands of transactions per second with latency rates that rival those of Frankfurt or London. This isn't just about speed—it's about reliability that our institutional clients demand. When milliseconds determine trading profitability, infrastructure quality directly translates to competitive advantage and customer retention."

The strategic importance of infrastructure investment in Cyprus is demonstrated through entrepreneurs' accounts of comprehensive government support and private sector collaboration. These efforts extend beyond mere connectivity to include ecosystem development and the fulfilment of industry-specific technical requirements. "Dimitris", the founder of a gaming company employing over 200 individuals and serving approximately 15 million users worldwide, explains, "The government recognised early on that infrastructure encompasses more than the installation of cables—it involves establishing conducive ecosystem conditions where digital enterprises can prosper. When our company required specialised connectivity for multiplayer gaming platforms, we engaged directly with telecommunications providers who comprehended our technical needs. This was not merely customer service; it represented strategic partnership development that enabled us to attain technical specifications that would be unachievable elsewhere."

The development of Cyprus's quantum communication infrastructure constitutes a notably significant competitive advantage for entrepreneurs centred on technology, thereby positioning the nation as an early adopter of next-generation communication technologies. Andreas, a cybersecurity entrepreneur whose company provides quantum-safe security solutions to European financial institutions, explains, "We're beta-testing quantum encryption protocols that most European countries won't have access to for years. These positions put our clients ahead of regulatory requirements, giving us early-mover advantages in quantum-safe security solutions. Our infrastructure advantage isn't just about current capabilities—it's about future-proofing our business models and creating defensible competitive positions."

Additional Cypriot perspectives highlight the broader impacts of infrastructure development excellence. "Theodoros", who operates a blockchain development company, notes, "Infrastructure quality enables us to attract international talent who expect seamless connectivity for remote collaboration. Our developers can participate in real-time coding sessions with teams across three continents without latency issues. This global collaboration capability would be impossible without world-class infrastructure." Similarly, Irene, a digital marketing entrepreneur serving clients across Europe and the Middle East, emphasises: "Our clients choose us because we can guarantee 99.9% uptime and immediate response capabilities. Infrastructure reliability becomes a service quality differentiator that justifies premium pricing and creates customer loyalty."

Greek Infrastructure Innovation: Resilience Through Creativity

Greek entrepreneurs exhibit exceptional creativity and adaptability in utilising existing infrastructure while devising innovative solutions to surmount constraints (Harkiolakis et al., 2025). They frequently transform limitations into competitive advantages through ingenuity and collaborative problem-solving approaches. "Melina," who leads a digital marketing agency in Athens serving clients across Southern Europe, describes sophisticated adaptive strategies: "We've learned to be infrastructure-agile. When our primary internet connection experiences issues, we have backup solutions through mobile networks, satellite connections, and partnerships with co-working spaces across the city. This redundancy makes us more resilient than competitors who rely on a single infrastructure source. Our clients appreciate that we never miss deadlines due to connectivity issues."

The division between urban and rural infrastructure in Greece presents both substantial challenges and unforeseen opportunities for entrepreneurial innovation, fostering the development of solutions that are valuable in international markets facing analogous connectivity constraints. "Alexandros," who operates a tourism technology startup from Thessaloniki that now serves clients across the Balkans and Eastern Europe, explains the transformation of constraints into opportunities: "We initially saw rural connectivity limitations as problems when developing applications for small hotels and restaurants in remote locations. But these constraints became opportunities when we developed offline-capable applications that sync when connectivity is available. This innovation now differentiates us in international markets where connectivity isn't guaranteed, giving us competitive advantages over competitors who assume perfect infrastructure."

Greek entrepreneurs similarly highlight that European Union infrastructure programs offer essential support in overcoming domestic limitations while facilitating access to resources that create a level playing field with larger international competitors. "Elena," a creative industry entrepreneur who produces digital content for global brands, notes, "EU funding for digital infrastructure has been transformative for small businesses like ours. We now have access to cloud computing resources and content delivery networks that would have been impossible to afford independently. This levels the playing field with larger competitors and enables us to serve international clients with professional-grade technical capabilities."

Additional Greek insights highlight advanced collaborative strategies for addressing infrastructure challenges. "Vassilis," who operates a software development company, describes: "We've formed infrastructure-sharing agreements with other local companies where we pool resources for backup connectivity, shared server capacity, and collective purchasing power for premium services. This collaboration enables us to achieve enterprise-grade infrastructure capabilities at startup prices." Despina", an e-commerce entrepreneur, adds, "Infrastructure constraints taught us to optimize everything—our applications run efficiently on limited bandwidth, and our websites load quickly even with slow connections. These optimisation skills become competitive advantages when we expand to markets with similar infrastructure limitations."

Comparative Infrastructure Impact: Divergent Paths to Success

The infrastructural disparities between Greece and Cyprus fundamentally influence diverse entrepreneurial opportunity frameworks, fostering various classes of ventures, business models, and competitive strategies (Harkiolakis & Komodromos, 2023). Cypriot entrepreneurs are more inclined to pursue ventures that require high-bandwidth, low-latency communication capabilities, whereas Greek entrepreneurs excel in developing solutions that are resilient to infrastructure variability (Zakzouk et al., 2023). "Ioannis," a Greek entrepreneur who established operations in both countries and now manages cross-border ventures, provides a comparative analysis: "Cyprus enables us to compete in global markets where infrastructure is assumed to be perfect—fintech, gaming, real-time trading platforms. Greece teaches us to build solutions that work everywhere—resilient applications, offline capabilities, and adaptive technologies. The combination gives us unique competitive advantages because we can serve both high-end markets that demand perfection and emerging markets that need reliability under constraints."

This infrastructure-driven specialisation engenders complementary rather than competitive advantages, as each nation cultivates unique entrepreneurial capabilities that are advantageous across various market segments and expansion strategies. The research indicates that the quality of infrastructure influences not only entrepreneurs' capabilities but also their cognitive approaches to problem-solving, solution development, and strategic positioning in international markets.

THEME 2: PLATFORM ECOSYSTEM NAVIGATION - STRATEGIC ORCHESTRATION IN DIGITAL ENVIRONMENTS

Entrepreneurs in both nations exhibit a sophisticated comprehension of platform dynamics that transcends mere platform utilisation to include strategic ecosystem orchestration, competitive positioning, and value creation via network effects. Their methodologies reflect varying strategic priorities and market positioning considerations, highlighting complex decision-making processes concerning platform selection, engagement strategies, and multi-platform orchestration, which are pivotal for long-term competitive sustainability and growth potential.

Subsequently, entrepreneurs and orchestrators—such as anchor organisations or platform owners—assume a crucial role in guiding the strategic direction of an ecosystem, aligning resources, and coordinating a heterogeneous array of participants (Yan et al., 2024). Effective orchestration necessitates meticulous management of institutional frameworks and the delineation of network boundaries to foster clarity and trust among actors. As ecosystems evolve, these leaders must adapt their roles to address changing dynamics, opportunities, and challenges (Hong et al., 2024). They also promote long-term resilience by balancing diversity and coherence within the network, ensuring that a multitude of ideas and capabilities coexist productively. Over time, decentralising orchestration responsibilities empowers stakeholders, distributes decision-making authority, and enhances the ecosystem's capacity to innovate and succeed (Orji, 2025).

Strategic Platform Selection: Risk Management vs. Value Optimization

Greek entrepreneurs emphasise platform diversification as a comprehensive risk management strategy designed to mitigate platform-specific vulnerabilities while expanding market reach across various geographical and demographic segments (Zakzouk et al., 2023). "Giorgos,", who operates an e-commerce business selling traditional Greek products internationally with annual revenues exceeding €2 million, explains his diversification approach: "We maintain presence on twelve different platforms because each serves different market segments and geographical regions with distinct preferences and behaviors. Amazon handles North American sales with their logistics excellence, Alibaba covers Asian markets with their payment systems, and local European platforms serve regional customers who prefer domestic service providers. This diversification protects us against algorithm changes, policy updates, or competitive threats on any single platform."

Cypriot entrepreneurs exhibit a more selective engagement with platforms, emphasising the maximisation of value derived from meticulously chosen platforms. This approach reflects increased resource availability and a strategic focus on in-depth optimisation rather than widespread diversification. "Christos," a fintech entrepreneur whose platform processes over €50 million in transactions monthly, describes his strategic approach: "We identified three critical platforms for our business model and became experts in their ecosystems. Rather than spreading resources across many platforms, we invested deeply in understanding algorithms, community dynamics, and partnership opportunities within our selected platforms. This focused approach enables us to achieve top-tier performance metrics and preferential treatment from platform providers."

The platform selection process involves an advanced analysis of network effects, competitive dynamics, and long-term strategic positioning, requiring entrepreneurs to balance current opportunities with the future evolution trajectories of the platform (Istrefi et al., 2025). "Anna," a Greek creative industry entrepreneur whose digital art reaches over 500,000 followers across platforms, provides insight into strategic decision-making: "Platform selection isn't just about where customers are today—it's about understanding platform evolution trajectories, policy changes, and emerging competitive threats" (Harkiolakis et al., 2025; Komodromos, 2025). We've seen platforms lose relevance quickly, so diversification protects against platform risk while positioning us for emerging opportunities. Instagram's algorithm changes,

TikTok's regulatory uncertainties, and LinkedIn's B2B focus shifts—we monitor these developments and adjust our strategies accordingly."

Additional perspectives on platform selection reveal nuanced strategic considerations. "Marinos," a Greek technology entrepreneur, notes, "We evaluate platforms based on five criteria: audience alignment, monetisation potential, competitive intensity, algorithm transparency, and long-term viability. This framework helps us make data-driven decisions rather than following trends." Meanwhile, Constantina, a Cypriot professional services entrepreneur, emphasises, "Platform selection requires understanding not just user numbers but engagement quality, conversion rates, and client lifetime value. We choose platforms where our audience is most receptive to our value proposition."

Algorithm Optimisation and Content Strategy: Technical Excellence vs. Collaborative Intelligence

Both Greek and Cypriot entrepreneurs allocate substantial resources towards understanding and optimising for platform-specific algorithms. Nonetheless, their methodologies exhibit variations rooted in differing resource availability and strategic priorities, thereby demonstrating distinct avenues to attain competitive advantage within algorithm-driven environments (Komodromos, 2024). "Omiros," a Cypriot gaming entrepreneur whose company achieves 48% organic reach rates compared to industry averages of 11%, explains his technical approach: "We employ dedicated algorithm specialists who analyse platform updates, test content variations, and optimise our visibility strategies through systematic A/B testing and performance analytics. This investment yields multiple returns through improved organic reach and reduced advertising costs. Our algorithm optimisation capabilities have become a core competitive differentiator that enables us to achieve superior performance metrics consistently."

Greek entrepreneurs exemplify resourceful strategies for algorithm optimisation by harnessing community expertise and fostering collaborative learning, thereby enabling effective competition with larger corporations endowed with specialised technical resources. "Stella," who operates a fashion e-commerce platform achieving 25% higher engagement rates than competitors, describes collaborative intelligence strategies: "We participate in entrepreneur groups where we share algorithm insights, test results, and optimisation strategies. This collective intelligence helps us compete with larger companies that have dedicated algorithm teams. We've created informal knowledge-sharing networks where successful strategies are shared, tested, and refined collectively. This collaboration multiplies our capabilities."

The development of content strategy demonstrates a sophisticated understanding of platform-specific audience expectations and engagement patterns, requiring entrepreneurs to cultivate multiple content competencies while consistently maintaining brand messaging. "Helena," a Cypriot professional services entrepreneur whose content reaches over 120,000 professionals monthly, notes the complexity of multi-platform content strategy: "LinkedIn requires thought leadership content that establishes expertise through detailed analysis and professional insights. Instagram needs visually compelling stories that humanise our brand through behind-the-scenes content and team personalities. TikTok demands authentic, entertaining content that showcases our company culture in creative, engaging formats. Each platform has distinct content languages, audience expectations, and engagement patterns that we must master while maintaining consistent brand messaging."

Additional insights into content strategy reveal sophisticated approaches to audience engagement. "Dimitrios," a Greek food entrepreneur, explains: "We develop content calendars that align platform-specific content with seasonal trends, cultural events, and industry discussions. Our content strategy balances educational value, entertainment appeal, and commercial messaging across different platforms." Evie," a Cypriot marketing consultant, adds, "Content optimisation requires understanding not just what performs well, but why it performs well. We analyse engagement patterns, comment sentiment, and conversion metrics to understand audience preferences and refine our content strategies continuously."

Multi-Platform Orchestration: Integration Strategies for Competitive Advantage

The research indicates that successful entrepreneurs formulate comprehensive multi-platform strategies that capitalise on the distinct advantages of various platforms while preserving consistent brand messaging and optimising cross-platform synergies. "Thodoris," a Greek technology entrepreneur whose integrated platform strategy generates 65% more leads than single-platform approaches, explains his orchestration methodology: "Our content strategy starts with core messages that we adapt for platform-specific formats and audience expectations while maintaining strategic consistency. We use LinkedIn for B2B relationship building and thought leadership, YouTube for educational content and product demonstrations, Twitter for real-time engagement with industry discussions and trending topics, and Instagram for visual storytelling that humanises our brand. Each platform serves distinct functions within our overall communication ecosystem."

Platform integration strategies necessitate advanced technical and strategic competencies that empower entrepreneurs to enhance efficiency while optimizing platform-specific performance indicators. "Demetra," a Cypriot marketing technology entrepreneur whose integration tools serve over 500 businesses, describes technical orchestration approaches: "We've developed internal tools that allow us to publish content simultaneously across multiple platforms while customising formats, hashtags, and engagement strategies for each platform's optimal performance patterns. This technical integration enables us to maintain consistent messaging while optimising for platform-specific algorithms and audience preferences. Our integration capabilities have become a service offering for other businesses."

Cross-platform analytics and optimisation exemplify sophisticated capabilities that empower entrepreneurs to comprehend audience behaviour patterns, optimise resource distribution, and maximise return on communication investments. "Antonios," a Greek digital marketing entrepreneur, provides insight into sophisticated analytics approaches:"We track audience journeys across platforms to understand how content consumption patterns influence conversion behaviors. LinkedIn followers who engage with our educational content are 3x more likely to convert on our website, while Instagram followers prefer direct messaging for initial contact. Understanding these cross-platform behaviours enables us to optimise our content strategies and resource allocation for maximum impact."

THEME 3: CULTURAL COMMUNICATION ADAPTATION—NAVIGATING GLOBAL MARKETS THROUGH CULTURAL INTELLIGENCE

The research reveals intricate cultural adaptation processes that extend beyond mere linguistic translation to encompass a profound understanding of cultural communication norms, relationship-building strategies, trust-building mechanisms, and the articulation of value propositions that resonate with a diverse international audience. Both Greek and Cypriot entrepreneurs exhibit advanced cultural competence, facilitating effective communication across various global markets while capitalising on their Mediterranean cultural heritage as a competitive edge in relationship-centric business environments.

Cultural Sensitivity in Digital Communication: Market-Specific Adaptation Strategies

Greek entrepreneurs emphasise the critical importance of understanding cultural communication preferences and behavioural patterns in different target markets, developing sophisticated frameworks for adapting their communication strategies to match cultural expectations and business practices. "Joseph," who exports Greek food products globally with distribution in 35 countries, explains his cultural adaptation methodology: "American customers want efficiency and convenience in their communication—quick responses, clear information, and minimal relationship building before transactions. German customers prefer detailed technical specifications, formal communication protocols, and systematic presentation of information. Middle Eastern customers expect extensive relationship development before business discussions, emphasising personal connections and demonstrating long-term commitment. Each market requires completely different communication approaches."

Cypriot entrepreneurs leverage their multicultural environment and strategic geographic location to cultivate a nuanced understanding of cross-cultural communication dynamics, which serves as a competitive advantage in the development of international markets. "Chloe," who provides professional services to global clients across four continents, notes the benefits of multicultural exposure: "Cyprus's position between Europe, Asia, and Africa gives us daily exposure to different communication styles and business cultures through our diverse client base, expatriate community, and international business environment. This experience helps us adapt our communication approaches for different client segments and market entry strategies more effectively than competitors from more homogeneous cultural environments."

Cultural adaptation necessitates a sophisticated understanding of both explicit and implicit communication norms that influence the development of business relationships, decision-making processes, and the sustainability of long-term partnerships. "Michalis," a Greek technology entrepreneur whose software serves clients in 23 countries, describes the complexity of cultural communication adaptation: "In Scandinavian markets, direct communication is appreciated and efficiency is prioritized—clients want straightforward information and quick decisions. In Mediterranean markets, relationship-building and indirect communication approaches are more effective, as clients prefer to understand the people behind the business before making commitments. In Asian markets, respect for hierarchy and formal communication protocols is essential. We train our team to recognise these cultural signals and adapt their communication styles accordingly."

Additional cultural insights highlight the extent of adaptation required to achieve international success (Komodromos, 2024). "Katerina," a Greek consulting entrepreneur, explains: "Cultural adaptation goes beyond communication style to include meeting structures, decision-making processes, and relationship maintenance approaches. American clients prefer structured, agenda-driven meetings with clear action items. Japanese clients require consensus-building processes and face-saving approaches to disagreement. Understanding these differences determines success or failure in international markets." "Maria," a Cypriot e-commerce entrepreneur, adds, "We customise not just our communication but our entire customer experience for different cultural contexts—website design, payment options, customer service approaches, and even product presentation reflect cultural preferences."

Trust Building Across Cultural Contexts: Relationship Development Strategies

Trust development emerges as a critical success factor that necessitates fundamentally different approaches across cultural contexts, with Mediterranean entrepreneurs required to adapt their inherently relationship-oriented strategies to attain success in both relationship-centric and efficiency-driven business cultures. "Irene," a Cypriot financial services entrepreneur whose company manages over €200 million in international client assets, explains cultural variations in trust building: "Trust building in Northern European markets relies on demonstrated competence, regulatory compliance, and technical excellence—clients want to see certifications, performance metrics, and systematic processes. In Middle Eastern markets, trust develops through personal relationships, mutual connections, and long-term commitment demonstrations—clients want to know our family backgrounds, personal values, and community involvement."

The research indicates that entrepreneurs in the Mediterranean region possess cultural advantages within relationship-based markets while encountering challenges in efficiency-oriented cultures. Consequently, they must cultivate dual competencies that utilise their inherent strengths and adapt to diverse business environments. "Agamemnon," a Greek entrepreneur operating in multiple international markets with revenues across four continents, notes the complexity of cultural positioning: "Our cultural background gives us advantages in markets that value relationship building and personal connection—Latin America, the Middle East, and parts of Asia. However, we've had to develop new competencies for markets that prioritise efficiency and transactional approaches—Northern Europe and parts of North America. The key is knowing when to emphasise relationships and when to emphasise efficiency."

Building digital trust necessitates distinct strategies compared to traditional face-to-face relationship development, compelling entrepreneurs to translate their relationship-building skills into digital formats while preserving authenticity and personal connection (Istrefi et al., 2025). "Anastasia," a Cypriot e-commerce entrepreneur whose online platforms achieve 95% customer satisfaction rates, describes digital trust translation: "Online trust building requires different signals—customer testimonials, security certifications, professional website design, responsive customer service, and transparent policies. We've learned to translate traditional relationship-building skills into digital formats while maintaining the personal touch that differentiates Mediterranean businesses. Video calls, personalised communications, and cultural sensitivity in digital interactions become crucial."

Language and Localization Strategies: Beyond Translation to Cultural Integration

Language adaptation transcends mere translation, encompassing comprehensive cultural localization of communication strategies, value propositions, and the development of business relationships that align with local market expectations and cultural norms. "Apollon," a Greek tourism entrepreneur whose platforms serve clients in 15 languages, explains the depth of localisation required: "We don't just translate our content—we localise our entire communication approach based on cultural research and market testing. Our German website emphasises efficiency, detailed information, and systematic organisation because German customers prefer comprehensive information before making decisions. Our Italian site focuses on emotional connection, experiential benefits, and family values because Italian customers respond to relationship-oriented messaging. Each market requires different persuasion strategies and communication approaches."

Cypriot entrepreneurs utilise multilingual capabilities as a competitive advantage in developing international markets, employing language skills as avenues to achieve deeper cultural understanding and more effective market penetration strategies. "Christopher," who operates a technology consultancy serving clients across multiple continents, notes the strategic value of multilingual capabilities: "Our team speaks Greek, English, Turkish, Arabic, and Russian, which allows us to communicate directly with clients across our target markets without translation barriers. This multilingual capability reduces communication barriers and fosters stronger client relationships, but more importantly, it enables us to understand cultural nuances that monolingual competitors often overlook. Language skills become competitive intelligence capabilities."

Cultural localization strategies necessitate comprehensive market research and ongoing adaptation informed by market feedback and cultural developments. "Aeneas," a Cypriot fintech entrepreneur, explains: "We conduct cultural research in each target market to understand not just language preferences but communication styles, decision-making processes, and relationship expectations. Our Russian clients prefer formal, detailed proposals with extensive documentation. Our British clients want concise, results-focused presentations with clear ROI demonstrations. Our UAE clients expect relationship building through face-to-face meetings and cultural sensitivity demonstrations."

THEME 4: COLLABORATIVE ECOSYSTEM DEVELOPMENT—STRATEGIC COMMUNITY BUILDING FOR COLLECTIVE SUCCESS

The research reveals advanced collaborative practices that extend beyond conventional networking activities to encompass strategic ecosystem management, systematic knowledge dissemination, and the development of collective competitive advantages, ultimately benefiting entire entrepreneurial communities. Entrepreneurs in both Greece and Cyprus exhibit a profound understanding that individual success is intrinsically linked to the overall health and development of the ecosystem, resulting in innovative collaboration models that generate shared value while preserving competitive positioning (Komodromos, 2025).

Strategic Partnership Development: Creating Win-Win Ecosystem Relationships

Greek entrepreneurs emphasise collaborative strategies to surmount resource limitations and market access challenges through strategic alliances that facilitate comprehensive service provision without necessitating substantial capital investment or full-time specialist recruitment. "Daphne," who operates a digital marketing agency that has grown 330% through partnership leverage, explains her ecosystem approach: "We've formed strategic partnerships with complementary service providers—web developers, graphic designers, content creators, and SEO specialists—that allow us to offer comprehensive digital solutions without hiring full-time specialists in each area. These partnerships create referral networks that benefit all participants while enabling us to compete with larger agencies that have in-house specialists. Our partnership network has become our primary competitive advantage."

Cypriot entrepreneurs leverage their advantageous geographic location to establish partnerships that connect diverse regional markets and cultural contexts, thereby generating distinctive value propositions that larger competitors struggle to emulate. "Achilles," a fintech entrepreneur whose partnership network spans three continents, describes cross-regional collaboration strategies: "Our partnerships extend across Europe, the Middle East, and Africa, leveraging Cyprus's position as a bridge between these regions with cultural understanding and regulatory knowledge. We provide European market access for Middle Eastern partners who require EU compliance expertise and Middle Eastern market knowledge, as well as support European clients in seeking expansion opportunities. Our geographic positioning becomes a partnership asset that creates value for all participants."

Partnership development necessitates advanced strategies in relationship management and value creation that effectively balance mutual advantages with individual competitive positioning (Komodromos, 2025). “Angelos,” a Greek technology entrepreneur whose partnership network includes over 60 active collaborations, notes the complexity of partnership orchestration: “Successful partnerships require clear value propositions for all participants, formal agreements that protect everyone's interests, and ongoing communication to ensure partnership objectives remain aligned as businesses evolve. We've learned that partnership success depends more on relationship management than initial agreements—markets change, businesses pivot, and partnerships must adapt accordingly.”

Further partnership insights uncover advanced strategies for creating collaborative advantages. “Elias,” a Greek software entrepreneur, explains, “We've created partner certification programs that ensure quality standards while providing partners with additional credentials and competitive advantages. This creates mutual value—partners get enhanced credibility, and we get reliable service delivery.” Meanwhile, “Zoe,” a Cypriot consulting entrepreneur, adds that, “Our partnerships include formal knowledge-sharing agreements where we exchange market intelligence, best practices, and client insights that help all partners serve clients more effectively.”

Knowledge Sharing and Community Building: Collective Intelligence Development

The research uncovers comprehensive knowledge-sharing practices that foster collective learning opportunities and confer competitive advantages through systematic information exchange, collaborative problem-solving, and peer mentorship programs (Komodromos et al., 2024). “Ariadne,” a Cypriot professional services entrepreneur whose community involvement has generated 45% of her client referrals, explains the value of collective intelligence: “We participate in entrepreneur groups where we share market intelligence, platform optimisation strategies, and customer acquisition approaches on a systematic basis. This collective knowledge enables all participants to compete more effectively against larger international competitors that have dedicated research teams. Our community becomes our research and development capability.”

Greek entrepreneurs exemplify robust community-building skills by leveraging cultural traditions of collective support and mutual assistance while adapting these practices to contemporary business challenges and digital communication environments (Komodromos et al., 2024). “Maria,” who operates a creative industry business that has achieved 290% growth through community support, describes the power of entrepreneurial community: “Our entrepreneur community functions like an extended family where we share resources, provide mutual support during challenges, and celebrate each other's successes collectively. This community strength helps individual businesses overcome obstacles that would be insurmountable on their own—financial challenges, technical problems, and market access barriers. The community becomes a collective resource that amplifies individual capabilities."

Knowledge sharing encompasses formal mentorship programs and structured learning initiatives that facilitate entrepreneurial development and contribute to reducing failure rates within entrepreneurial ecosystems (Harkiolakis & Komodromos, 2023). "Marcos," a Cypriot gaming entrepreneur who has mentored over 40 new ventures, notes the systematic nature of knowledge transfer: "We've established mentorship programs where successful entrepreneurs guide newer ventures through common challenges based on structured curricula and regular progress meetings. This knowledge transfer accelerates entrepreneurial learning and reduces failure rates within our ecosystem while creating networks that benefit all participants long-term."

Collective Market Development: Rising Tide Strategies

Entrepreneurs in both countries participate in collective market development initiatives that generate opportunities for entire industry sectors, rather than concentrating exclusively on individual competitive advantages. This approach acknowledges that market expansion benefits entire ecosystems more significantly than zero-sum competition within static markets. "Theo," a Greek food export entrepreneur whose collaborative marketing has opened three new international markets, explains sector-wide collaboration: "We work with competitor companies to promote Greek cuisine internationally, recognising that market expansion benefits everyone involved. Individual company competition occurs within growing markets rather than fighting for shares of static markets. When we collectively promote Greek food culture, all Greek food companies benefit from increased international awareness and demand."

Cypriot entrepreneurs employ collective positioning strategies to promote Cyprus as a regional business hub. Such initiatives generate ecosystem-wide benefits that enhance individual company opportunities by improving the national reputation and market positioning. "Tatiana," who provides business development services and has participated in five collective market development initiatives, describes collaborative positioning: "We collaborate with government agencies and other businesses to promote Cyprus as a gateway to European, Middle Eastern, and African markets through joint trade missions, collective marketing campaigns, and shared participation in international trade shows. This collective positioning creates opportunities that individual companies couldn't achieve independently while building Cyprus's reputation as a business-friendly environment."

Further collaborative development insights reveal advanced strategies for ecosystem building. "Constantinos," a Greek technology entrepreneur, notes, "We've organised sector-specific working groups that address common challenges like regulatory compliance, talent recruitment, and technology standards. These collaborative solutions benefit entire industry sectors. "Evagneline," a Cypriot marketing entrepreneur, adds, "Our collective market research initiatives allow small companies to access market intelligence that would be too expensive for individual businesses, creating shared competitive advantages."

THEME 5: ADAPTIVE COMMUNICATION STRATEGIES—RESILIENCE AND INNOVATION IN DYNAMIC ENVIRONMENTS

The research uncovers advanced adaptive capabilities that empower entrepreneurs to respond effectively to evolving communication technologies, shifting market conditions, unforeseen disruptions, and competitive threats through systematic innovation, strategic flexibility, and ongoing learning approaches. The COVID-19 pandemic exemplifies a notable instance of entrepreneurial communication adaptation, highlighting significant variations in adaptive capabilities and strategic responses that influence long-term competitive sustainability.

Technology Adaptation and Innovation: Embracing Digital Transformation

Greek entrepreneurs exhibit remarkable adaptability in adopting new communication technologies and platforms, frequently transforming traditional business models into sophisticated digital operations within compressed timeframes through resourceful innovation and collaborative learning. "Socrater," who operates a conventional manufacturing business that achieved 480% revenue growth through digital transformation, explains his adaptation journey: "We transformed from primarily offline operations to sophisticated digital communication strategies within six months during the pandemic. This transformation included developing e-commerce platforms, implementing video conferencing for customer consultations, creating virtual product demonstrations, and establishing social media presence across multiple platforms. The digital transformation wasn't just about survival—it became our primary growth engine."

Cypriot entrepreneurs demonstrate a commendable tendency to adopt emerging communication technologies at an early stage, frequently serving as beta testers and initial implementers before widespread adoption. This proactive approach fosters competitive advantages through technological leadership and platform expertise. "Maya," a marketing technology entrepreneur whose early adoption strategies have generated 250% year-over-year growth, describes her innovation approach: "We continuously experiment with new platforms and communication tools, often becoming early adopters before they reach mainstream adoption through systematic testing and performance evaluation. This early adoption provides us with competitive advantages and enables us to understand platform evolution trends before our competitors. We have been early adopters of TikTok for B2B marketing, LinkedIn video features, and AI-powered content creation tools.

The enhancement of technological adaptation necessitates ongoing learning and the development of capabilities, which entrepreneurs must systematically incorporate into their operations and strategic planning procedures. "Michalis," a Greek technology entrepreneur who has successfully navigated five major platform transitions, notes the systematic nature of adaptation requirements: "Communication technology evolution requires constant education and experimentation through dedicated resource allocation. We allocate specific budgets for testing new platforms, tools, and approaches, recognising that communication innovation drives business growth more than traditional marketing investments. Our technology adaptation capabilities have become our core competitive differentiator."

Further insights into technology adaptation uncover advanced methods in managing innovation. "Panagiotis," a Greek e-commerce entrepreneur, explains, "We maintain innovation pipelines where we test emerging technologies in controlled environments before full implementation, reducing adoption risks while maintaining competitive positioning." Additionally, "Michael," a Cypriot fintech entrepreneur, adds, "We participate in technology preview programs with major platforms, giving us early access to new features and the ability to provide feedback that influences platform development."

Crisis Response and Resilience: Learning from Disruption

The COVID-19 pandemic served as a natural experiment in the adaptation of entrepreneurial communication, highlighting notable disparities in adaptive capacities, strategic flexibility, and resilience strategies that differentiate successful entrepreneurs from those who are unable to navigate disruptions effectively. "Penelope," a Cypriot tourism entrepreneur whose virtual pivot generated 80% of pre-pandemic revenues during lockdowns, explains crisis adaptation strategies: "When travel restrictions eliminated traditional tourism marketing approaches, we pivoted to virtual experiences, online cultural events, and digital relationship building that maintained customer engagement and generated revenue during lockdowns. We created virtual tours, online cooking classes featuring traditional Cypriot cuisine, and digital cultural experiences that attracted international audiences. The crisis forced innovation that expanded our market reach beyond geographical limitations."

Greek entrepreneurs demonstrated a notably robust resilience through collaborative adaptation strategies that leveraged community support and collective problem-solving to overcome individual resource limitations and market challenges.

"Armagedon," who operates a restaurant chain that maintained profitability throughout the pandemic, describes collaborative crisis response: "We worked with other local businesses to develop collective delivery platforms, shared marketing campaigns, and joint customer loyalty programs that helped everyone survive during the crisis while building stronger community relationships. Our collaborative approach enabled us to maintain operations with reduced individual investment while creating customer experiences that individual businesses couldn't provide independently."

Crisis adaptation frequently leads to enduring enhancements in communication strategies and capabilities, thereby strengthening long-term competitive positioning and operational efficiency beyond mere crisis recovery. "Natasha," a Cypriot professional services entrepreneur whose pandemic innovations increased efficiency by 65%, notes the lasting impact of crisis-driven innovation: "Pandemic-driven communication innovations—virtual consultations, digital document management, and online collaboration tools—proved more efficient than traditional approaches in many contexts. We've maintained these innovations even after returning to normal operations because they improve service delivery, reduce costs, and enhance client satisfaction. Crisis adaptation became permanent competitive advantage development."

Continuous Learning and Development: Systematic Capability Building

Successful entrepreneurs consistently commit to ongoing learning and continually improve their communication skills. They employ structured methods for skill development, gathering market intelligence, and enhancing strategic capabilities, all of which enable them to maintain a competitive edge in dynamic environments. "Maria," a Greek digital marketing entrepreneur whose continuous learning investments have generated 350% team productivity improvements, explains her development approach: "Communication strategy evolution requires constant learning through industry publications, conference participation, online courses, and peer networking activities. We invest significantly in team education because communication competence has a direct impact on business performance that exceeds that of traditional operational investments. Our learning investments have become our primary competitive advantage development strategy." (Komodromos et al., 2026).

The research highlights systematic methods for developing and optimising communication strategies that successful entrepreneurs employ to stay competitive and adapt to changing market conditions and technological advances. "Sofronis," a Cypriot cybersecurity entrepreneur whose systematic optimisation approach has achieved industry-leading performance metrics, describes strategic development processes: "We conduct quarterly communication strategy reviews that analyse platform performance, audience engagement patterns, and competitive positioning through systematic data analysis and performance evaluation. These reviews inform resource allocation and strategic adjustments for the upcoming quarter, identifying emerging opportunities and competitive threats. Our systematic approach to strategy development has become a core competitive capability."

Additional insights into learning and development illustrate advanced methodologies for enhancing capabilities. "Aeneas," a Greek software entrepreneur, notes, "We maintain individual development plans for each team member that include communication skills, platform expertise, and cultural competency development aligned with business objectives." "Cassandra," a Cypriot consulting entrepreneur, explains, "We've established learning partnerships with other companies where we share training resources, expertise, and development costs while maintaining competitive advantages through collaborative capability building."

These five themes collectively elucidate the sophisticated nature of communication-mediated entrepreneurship within Mediterranean contexts, demonstrating that successful entrepreneurs cultivate complex, interconnected capabilities that enable them to navigate digital ecosystems while leveraging cultural advantages and overcoming structural constraints through innovation, collaboration, and adaptive strategies.

DISCUSSION AND ANALYSIS

Cross-Country Comparative Analysis

The comparative analysis between Greek and Cypriot entrepreneurs reveals both convergent patterns and divergent strategies, reflecting different structural conditions, resource availability, and market positioning approaches. While both countries share Mediterranean cultural foundations and the benefits of European Union membership, their approaches to communication-mediated entrepreneurship exhibit significant variations, offering insights into how national contexts influence entrepreneurial strategies.

Infrastructure-Driven Opportunity Structures

The research demonstrates that the quality of communication infrastructure fundamentally alters the opportunity structures available to entrepreneurs in Greece and Cyprus. Cyprus's investment in advanced digital infrastructure enables entrepreneurs to pursue ventures that require high-performance communication capabilities, such as fintech platforms processing real-time transactions or gaming companies delivering low-latency multiplayer experiences. These infrastructural capabilities position Cypriot entrepreneurs to compete directly with ventures from larger European economies within technology-intensive sectors.

Greek entrepreneurs, navigating variable infrastructure environments, adopt resilience-driven strategies to gain competitive edges in markets with uncertain infrastructure reliability. This flexibility is especially beneficial for expanding internationally into emerging or developing regions where communication infrastructure might be less dependable. The research indicates that varied infrastructure settings cultivate unique entrepreneurial skills, which lead to advantages that complement rather than compete with one another.

Cultural Capital and Communication Competence

Both Greek and Cypriot entrepreneurs draw upon Mediterranean cultural traditions, emphasising relationship cultivation and trust development. However, they adapt these capabilities distinctively for digital communication contexts. Greek entrepreneurs typically emphasize collaborative strategies and community engagement, which mirror traditional collective support systems. Conversely, Cypriot entrepreneurs utilise their multicultural exposure derived from their geographic location to cultivate advanced cross-cultural communication skills.

The research demonstrates that cultural communication competence offers Mediterranean entrepreneurs competitive advantages in relationship-intensive markets, while necessitating supplementary capability development for efficiency-oriented business cultures. This discovery suggests that cultural background presents both opportunities and limitations that entrepreneurs must strategically manage through the development of effective communication strategies.

Resource Optimisation and Strategic Focus

Differences in resource availability between Greece and Cyprus lead to unique strategic approaches regarding the utilisation of communication platforms. Cypriot entrepreneurs, endowed with superior financial resources and infrastructure, tend to adopt a selective approach to platforms, accompanied by comprehensive optimisation strategies. Conversely, Greek entrepreneurs exhibit considerable resourcefulness in crafting diversified platform strategies aimed at mitigating platform-specific risks and optimising outreach within their resource limitations.

These various approaches suggest that resource limitations can stimulate innovation and adaptive abilities, which may serve as competitive advantages in volatile or uncertain business environments. The research demonstrates that both the availability of resources and constraints can generate opportunities for entrepreneurial success; however, they necessitate distinct strategic approaches and prioritisation of capability development.

THEORETICAL IMPLICATIONS

Extending Platform Ecosystem Theory

The research advances platform ecosystem theory by illustrating how entrepreneurs in smaller economies cultivate refined orchestration capabilities that allow them to harness network effects, despite constrained domestic market sizes. The findings indicate that successful entrepreneurs develop what we refer to as "ecosystem bridging capabilities"—the capacity to connect multiple platform ecosystems and generate value through cross-platform integration and coordination.

This theoretical contribution proposes that platform ecosystem theory should incorporate the role of individual entrepreneurs as ecosystem orchestrators, who generate value not solely through platform participation but also via strategic ecosystem design and management. The study illustrates that entrepreneurs are capable of impacting platform ecosystem dynamics through collective action, the development of strategic partnerships, and innovative approaches to utilisation that transcend mere platform optimization.

Communication-Mediated Entrepreneurship Theory

The research advances theoretical insights into communication-mediated entrepreneurship as a distinct phenomenon necessitating specialised frameworks for comprehension and analysis. The findings imply that communication competence has become a primary determinant of entrepreneurial success, requiring the integration of traditional entrepreneurship theory with perspectives from communication and media studies.

The emerging theoretical framework from this research positions communication capabilities as fundamental entrepreneurial competencies rather than ancillary functions. This perspective calls for a reconceptualisation of entrepreneurial education, support programs, and evaluation metrics to emphasise communication strategy development as central to entrepreneurial achievement, rather than as a peripheral element of core business activities (Komodromos et al., 2026).

Cultural Adaptation in Digital Contexts

The research advances the understanding of how cultural communication competencies are applied within digital entrepreneurship environments. The findings illustrate that cultural background presents both opportunities and challenges for effective digital communication, necessitating entrepreneurs to cultivate advanced adaptation skills that harness cultural strengths while addressing cultural limitations. This theoretical contribution posits that cultural competence in digital entrepreneurship encompasses not only the comprehension of diverse cultural communication norms but also the development of capabilities to adapt these norms to digital platform contexts and multicultural audiences. The research further suggests that successful cultural adaptation demands both profound cultural understanding and technological communication proficiency.

PRACTICAL APPLICATIONS AND POLICY IMPLICATIONS

Entrepreneurial Education and Development Programs

The research findings have significant implications for the design of entrepreneurial education and support programs. The emphasis on communication competence as a core entrepreneurial capability suggests that education programs should integrate communication strategy development, platform utilisation skills, and cultural adaptation competencies as central rather than supplementary curriculum elements.

Support programs for entrepreneurs should include specialised training in platform ecosystem navigation, multi-channel communication orchestration, and cross-cultural digital communication strategies. The research suggests that these competencies are learnable through structured education and mentorship programs but require a dedicated focus and allocation of resources rather than informal development approaches.

Infrastructure Investment and Policy Development

The research demonstrates clear relationships between the quality of communication infrastructure and entrepreneurial opportunity creation, supporting arguments for continued public investment in digital infrastructure development. However, the findings also suggest that infrastructure policy should consider not only technical capabilities but also ecosystem development, platform access, and international connectivity as critical components of entrepreneurial support infrastructure.

Policy frameworks should recognise that communication infrastructure investment creates multiplier effects through entrepreneurial ecosystem development that extend beyond direct economic impact measurements. The research suggests that infrastructure policy should be integrated with entrepreneurship support policies to maximise economic development outcomes and competitive positioning.

International Market Development Strategies

The research offers valuable insights for government agencies and economic development organisations aiming to facilitate international market expansion for domestic entrepreneurs. The findings recommend that support for market development should prioritise the development of communication strategies, cultural adaptation skills, and platform ecosystem integration rather than concentrating solely on conventional trade promotion methods.

International market development programs should provide entrepreneurs with resources to understand platform dynamics within target markets, adapt to cultural communication norms, and implement multi-platform integration strategies. The research highlights that these capabilities are crucial for achieving sustainable international growth and maintaining a competitive edge in global markets.

CONCLUSION AND FUTURE RESEARCH DIRECTIONS

This in-depth qualitative study shows that communication networks, media systems, and platform behaviours are now crucial factors for entrepreneurial success in today's Mediterranean economies. It highlights how Greek and Cypriot entrepreneurs use advanced communication strategies to harness available technologies and platforms. These efforts help them break through traditional market barriers and expand their economic prospects through digital connectivity and network effects.

KEY RESEARCH CONTRIBUTIONS

The study makes a significant contribution to both academic knowledge and practical understanding of communication-mediated entrepreneurship. It advances the theoretical understanding of how communication processes influence entrepreneurial success in the digital economy by integrating systems theory, network effects theory, and strategic communication theory into a comprehensive framework that analyses entrepreneurial ecosystem dynamics.

Methodologically, the cross-national comparative design facilitates the identification of common patterns and local variations in communication-driven entrepreneurship, thereby offering insights that are challenging to attain through single-country or purely quantitative studies. The qualitative approach affords detailed insights into entrepreneurial sense-making and adaptive capabilities—dimensions that quantitative methods often fail to capture sufficiently.

Practically, the research provides actionable insights for entrepreneurs seeking to enhance their communication effectiveness, policymakers interested in supporting the development of entrepreneurial ecosystems, and educators developing programs for digital entrepreneurship. The findings suggest that communication competencies are trainable through structured approaches, albeit requiring a dedicated focus and a significant commitment of resources.

POLICY AND PRACTICE IMPLICATIONS

The research findings endorse several policy recommendations aimed at enhancing the development of entrepreneurial ecosystems within Mediterranean economies and analogous contexts. Infrastructure investments should prioritise not only technical capabilities but also the development of ecosystems, platform access, and international connectivity as integrated components of entrepreneurial support systems. Educational institutions are advised to redesign entrepreneurship programs to emphasise the development of communication strategies, navigation of platform ecosystems, and cross-cultural digital communication as core competencies rather than auxiliary skills.

Support organisations should offer specialised training and mentorship in areas identified as critical for entrepreneurial success, including algorithm optimisation, multi-platform orchestration, and cultural adaptation strategies.

Government agencies should recognise the development of communication infrastructure and ecosystems as strategic investments that generate multiplicative economic benefits through entrepreneurial activities, rather than viewing them solely as technical projects. International market development initiatives should prioritise the advancement of communication capabilities and the integration of platform ecosystems as a means to achieve sustainable competitive positioning.

FUTURE RESEARCH DIRECTIONS

The research suggests promising areas for future studies better to understand communication-mediated entrepreneurship and its role in economic growth. Long-term studies can track entrepreneurial communication and business results to uncover development patterns, adaptive strategies, and the lasting effectiveness of different communication methods. Cross-cultural research that extends beyond Mediterranean countries to include other European economies, emerging markets, and diverse cultures could reveal broader patterns and unique variations. Investigating emerging technologies, such as AI, virtual reality, and blockchain, can deepen our understanding of how innovation creates new opportunities and challenges for entrepreneurs.

Policy intervention studies on government support for communication-mediated entrepreneurship can inform investment and program design to achieve a better economic impact. They compare policies across countries, measuring their success in developing entrepreneurial ecosystems and boosting international competitiveness.

FINAL REFLECTIONS

As digital technologies continue to reshape entrepreneurial processes and outcomes, the ability to effectively navigate and leverage communication ecosystems becomes increasingly critical for entrepreneurial success and broader economic transformation (Halkias et al., 2024). The Mediterranean context examined in this research offers valuable insights into how smaller economies can leverage communication technologies and strategic positioning to compete effectively in global markets whilst preserving cultural distinctiveness and competitive advantages.

The research emphasises that entrepreneurial success in digital contexts necessitates a sophisticated integration of technological capabilities, cultural competencies, and strategic reasoning that transcends traditional entrepreneurship frameworks (Halkias et al., 2024; Komodromos et al., 2024). For academics, practitioners, and policymakers, this study provides compelling evidence that communication-mediated entrepreneurship constitutes a unique phenomenon requiring specialised approaches to education, support, and policy development.

The findings indicate that ongoing investment in communication infrastructure, capability development, and ecosystem support constitutes a vital avenue for sustained economic growth and maintaining international competitiveness in an increasingly interconnected global economy. As Mediterranean economies advance their transformation initiatives, the insights gained from this research offer essential guidance for optimising entrepreneurial potential while fostering resilient and inclusive economic development strategies (Komodromos et al., 2024).

REFERENCES

Abaci, N. (2022). Relationship between entrepreneurship perception and communication skill: A structural equation model. *International Journal of Management Education*. Advance online publication. DOI: 10.1016/j.ijme.2022.100725

Agustina, T., Isnaini, S., & Said, I. (2025). Examining the Importance of Developing Entrepreneurial Communication for Students. Teumulong: Journal of Community Service. https://doi.org/.DOI: 10.62568/jocs.v3i3.35

Bachmann, N., Rose, R., Maul, V., & Hölzle, K. (2024). What makes for future entrepreneurs? The role of digital competencies for entrepreneurial intention. *Journal of Business Research*. Advance online publication. DOI: 10.1016/j.jbusres.2023.114481

Baig, U., Hussain, B., Meidutė-Kavaliauskienė, I., & Davidavičius, S. (2022). Digital Entrepreneurship: Future Research Directions and Opportunities for New Business Model. *Sustainability*. Advance online publication. DOI: 10.3390/su14095004

Bejjani, M., Göcke, L., & Menter, M. (2023). Digital entrepreneurial ecosystems: A systematic literature review. *Technological Forecasting and Social Change*. Advance online publication. DOI: 10.1016/j.techfore.2023.122372

Boichenko, K., & Sierhieiev, M. (2024). ENTREPRENEURIAL ECOSYSTEMS IN THE CONTEXT OF DIGITALISATION. Economic scope. https://doi.org/.DOI: 10.32782/2224-6282/191-61

Chatzinikolaou, D. (2025). Integrating Sustainable Energy Development with Energy Ecosystems: Trends and Future Prospects in Greece. *Sustainability*. Advance online publication. DOI: 10.3390/su17041487

Chen, A., Lin, Y., Mariani, M., Shou, Y., & Zhang, Y. (2023). Entrepreneurial growth in digital business ecosystems: An integrated framework blending the knowledge-based view of the firm and business ecosystems. *The Journal of Technology Transfer*, *48*, 1628–1653. DOI: 10.1007/s10961-023-10027-9

Constantinou, S., Pamboris, A., Alexandrou, R., Kronis, C., Zeinalipour-Yazti, D., Papadopoulos, H., & Konstantinidis, A. (2022). EnterCY: A Virtual and Augmented Reality Tourism Platform for Cyprus. 2022 23rd IEEE International Conference on Mobile Data Management (MDM), 314-317. https://doi.org/DOI: 10.1109/MDM55031.2022.00069

Dahwa, C. (2023). Adapting and blending grounded theory with case study: A practical guide. *Quality & Quantity*. Advance online publication. DOI: 10.1007/s11135-023-01783-9

Diaz, J., Halkias, D., & Harkiolakis, T. (2025). *Digital Entrepreneurship and Disruptive Innovation: Intersections of Regional Development, Innovation Systems, and Sustainability*. Routledge.

Felicetti, A., Corvello, V., & Ammirato, S. (2023). Digital innovation in entrepreneurial firms: A systematic literature review. *Review of Managerial Science*, •••, 1–48. DOI: 10.1007/s11846-023-00638-9

Fernandes, C., Ferreira, J., Veiga, P., Kraus, S., & Dabić, M. (2022). Digital entrepreneurship platforms: Mapping the field and looking towards a holistic approach. *Technology in Society*. Advance online publication. DOI: 10.1016/j.techsoc.2022.101979

Giotopoulos, K. (2024). *A Systematic Review in Digital Entrepreneurship Ecosystems*. Technium Social Sciences Journal., DOI: 10.47577/tssj.v61i1.11562

Goertler, T., Papert, M., Fischer, I., & Schmidt, M. (2025). *Building digital platform ecosystems: A synthetization of fundamental design topics from a literature review*. Digital Business., DOI: 10.1016/j.digbus.2025.100109

Gudi, A., Chinta, R., & Jin, A. (2024). Competencies and capabilities as determinants of digital entrepreneurship: An empirical validation. *Journal of International Technology and Information Management*, *32*(1), 271–302.

Hadjimitsis, D., Themistocleous, K., Michaelides, S., Neocleous, K., Danezis, C., Kyriakides, N., . . . Philimis, P. (2024, July). Emerging Prospects in Earth Observation for Cyprus and the Eastern Mediterranean, Middle East and North Africa (EMMENA) Region Through the Eratosthenes Centre of Excellence and Excelsior H2020 Teaming Project. In *IGARSS 2024-2024 IEEE International Geoscience and Remote Sensing Symposium* (pp. 2863-2866). IEEE.

Halkias, D., Esposito, M., Diaz, J., & Harkiolakis, T. (2024). Young Digital Entrepreneurs Innovate the Greek Maritime Industry: A Case Study. In Ramadani, V., Kjosev, S., & Sergi, B. (Eds.), *Entrepreneurship Development in the Balkans Perspectives from Diverse Contexts*. Emerald.

Halkias, D., & George-Ufot, D. (2021). Challenges of women micro-entrepreneurs in the Niger delta region of Nigeria for business survival: A narrative inquiry. The ISM Journal of International Business, 4(1), 11-22. https://www.proquest.com/scholarly journals/challenges-women-micro-entrepreneurs-niger-delta/docview/2609119503/se 2?accountid=14872

Harkiolakis, T., Alserhan, B. A. A., & Qahoush, R. (2025). Ethnic Fashion and Sustainability in the MENA Region: The Case of Sanaa Mohammad in Jordan. In Diaz, J., Halkias, D., & Harkiolakis, T. (Eds.), *Digital Entrepreneurship and Disruptive Innovation: Intersections of Regional Development, Innovation Systems, and Sustainability*. Routledge.

Harkiolakis, T., & Komodromos, M. (2023). Supporting Knowledge Workers' Health and Well-being in the Post-Lockdown Era. Administrative Sciences - Special Issue "Innovative Leaders and Empowered Employees. *Insights into the Future of Work*, *13*(2), 49. Advance online publication. DOI: 10.3390/admsci13020049

Hawlitschek, F., & Hodapp, D. (2024). Challenges to Co-Creating Value in Nascent Platform Ecosystems: Gaps Between Theory and Practice. *International Journal of Electronic Commerce*, *28*, 156–185. DOI: 10.1080/10864415.2024.2332046

Hong, Z., Zhang, Q., Xu, X., & Lyu, Z. (2024). Competition and cooperation in a platform-based business ecosystem within complementary partners. *International Journal of Production Economics*. Advance online publication. DOI: 10.1016/j.ijpe.2024.109337

Istrefi, A., Zeqiri, J., Hasani, V., & Komodromos, M. (2025). The role of entrepreneurial marketing and digital transformation on women's entrepreneurial intentions in Kosovo, Int. *J. of Technology Enhanced Learning*, *17*(2), 174–196. DOI: 10.1504/IJTEL.2024.10063470

Jacobides, M., Cennamo, C., & Gawer, A. (2024). Externalities and complementarities in platforms and ecosystems: From structural solutions to endogenous failures. *Research Policy*. Advance online publication. DOI: 10.1016/j.respol.2023.104906

Koleva, P. (2023). Critical analysis of grounded theory strategy in organisational and management research. *Qualitative Research Journal*. Advance online publication. DOI: 10.1108/qrj-08-2022-0118

Komodromos, M. (2024). Gender Barriers Experienced by Women Entrepreneurs in Cyprus. In Dana, L. P., & Chhabra, M. (Eds.), *Women Entrepreneurship Policy*. Springer., DOI: 10.1007/978-981-97-3607-2_9

Komodromos, M. (2025). A Literature Review on The Emergence of Entrepreneurship Education: Development, Trends, and Challenges in Masouras, A., Anastasiadou, S., Constantelou, A., & Apostolopoulos, S., pp. 21-46 (Eds.). (2025). Real-World Tools and Scenarios for Entrepreneurship Exploration. IGI Global. DOI: 10.4018/979-8-3693-9040-5

Komodromos, M., Anastasiadou, S., & Seremeti, L. (2026). Transforming Corporate Sustainability: Integrating Advanced Analytics and Business Intelligence Into ESG Implementation. In Komodromos, M., Seremeti, L., Anastasiadis, L., Liargovas, P., & Anastasiadou, S. (Eds.), *Data-Driven ESG Strategy Implementation Through Business Intelligence* (pp. 1–36). IGI Global Scientific Publishing., DOI: 10.4018/979-8-3373-5142-1.ch001

Komodromos, M., Harkiolakis, T., Khoshtaria, T., & Vassiliou, M. (2024). Adoption of digital technology and investigation of the impact in the marketing industry for marketing communication professionals: The case study of Cyprus and Greece. *International Journal of Technology Enhanced Learning*, *16*(2), 173–191.

Kreiterling, C. (2023). Digital innovation and entrepreneurship: A review of challenges in competitive markets. *Journal of Innovation and Entrepreneurship*, *12*, 1–13. DOI: 10.1186/s13731-023-00320-0

Kusbandono, D., Zaakiyyah, H., & Junaidi, A. (2024). *THE ROLE OF INTERPERSONAL COMMUNICATION IN BUILDING LEADERSHIP THAT POSITIVELY IMPACTS BUSINESS PERFORMANCE IN THE ENTREPRENEURIAL CONTEXT. Technopreneurship and Educational Development Review*. TENDER., DOI: 10.61100/tender.v1i2.187

Makri, C., & Neely, A. (2021). Grounded Theory: A Guide for Exploratory Studies in Management Research. *International Journal of Qualitative Methods*, *20*. Advance online publication. DOI: 10.1177/16094069211013654

Malthouse, E., & Copulsky, J. (2022). Artificial intelligence ecosystems for marketing communications. *International Journal of Advertising*, *42*, 128–140. DOI: 10.1080/02650487.2022.2122249

Manafe, M., Ohara, M., Gadzali, S., Harahap, M., & Ausat, A. (2023). Exploring the Relationship Between Entrepreneurial Mindsets and Business Success: Implications for Entrepreneurship Education. *Journal of Education*. Advance online publication. DOI: 10.31004/joe.v5i4.2238

McGloin, R., Saxton, A., Rao, A., Hintz, E., Coletti, A., Hamlin, E., Turner, M., & Mathieu, J. (2023). Examining the Importance of Developing Entrepreneurial Communication Skills in Accelerator Programs: A Focus Group Based Approach. *Entrepreneurship Education and Pedagogy*, *7*, 468–493. DOI: 10.1177/25151274231183382

Naeem, M., Ozuem, W., Howell, K., & Ranfagni, S. (2023). A Step-by-Step Process of Thematic Analysis to Develop a Conceptual Model in Qualitative Research. *International Journal of Qualitative Methods*, *22*. Advance online publication. DOI: 10.1177/16094069231205789

Nicoli, N., & Komodromos, M. (2019). CSR Communication in the Digital Age. The Case of The Bank of Cyprus. In Alexandros, A., & Evi, D. (Eds.), *Cases on Corporate Social Responsibility and Contemporary Issues in Organizations* (pp. 71–89). IGI Global.

O'Callaghan, C., Dwyer, J., & Schofield, P. (2024). Thematic analysis informed by grounded theory (TAG) in healthcare research: Foundations and applications. *Qualitative Research in Psychology*, *21*, 279–306. DOI: 10.1080/14780887.2024.2347580

Orji, G. (2025). *Influence of Digital Ecosystems on Value Creation and Competitive Positioning in the E-Commerce Industry*. European Journal of Business and Strategic Management., DOI: 10.47604/ejbsm.3330

Pascua, D. (2024). *Exploring the Application of Grounded Theory: A Systematic Review of Methodological Approaches*. International Journal of Multidisciplinary Research and Growth Evaluation., DOI: 10.54660/.ijmrge.2024.5.4.1362-1366

Ratnasari, A., Sofyan, A., & Yuniati, Y. (2023). *Business Communication in Improving the Entrepreneurial Skills of SME Members*. KnE Social Sciences., DOI: 10.18502/kss.v8i18.14252

Ritala, P. (2023). Grand challenges and platform ecosystems: Scaling solutions for wicked ecological and societal problems. *Journal of Product Innovation Management*. Advance online publication. DOI: 10.1111/jpim.12682

Satar, M., Alshibani, S., & Alarifi, G. (2024). Effects of Firm-Level Entrepreneurship Orientation on Digital Transformation in SMEs: The Moderating Role of Strategic Agility. *Entrepreneurship Research Journal*, *15*, 91–124. DOI: 10.1515/erj-2023-0267

Scott, S., Hughes, M., & Ribeiro-Soriano, D. (2021). Towards a network-based view of effective entrepreneurial ecosystems. *Review of Managerial Science*, *16*, 157–187. DOI: 10.1007/S11846-021-00440-5

Seremeti, L., Anastasiadis, L., & Komodromos, M. (2026). ESG Strategy and Business Intelligence Semantic Network. In Komodromos, M., Seremeti, L., Anastasiadis, L., Liargovas, P., & Anastasiadou, S. (Eds.), *Data-Driven ESG Strategy Implementation Through Business Intelligence* (pp. 37–48). IGI Global Scientific Publishing., DOI: 10.4018/979-8-3373-5142-1.ch002

Venizelou, V., & Poullikkas, A. (2025). Navigating the Evolution of Cyprus' Electricity Landscape: Drivers, Challenges and Future Prospects. *Energies*. Advance online publication. DOI: 10.3390/en18051199

Yakhshiboeva, L., & Eshbayev, O. (2023). *The Synergistic Impact of Artificial Intelligence and Big Data Analytics on Marketing Communication Strategies in Entrepreneurial Ecosystems: A Data-Driven Approach*. Financial Technology and Innovation., DOI: 10.54216/fintech-i.030202

Yan, J., Wang, S., Xiong, J., Scaringella, L., & Chen, X. (2024). Value Creation Reflecting CVC Strategic Orientations in Internet Platform Business Ecosystems: The Case of Tencent. *IEEE Transactions on Engineering Management*, *71*, 1531–1541. DOI: 10.1109/TEM.2023.3336407

Zakzouk, F.A., Sumadi, A. M., & Khadra, T, & Komodromos.m. (2023). Factors affecting women entrepreneurs in Jordan. International Journal of Environment. *Workplace and Employment*, *7*(2), 164–185. DOI: 10.1504/IJEWE.2023.134532

Zhang, Z., Khan, R., & Nandhakumar, J. (2024). Exploring liminal experience of entrepreneurs in information and communication technology innovation in underdeveloped regions. *Information & Management*, *62*, 104067. DOI: 10.1016/j.im.2024.104067

Zhou, F. (2024). Utilizing Social Media Platforms to Foster Entrepreneurial Communication among Start-up Communities. *El Profesional de la Información*. Advance online publication. DOI: 10.3145/epi.2024.ene.0525

Chapter 10
Educational Infrastructures and Pluriversal Imagination for Equitable Economic Futures:
Colonial Curricula to Post-Growth Pedagogies

Ashraf Alam
https://orcid.org/0000-0001-6178-1187
Alliance University, Bengaluru, India

ABSTRACT

This chapter interrogates the constitutive role of education in shaping economic futures in the Global South, arguing that educational systems function both as narrative infrastructures, i.e., constructing societal grammars of value, success, and belonging, and as systemic design spaces, structuring the material conditions under which these narratives are enacted. Integrating insights from narrative economics, systemic design theory, and decolonial pedagogy, it examines how colonial legacies, neoliberal reforms, and alternative pedagogies influence economic subject formation. Drawing on case studies from Asia, Africa, and Latin America, including Bhutan's GNH schools, Brazil's Landless Workers' Movement, and innovation hubs across Africa, it demonstrates that plural, justice-oriented models are operational realities. The chapter advances a framework for scaling such innovations through

DOI: 10.4018/979-8-3373-4566-6.ch010

policy reform, participatory governance, and post-growth valuation systems, positioning education as a generative commons for building equitable, sustainable, and pluriversal economies.

1. INTRODUCTION

The early decades of the twenty-first century constitute a historical inflection point defined by an unprecedented simultaneity of macrostructural perturbations, i.e., hyper-volatility in global political economy, biospheric destabilisation through anthropogenic climate perturbations, accelerated demographic transitions, and a tectonic reconfiguration of geopolitical hegemonies (Reynolds-Cuéllar et al., 2024; Novoa & Vasudevan, 2025; Agasisti & Bertoletti, 2022). The synergistic interplay of exponential automation, algorithmic governance, and platform-mediated capital accumulation with intensifying climatological disruptions has rendered the axioms of linear, growth-maximising developmentalism, bequeathed from the industrial and colonial epochs, empirically untenable and normatively indefensible. The asymmetric concentration of wealth, evidenced in Oxfam's 2024 analysis wherein the richest 1 percent command over two-thirds of global asset expansion, intersects with planetary boundary overshoot to expose the structural bankruptcy of GDPism as the cardinal metric of collective prosperity. Within this reconstituted terrain, education emerges not as a peripheral adjunct to economic strategy but as a generative architecture of economic ontology itself (Castro-Sotomayor & Minoia, 2023; Chan et al., 2023; Habibi & Zabardast, 2020). Formal and non-formal learning ecologies, spanning state-administered schools, polytechnic institutes, universities, community academies, and vernacular knowledge networks, are not merely provisioning sites for labour-market insertion; they are primary arenas where the semiotic, epistemic, and algorithmic scaffolding of "*the economy*" is fabricated, rehearsed, and normalised. The relationship between education and economic transformation is, therefore, constitutive rather than instrumental, i.e., pedagogical architectures actively mediate the formation of economic imaginaries, delimit the taxonomies of legitimate economic activity, and calibrate the moral economies of productivity, merit, and distributive justice.

The Global South, encompassing over 6.5 billion inhabitants and projected to host approximately 90 percent of the world's youth cohort by mid-century, represents a critical locus for interrogating this dynamic. Here, economic trajectories are bifurcated between the longue durée of extractive colonial political economies, structured by dependent trade regimes, monoculture export specialisation, and fiscal conditionalities, and a proliferating topography of insurgent innovation, epistemic disobedience, and post-capitalist experimentation (Huaman, 2022; Lazarus et al.,

2025; Hanushek & Woessmann, 2021). From sub-Saharan African municipalities leveraging municipal cooperatives to Pacific island nations embedding climate adaptation into vocational training, education functions as both a battleground and a crucible for competing teleologies of economic life. The stakes are quantifiable. UNESCO's *Global Education Monitoring Report* (2023) documented 244 million out-of-school children and youth worldwide, with over 40 percent concentrated in sub-Saharan Africa, a demographic exclusion that is not simply developmental lag but the foreclosure of entire socio-economic futurities for rural, Indigenous, and subaltern populations. In low- and middle-income countries, the World Bank's *Learning Poverty Index* (2022) registers that 70 percent of ten-year-olds fail to attain basic reading comprehension, a structural predictor of exclusion from high-value innovation sectors. Such deficits are not reducible to cognitive skill gaps; they index an absence of pedagogical ecologies capable of fostering anticipatory governance, collective problem-solving, and the capacity to articulate heterodox economic futures.

In this analytic schema, education operates as both medium and mechanism for the production of economic imagination. As a medium, it circulates hegemonic narratives, i.e., productivity fetishism, competitive individualism, and perpetual growth, embedding them in the collective cognitive infrastructure (Braga Campos, 2020; Hill et al., 2025; Hanushek & Woessmann, 2020). As a mechanism, it codifies these narratives into curricular taxonomies, pedagogical repertoires, and institutional architectures, thus materialising them as competencies, dispositions, and decision-making heuristics in emergent economic actors. Robert Shiller's theoretical corpus on narrative economics demonstrates that such socio-economic mythologies, whether technophilic disruptionism, austerity rationalism, or heroic entrepreneurialism, can exert allocative and policy effects as potent as econometric data. Educational apparatuses, through curricular canon formation, assessment regimes, and role-modelling, function as key sites of narrative naturalisation (Perry, 2023; Garcia & Mirra, 2023). The strategic question for the Global South is whether these narratives will remain tethered to neoliberal developmentalism or pivot toward solidaristic, ecologically consonant, and distributively just economic architectures. Policy praxis reflects this ambivalence. India's *National Education Policy* (2020) hybridises vernacularisation, via multilingualism and interdisciplinarity, with explicit benchmarking against global competitiveness indices. Kenya's Competency-Based Curriculum (CBC), incubated in 2017 and scaled nationally in 2019, enshrines creativity and digital literacy yet evidences stratified uptake due to urban-rural infrastructural asymmetries. Brazil's *Movimento dos Trabalhadores Rurais Sem Terra* (MST) constitutes a counter-hegemonic intervention, embedding agroecology, cooperative governance, and political economy into agrarian reform settlements, thus operationalising education as a platform for economic counter-design against extractivist orthodoxy.

The methodological architecture of this chapter is deliberately transdisciplinary. Educational theory elucidates the micro-processes of identity and disposition formation; development studies situates pedagogical reform within global political economy; narrative economics decodes the performative power of economic storytelling; and decolonial thought unmasks the epistemic hierarchies constraining economic imagination (Qureshi, 2025; Perry, 2021). Synthesised, these frameworks enable a granular analysis of how educational ecologies in the Global South operate as both reproduction mechanisms for inherited economic orders and as generative spaces for systemic transformation. This inquiry is concretised through empirical exemplars, i.e., Lagos-based graduates of the University of Nigeria populating Yaba's "*Silicon Lagoon*" fintech clusters to advance financial inclusion for unbanked populations; rural Bangladeshi girls, through BRAC's non-formal education programmes, deploying solar micro-grid entrepreneurship for off-grid electrification; Cape Town's community innovation labs enabling township youth to prototype water-harvesting systems in response to hydrological crises. These are not idiosyncratic anomalies but instantiations of a pluriversal economic metamorphosis in which education is an active design variable (Becerra Posada, 2025; Garcia-Arias et al., 2023). Accordingly, the chapter theorises education as *narrative infrastructure* and *systemic design space*, i.e., arenas where economic futurities are imagined, contested, and enacted. The learner is positioned as an *economic actor-in-formation*, whose imaginative latitude, critical reflexivity, and cooperative problem-solving repertoires will decisively contour the morphology of emergent economic orders. The analytical trajectory proceeds through an exposition of the theoretical foundations, an excavation of colonial continuities, an exploration of entrepreneurial imaginaries, an interrogation of valuation regimes, a dissection of identity-market entanglements, and a proposition of systemic design principles for embedding economic justice at the heart of educational transformation.

2. THEORETICAL FRAMEWORK

The conventional conceptualisation of education as a linearised technocratic conduit for producing labour-market-ready human capital is epistemologically impoverished and analytically myopic when interrogated against the exigencies of systemic economic transformation in the Global South. To apprehend education as a constitutive driver of socio-economic reconfiguration, it must be situated within two interlinked theoretical strata, i.e., *narrative infrastructure* and *systemic design*. This reconceptualisation foregrounds education not as a passive skill-dispensing mechanism but as a contested socio-institutional arena wherein collective imaginaries of value, temporality, progress, and futurity are authored, rehearsed, and institutionalised (Minoia & Castro-Sotomayor, 2024). The construct of *narrative infrastructure* derives from the recognition, elaborated by economists such as Robert Shiller and historiographers like Deirdre McCloskey, that economic systems are sustained, legitimated, and periodically reconstituted as much by semiotic architectures and discursive repertoires as by econometric indicators or formal models. McCloskey's analysis of the rhetoric of economics demonstrates that policy regimes and developmental trajectories are narratively scaffolded through persuasive tropes, while Shiller's formulation of narrative economics elucidates how memetically potent economic storylines, whether the euphoria of the 1990s dot-com boom or the pessimism preceding the 2008 global financial contraction, induce macroeconomic shifts through contagion in collective expectation-formation. Within educational ecologies, narrative infrastructure is instantiated in textbook canon formation, the curated pantheon of historical exemplars, the symbolic hierarchisation of career pathways, and the ceremonial rites of merit recognition (Sandoval-Llanos et al., 2025; Feukeu, 2024). These narratives are not inert reflections of economic life; they are prescriptive grammars that delimit the spectrum of legitimate economic subjectivities, frequently entrenching structural asymmetries.

In the Global South, narrative infrastructures are profoundly inflected by the longue durée of coloniality. Achille Mbembe's theorisation of *necropolitics* clarifies how colonial education systems were architected not to engender autonomous, rights-bearing citizens but to manufacture compliant intermediaries for extractive economic regimes (Iwu et al., 2021; Letkemann, 2023). Walter Mignolo's analytic of the *coloniality of knowledge* exposes how Euro-American epistemic frames are canonised as universal benchmarks, delegitimising endogenous economic ontologies. Paulo Freire's *pedagogy of the oppressed* furnishes a counter-epistemology, recasting education as dialogical praxis through which learners interrogate structural determinants and prototype alternative socio-economic orders. Operationalising this in contemporary curricula necessitates dismantling the implicit syllabus that equates economic legitimacy with assimilation into globalised capitalist circuits, replacing it with pluralised pathways grounded in community-defined metrics of prosperity. If narrative infrastructure explicates *what* economic futures education habitually imagines, systemic design theorises *how* educational architectures can be reconfigured to materialise different futures. Systemic design, as articulated by Bela Banathy and expanded by Peter Jones, applies design science to complex, adaptive socio-technical systems, privileging participatory co-creation, iterative prototyping, and holism over reductionist optimisation. In educational governance, this translates into reframing curriculum, pedagogy, and institutional regulation as co-dependent subsystems nested within a broader socio-ecological meta-system. Such an approach repudiates the mechanistic input–output orthodoxy dominating policy discourse, i.e., where fiscal allocations are presumed to yield predictable labour-market outputs, and instead embraces the non-linearity, feedback loops, and emergent properties characteristic of complex adaptive systems.

Applying systemic design in Global South contexts necessitates direct engagement with structural political economy constraints. UNESCO's *Global Education Monitoring* data (2022) reveal that in low-income economies, an average of 43 percent of education budgets is externally financed, structurally exposing policy agendas to donor-driven epistemic governance that privileges quantifiable, short-cycle deliverables over endogenous resilience-building. This conditionality often truncates curricular innovation to donor-sanctioned metrics, thereby constraining the incubation of heterodox economic imaginaries (Tao et al., 2022; Asher & Novosad, 2020). A systemic design paradigm inverts these priorities, institutionalising adaptability, multi-stakeholder governance, and the capacity for continuous pedagogical prototyping responsive to shifting socio-economic and ecological contingencies. Integrating decolonial epistemologies into systemic design entails more than the tokenistic inclusion of Indigenous knowledge as a curricular annex; it requires reconstitution of the epistemic operating system. Boaventura de Sousa Santos' *epistemologies of the South* assert the ontological and methodological validity of knowledge systems sedimented in the lived praxis of subaltern communities. Operational exemplars might include vocational curricula in Pacific Island states embedding artisanal climate-adaptation logics of fisherfolk, or entrepreneurship modules in East African polytechnics co-delivered by informal economy actors whose adaptive market ecologies remain invisible to GDP calculus but are vital to socio-economic reproduction.

Theories of *economic imagination*, as formulated by Jens Beckert and Arjun Appadurai, further complexify the analytic. Beckert's construct of *fictional expectations* foregrounds how economic agency is predicated on socially constructed projections of the future, while Appadurai's *capacity to aspire* situates imagination as a culturally mediated navigational skill, asymmetrically distributed across socio-economic strata (McGinnv et al., 2020; Leigh, 2024). Education systems are thus implicated in expanding or contracting aspirational horizons. Centralised, inflexible curricula, detached from vernacular economic ecologies, compress these horizons into narrow, sanctioned occupational trajectories; conversely, participatory, locally embedded curricula amplify the cognitive bandwidth for envisioning heterodox futures that subvert entrenched hierarchies. The learner, within this synthesis, is reframed as an *economic actor-in-formation*, i.e., a locus where narrative inscription and systemic affordances converge. Human capital theory, canonised by Theodore Schultz and Gary Becker in the 1960s, cast the learner as a rational investment object whose productivity-enhancing attributes can be augmented through educational expenditure. While influential, this economistic ontology collapses education into a factor of production and the learner into commodified labour power (Abad-

Segura & González-Zamar, 2021; Jorgenson & Fraumeni, 2020). In the Global South, where, per ILO statistics, informal economies encompass up to 85 percent of total employment in sub-Saharan Africa, where this reductive framing elides the multiplicity of livelihood strategies that traverse formal, informal, subsistence, and hybrid economic modalities.

The implications for design and evaluation are profound. In a narrative infrastructure lens, the critical interrogation is: *Which economic narratives are learners being socialised to inhabit?* In a systemic design lens, it is: *How do institutional architectures scaffold or constrain the co-production of alternative futures?* Synthesised, these frameworks demand a paradigmatic shift in educational assessment, i.e., from the narrow metrics of standardised test scores, credential attainment, and formal sector placement to multi-dimensional indicators encompassing imaginative capacity, socio-ecological resilience, and distributive equity. Empirical instantiations across the Global South attest to the operability of this dual framework. Vietnam's 2018 general education reform inscribed *entrepreneurial competence* as a transversal competency, embedding it in project-based learning linked to community economic engagement, an intervention co-designed with industry, academia, and civil society (Hanushek & Woessmann, 2023; Habibi & Zabardast, 2020). Colombia's *Escuela Nueva* model, transposed to over 20 national contexts, reconfigures rural schooling through multi-grade autonomy, learner self-governance, and community co-production, simultaneously diversifying economic narratives and enhancing systemic adaptivity. Tanzania's *Building a Future* initiative integrates agronomic science with indigenous soil stewardship praxis, directly contesting the colonial developmentalist narrative that equates modernity with epistemic displacement of traditional agroecologies.

By construing education simultaneously as narrative infrastructure and as systemic design ecology, latent leverage points for structural transformation become perceptible. Teacher education can be re-engineered to cultivate *narrative literacy*, enabling practitioners to deconstruct and reconstruct economic storylines embedded in disciplinary canons (McGinnv et al., 2020; Hanushek & Woessmann, 2020). Curriculum architecture can be retooled for iterative, multi-stakeholder feedback integration, ensuring congruence between classroom economic imaginaries and lived socio-economic realities. Policy evaluation protocols can be expanded to include indices of cultural relevance, community integration, and ecological stewardship as co-equal with conventional economic outputs (León-Gómez et al., 2021; Ozturk & Ullah, 2022). This integrated theoretical armature undergirds the subsequent interrogation of colonial legacies in economic subject formation. It clarifies why ostensibly sophisticated reforms often fail to rewire the economic logic of educational systems, and why insurgent, community-rooted innovations, despite operating at the margins of formal policy, can catalyse profound reimaginings of economic participation that are both inclusively distributive and structurally transformative.

3. COLONIAL LEGACIES AND THE FORMATION OF ECONOMIC SUBJECTS

The educational topographies of the Global South are deeply sedimented within the longue durée architectures of colonial domination, wherein schooling was engineered not as a locus of emancipatory epistemogenesis but as an apparatus for fabricating economic subjectivities calibrated to the imperatives of imperial extraction. The historical genealogy of these formations reveals how pedagogical institutions were instrumentalised to sustain a stratified international division of labour that structurally privileged metropolitan economies while rendering colonised polities into peripheral suppliers of primary commodities, raw materials, and low-cost labour power (Etim & Daramola, 2020; Tarkar, 2020). In British India, the paradigmatic crystallisation of this logic is epitomised in Thomas Babington Macaulay's 1835 *Minute on Education*, wherein he explicitly advocated the cultivation of an intermediary elite "*Indian in blood and colour, but English in tastes, in opinions, in morals, and in intellect.*" The curricular architecture institutionalised under this regime was explicitly designed to produce clerks, translators, and minor bureaucratic functionaries to service the colonial administrative machine, thereby foreclosing indigenous innovation trajectories. Analogous pedagogical blueprints proliferated in French West Africa, where colonial functionaries such as Jules Harmand codified vocational instruction to emphasise agricultural and artisanal competencies optimised for export production, while systematically suppressing endogenous knowledge systems and innovation ecosystems.

The epistemic and institutional residues of this colonial educational economism endured with remarkable tenacity into the post-independence period (Stoica et al., 2020; Hardi et al., 2023). Newly sovereign states frequently inherited not only the material infrastructures, e.g., school buildings, inspection regimes, and examination boards, but also the epistemological frameworks and developmentalist policy logics embedded within them. Ghana's 1951 *Accelerated Development Plan for Education* radically expanded access to primary schooling, yet retained curricular architectures tethered to British examination standards. In Kenya, the 1964 Ominde Commission sought to "*Africanise*" content, yet continued to benchmark educational outcomes against labour-market integration into economic sectors structurally configured under colonial economic planning. The onset of International Monetary Fund and World Bank–driven structural adjustment programmes in the 1980s compounded these constraints. Austerity regimes and conditional lending precipitated severe contractions in public expenditure on education, with Zambia registering a greater than 50 percent decline in per-capita educational investment between 1980 and 1990. These fiscal compressions crippled curricular innovation capacity, degraded teacher professional development pipelines, and reinforced dependency on donor-scripted education reform templates.

Such dynamics consolidated what critical development economists like Ha-Joon Chang have described as a "*race to the bottom*" in educational design, i.e., systems optimised to produce cost-efficient human capital for insertion into low-wage segments of global commodity chains, rather than cultivating autonomous, contextually adaptive economic capacities (Myovella et al., 2020; Erlando et al., 2020; Alshurafat et al., 2021). The human capital paradigm, canonised in the early 1960s by Theodore Schultz, reframed education as an investment calculus to enhance labour productivity. While this economistic framing facilitated expansions in enrolment, it simultaneously reinforced the reduction of learners to fungible units of economic input, thereby obscuring education's role as a narrative and structural architect of collective economic futures. In many postcolonial polities, the aspirational architectures embedded within education remain tethered to Euro-Atlantic metrics of success. Arjun Appadurai's formulation of the *capacity to aspire* illuminates the uneven distribution of navigational competence for articulating and realising desired futures. Elite metropolitan schools cultivate fluency in globalised economic scripts, while rural and marginalised institutions often constrain learners to parochial visions of subsistence livelihoods or precarious wage labour. In South Africa, the post-apartheid persistence of *Model C* schools as gateways to elite university placements and high-remuneration employment starkly contrasts with the infrastructural and pedagogical deficits of township schools, producing systemic stratifications in life-course trajectories.

Colonial epistemologies are further perpetuated through linguistic hierarchies embedded in education. In Mozambique, Portuguese remains the dominant medium of instruction despite UNESCO data indicating that fewer than 15 percent of citizens speak it as a first language. This *linguistic coloniality* alienates learners from vernacular socio-economic contexts and codifies economic participation as contingent upon assimilation into imported linguistic-cultural codes. Francophone West Africa replicates these dynamics, with mastery of French functioning simultaneously as a prerequisite for formal-sector mobility and as a symbolic marker of social capital. Nominally progressive reforms often reinscribe these colonial logics. India's *National Education Policy* (2020) foregrounds multilingualism and multidisciplinary integration but is also deeply aligned with global competitiveness indices, thereby reproducing neoliberal development imaginaries. Kenya's Competency-Based Curriculum, rolled out nationally in 2019, aspires to inculcate critical thinking and entrepreneurial agency, yet implementation assessments by the Kenya Institute of

Curriculum Development (2022) reveal stark disparities in access to requisite pedagogical resources, disproportionately privileging urban, capital-intensive schools.

Some national experiments, however, deliberately attempt to disarticulate education from colonial economic scripts. Bolivia's 2010 *Avelino Siñani–Elizardo Pérez Education Law* institutionalises an intracultural, intercultural, and multilingual curriculum embedding Indigenous epistemologies, agroecological practices, and community-based economic logics into the national framework, drawing intellectual lineage from the 1930s Warisata School. Brazil's *Movimento dos Trabalhadores Rurais Sem Terra* (MST), operating over 2,000 schools in agrarian reform settlements, integrates literacy with cooperative governance and political economy, positioning agricultural labour as a locus of economic sovereignty rather than subsumption under export-oriented monocultures (Fahad et al., 2023; Liu et al., 2020). Yet the homogenising pressures of globalisation constrain such counter-hegemonic initiatives. University ranking regimes administered by *Times Higher Education* and QS, alongside OECD's PISA benchmarking, exert centripetal forces toward STEM privileging, English-language dominance, and marketised research outputs. Malaysia's 2015 pivot to English-medium STEM instruction exemplifies this dynamic, justified as a competitiveness strategy, despite Ministry of Education data (2018) evidencing declining rural student performance due to linguistic alienation. The resultant economic subject is thus a hybrid formation, i.e., formally postcolonial yet epistemically and aspirationally structured by the interpenetration of coloniality and neoliberalism. This subject is normatively oriented toward insertion into global supply chains, valorisation of urban white-collar labour, and conflation of innovation with high-tech entrepreneurialism, while informal economies, subsistence production, and cooperative modes of livelihood are relegated to residual or "*alternative*" status.

Dismantling these entrenched formations necessitates interrogating not only curricular content but also the institutional architectures and policy logics that govern educational reproduction. Policy innovations in Rwanda and Ethiopia exemplify embryonic moves toward integration of community service, environmental stewardship, and endogenous enterprise into formal learning outcomes. Rwanda's 2015 Competence-Based Curriculum embeds entrepreneurship and citizenship as transversal competencies across all schooling levels, explicitly aligning them with self-reliance objectives in national development planning. Nevertheless, such reforms operate within a global economic regime that structurally rewards compliance with neoliberal templates. This critical excavation of colonial legacies reveals a constitutive tension, i.e., educational systems in the Global South are simultaneously tasked with catalysing sovereign national development and operating within an international economic order whose definitional metrics of "*development*" reproduce dependency and asymmetry. Addressing this contradiction demands a dual reorientation, i.e., transforming both the narratives that educational systems reproduce and the systemic designs through which those narratives are operationalised. The following section turns to one of the most contested terrains in this struggle, i.e., the cultivation, or foreclosure, of entrepreneurial imaginaries within educational praxis.

4. ENTREPRENEURIAL IMAGINATIONS IN EDUCATIONAL PRACTICE

The cultivation of entrepreneurial capacity within educational ecosystems of the Global South constitutes a complex and ideologically contested domain wherein multiple, and often antagonistic, logics converge. On one axis lies the neoliberal teleology of market competitiveness, hyper-individualised self-reliance, and accelerated capital circulation; on another, the developmentalist imperative to catalyse GDP expansion through technological innovation and commercial modernisation; and, counterpoised to both, insurgent paradigms that reframe entrepreneurship as a praxis of social transformation, ecological stewardship, and collective well-being. To apprehend how educational institutions configure entrepreneurial subjectivities, one must interrogate not merely the explicit curricular architectures that codify entrepreneurial competencies but also the latent pedagogical dispositifs that shape learners' aspirational economies, normative value orientations, and modalities of risk negotiation (Saleh et al., 2020; Zhao et al., 2022). Policy discourses across numerous postcolonial polities position entrepreneurship as a palliative to structurally embedded unemployment and underemployment, particularly among burgeoning youth cohorts. The International Labour Organization's 2023 data indicate that youth unemployment in sub-Saharan Africa persists above 12 percent, with underemployment figures markedly higher in precarious, informal, and low-wage labour niches. Constrained by limited absorptive capacity in formal employment sectors, governments have recalibrated educational mandates to embed entrepreneurial training as a labour-market pressure valve. Nigeria's *National Policy on Education* (revised 2013) institutionalises compulsory entrepreneurship modules across tertiary institutions, encompassing business plan design, financial literacy, and market analytics. This policy trajectory is anchored in the neoliberal postulate that joblessness is predominantly a skills-mismatch phenomenon, thereby occluding the structural dynamics of capital concentration, asymmetrical market access, and ownership inequities.

Within institutional settings, the dominant framing of entrepreneurship tends toward a market-centred paradigm. Business incubators embedded in universities, hackathons in technical institutes, and corporate-sponsored pitch competitions valorise "*innovation*" as the capacity to generate scalable, high-margin enterprises (Ali, 2022;

Ben-Salha et al., 2021). The African Leadership University's School of Business, inaugurated in Mauritius in 2015, explicitly models its curricular architecture on Silicon Valley's start-up epistemology, privileging rapid prototyping, competitive pitching, and venture capital engagement. While such modalities transmit operational skills in business development, they also sediment an entrepreneurial imaginary disproportionately weighted toward individual success narratives and high-growth models, often at the expense of cooperative, territorially embedded, and socially accountable forms of enterprise. The neoliberalisation of entrepreneurial discourse manifests further in the axiological elevation of risk-taking, market disruption, and competitive aggression as inherently valorous attributes. Pedagogical programmes influenced by this discourse frequently import the idioms of "*fail fast*" and "*pivot*" from the lean start-up methodology canonised in U.S. entrepreneurial culture. In socio-economic contexts lacking robust safety nets or equitable capital markets, the romanticisation of failure is an aspirational fiction accessible primarily to a minority insulated by pre-existing resource endowments. A 2022 impact evaluation of entrepreneurship education within South Africa's Technical and Vocational Education and Training (TVET) colleges revealed that while graduates exhibited enhanced ideational confidence, fewer than 10 percent accessed start-up capital within two years, highlighting the disjuncture between entrepreneurial formation and the systemic conditions for business viability.

In contraposition, emergent models of entrepreneurial pedagogy from within the Global South foreground logics of cooperation, ecological reciprocity, and communal well-being. The Honey Bee Network, established in India by Anil Gupta in 1988, operates as a grassroots innovation clearinghouse, cataloguing and supporting rural inventors while preserving intellectual property within the originating communities. Its educational arm collaborates with rural schools in Gujarat and Rajasthan, embedding innovation scouting into science curricula and cultivating entrepreneurial imaginaries anchored in endogenous problem-solving. In Uganda, the Junior Achievement Africa programme has re-engineered its model to institutionalise cooperative agribusiness ventures among secondary learners, thereby shifting ownership structures from atomised entrepreneurship toward collective enterprise. Latin American exemplars illuminate the solidaristic reorientation of entrepreneurial education. In Brazil's southern state of Rio Grande do Sul, the *Rede de Escolas de Economia Solidária* (Solidarity Economy School Network) collaborates with secondary and vocational institutions to integrate cooperative governance, ethical finance, and fair-trade protocols into business curricula. A 2021 evaluation by the state Secretariat of Education demonstrated elevated graduate participation in cooperative enterprises and civic associations, signalling that when entrepreneurial education is grounded in alternative economic logics, it can function as a vector for economic democratisation.

Sustainability-infused entrepreneurial curricula have similarly gained salience. In the Philippines, the Department of Education's K–12 curriculum embeds an "*Entrepreneurship*" strand at the senior high school level incorporating environmental accountability and social enterprise models. Piloted in 2016 in climate-vulnerable Eastern Visayas, student-led ventures developed flood-resistant housing prototypes and community disaster-preparedness services, evidencing the mobilisation of entrepreneurial imagination toward climate resilience and distributive social equity. Innovation hubs and makerspaces across the Global South have emerged as experimental pedagogical laboratories. AfriLabs, a continental network founded in 2011, interlinks over 400 hubs in 53 countries, provisioning mentorship, technical capacity-building, and seed capital. Member nodes such as Nairobi's iHub and Lusaka's BongoHive operationalise curricula integrating coding, design thinking, and social impact assessment, frequently eschewing didactic instruction in favour of project-based, peer-mediated, and open-innovation methodologies. Yet, their sustainability remains precariously tethered to donor funding cycles, raising concerns regarding institutionalisation within national education systems.

Rural innovation schools provide further evidence of entrepreneurship as a collective adaptive strategy. Nepal's Rural Education and Development (READ) model amalgamates community libraries with vocational instruction in organic agriculture, livestock husbandry, and eco-tourism, operated in partnership with cooperatives to align entrepreneurial outputs with local developmental priorities. In Malawi, the Nkhata Bay Skills for Youth Employment project imparts technical skills in carpentry, tailoring, and fish processing, coupled with cooperative business governance and micro-savings mechanisms. Nevertheless, the proliferation of entrepreneurial education is not without regressive potential. Absent critical reflexivity, such initiatives risk entrenching existing socio-economic asymmetries by disproportionately benefitting those with pre-existing network capital, financial resources, and cultural capital. The valorisation of exceptionalised "*success stories*" as programme efficacy proxies can occlude systemic barriers facing the majority. Kenya's Youth Enterprise Development Fund, operational since 2007, disbursed loans to over 500,000 youth by 2020; repayment patterns, however, skewed heavily toward urban graduates with familial business linkages, while rural beneficiaries lacked the collateral, mentorship, and infrastructural access necessary for enterprise sustainability.

A genuinely transformative cultivation of entrepreneurial imagination necessitates more than the additive inclusion of business modules or start-up competitions. It demands pedagogical ecologies that pluralise exposure to economic models, embed critical interrogation of socio-environmental consequences, and democratise access to the infrastructural and financial resources essential for enterprise realisation (Abad-Segura & González-Zamar, 2021). This entails bridging formal and informal learning architectures, valorising community knowledge economies, and scaling proven local innovations through enabling policy frameworks without eroding contextual specificity. In this reconstituted schema, the entrepreneur is reframed not solely as a market disruptor but as a civic-economic agent co-producing just, resilient, and sustainable socio-economic systems. Educational institutions, by actively scripting the narratives, competencies, and cooperative praxis underpinning such roles, hold the epistemic and structural capacity to redefine the ontological parameters of economic participation for emergent generations. The subsequent section interrogates the valuation regimes through which educational outputs are measured, and whether these regimes reinforce or destabilise the GDP-aligned logics that currently circumscribe economic imagination in the Global South.

5. EDUCATION, MEASUREMENT, AND POST-GROWTH LOGICS

The evaluative architectures through which educational systems in the Global South are appraised remain profoundly tethered to productivist and GDPist epistemologies of value. State bureaucracies, multilateral donor agencies, and global benchmarking regimes habitually mobilise a narrow suite of quantitative indicators, i.e., standardised literacy and numeracy proficiency scores, gross and net enrolment ratios, graduate employability indices, and ordinal positioning in transnational comparative assessments such as the OECD's Programme for International Student Assessment (PISA). While ostensibly offering ostensibly objective snapshots of system performance, these metrics instantiate a reductive telos, i.e., the calibration of human capital stocks to maximise throughput into growth-centric macroeconomic circuits. In privileging standardisation, efficiency optimisation, and inter-jurisdictional competitiveness, such regimes systematically marginalise equity, ecological sustainability, and integrative human flourishing as secondary or non-essential outcomes. The hegemony of GDP-aligned valuation modalities is historically sedimented. Post-independence development blueprints in states such as India, Ghana, and Malaysia were grafted onto the growth orthodoxies propagated by Bretton Woods institutions, positioning mass education as a primary productivity vector in the Keynesian–developmentalist synthesis. World Bank sectoral reviews in the 1970s and 1980s operationalised curriculum reform as a function of projected labour-market demand curves, thereby subsuming the telos of education under the calculus of GDP augmentation. This instrumentalist logic remains pervasive, e.g., Kenya's *Education Sector Analysis* (2022) reported secondary transition rates in the same statistical tableau as GDP growth forecasts, rendering the former an input variable to the latter without interrogating the substantive quality, inclusivity, or civic-ecological orientation of the learning experience.

The structural inadequacies of this evaluative paradigm are increasingly empirically irrefutable. UNESCO's *Global Education Monitoring Report* (2023) documents that in sub-Saharan Africa, notwithstanding significant expansion in school access post-2000, youth unemployment and underemployment persist at intransigently high levels, with informal-sector livelihoods constituting the modal form of economic participation. Such patterns disclose the inability of conventional metrics to capture the disjuncture between formal system outputs and the plural, often precarious, realities of economic life. Moreover, the fixation on employability as a principal performance indicator occludes education's constitutive role in cultivating civic republicanism, ecological stewardship, and cultural continuity, i.e., domains critical to long-term socio-ecological viability yet non-fungible within GDP accounting frameworks. Post-growth economic epistemologies articulate alternative grammars of valuation that contest productivist metric hegemony. Kate Raworth's *Doughnut Economics* model proposes recalibrating economic performance assessment to a dual threshold, i.e., meeting universal human needs while remaining within biophysical planetary boundaries. Transposed to education, this entails evaluating whether pedagogical systems equip individuals and collectivities with the cognitive repertoires, ethical dispositions, and praxis-oriented skills necessary for sustaining socio-ecological equilibrium and distributive justice (Acquah & Ibrahim, 2020; Surya et al., 2021). Tim Jackson's theorisation of "*prosperity without growth*" reframes prosperity as the capacity for human flourishing under ecological constraint, thereby implicating education in the cultivation of post-materialist competencies, i.e., relational intelligence, adaptive resilience, and moral imagination, indispensable to such flourishing.

Operational exemplars from the Global South evidence the viability of such paradigms. Bhutan's *Gross National Happiness* (GNH) index, codified into educational governance in 2008, institutionalises multidimensional indicators encompassing psychological well-being, community vitality, and ecological resilience. Schools are evaluated not solely on academic attainment but also on student engagement in cultural praxis, curricular integration of environmental ethics, and the cultivation of prosocial dispositions. The Ministry of Education's 2021 evaluations found GNH-aligned schools demonstrating statistically significant gains in civic participation and reduced dropout rates relative to control cohorts, highlighting the measurability of non-GDP-aligned benefits. In India's capital territory, the *Happiness Curriculum*, initiated in 2018 across 1,000+ schools, integrates mindfulness training, socio-emotional literacy, and moral reasoning into daily pedagogical cycles. The State Council of Educational Research and Training's 2022 report documented marked improvements in student comportment and self-reported life satisfaction, i.e., outcomes absent from standardised test matrices yet integral to the formation of ethically reflexive, cooperatively inclined economic actors.

Colombia's *Escuela Nueva* model, originating in the 1970s as a rural education reform, operationalises participatory pedagogy, flexible promotion regimes, and embedded community engagement. Evaluation metrics encompass community governance participation rates, incidence of student-led local development initiatives, and alignment of school projects with ecological and economic priorities. Longitudinal studies by Fundación Escuela Nueva demonstrate that alumni exhibit higher propensities toward cooperative enterprise participation and local governance roles, exemplifying a broadened value horizon linking learning directly to socio-economic resilience. Decolonial valuation frameworks extend this reimagination further. Boaventura de Sousa Santos' *epistemologies of the South* foreground valuation systems anchored in the lived priorities of subaltern collectivities rather than the normative priorities of global capital. In praxis, this could entail, for example, the co-design of assessment rubrics in Amazonian Indigenous schools that confer equal legitimacy to competencies in ethnobotanical knowledge, fluvial navigation, and restorative conflict mediation alongside literacy and numeracy. In South Africa's Eastern Cape, Rural Education Access Programme pilots have embedded community-defined metrics, such as the quantum of school projects mitigating local food insecurity, into evaluative protocols, thus recalibrating education to align with endogenous development agendas.

Transitioning to post-growth and pluriversal valuation regimes necessitates structural re-engineering at multiple governance scales. Ministries of education must decouple performance audits from GDP growth targets, embedding multidimensional well-being indicators as core evaluative criteria. Teacher education institutions require curricular integration of ecological literacy, systems thinking, and community co-design methodologies. Multilateral donors and philanthropic funders must recalibrate financing logics to reward systemic resilience, distributive equity, and ecological sustainability rather than throughput efficiency and labour-market absorption rates alone (Rao et al., 2023; Ahmad et al., 2021). Yet structural impediments persist. Global university ranking systems, dominated by variables such as citation indices, international faculty ratios, and graduate median salaries, exert a homogenising pressure that disincentivises epistemic innovation. The University of Cape Town's 2021 strategic plan candidly acknowledged the tension between sustaining its *Times Higher Education* ranking and deepening its community-engaged research mandate, with ranking algorithms disproportionately weighting English-language publication output over measurable local development impact.

The strategic challenge, therefore, is the construction of hybridised evaluative architectures that retain cross-institutional comparability while embedding contextually grounded, culturally resonant indicators of educational efficacy. The African Union's

Continental Education Strategy for Africa (2016–2025) represents an incipient step toward this, proposing continental benchmarking instruments that integrate metrics of social cohesion, cultural preservation, and ecological sustainability alongside cognitive achievement indices. Ultimately, educational value cannot be algorithmically condensed into test-score aggregates or formal-sector placement percentages. Within a post-growth paradigm, the metric of success must be the system's capacity to engender equitable, adaptive, and ecologically regenerative economic orders. Achieving this requires not only alternative indicators but also a paradigmatic re-orientation of pedagogy, institutional governance, and policy architectures toward an expanded, justice-oriented conception of value. The ensuing section interrogates how the intersection of cultural markets and identity formation shapes the economic dispositions, aspirations, and positionalities of learners in the Global South.

6. CULTURAL MARKETS AND IDENTITY FORMATION

Markets function not solely as arenas of commodity exchange but as dense socio-cultural ecologies wherein identities, aspirations, and symbolic capital are incessantly produced, negotiated, and performed. Education, as one of the principal institutional mechanisms for the transmission and legitimation of the dispositions, linguistic repertoires, and competencies valorised in market participation, occupies a pivotal role in scripting how individuals imagine, inhabit, and perform economic subjectivity (Mahalik et al., 2021; Thathsarani et al., 2021). In the Global South, this process of economic subject formation is inextricably imbricated with the sedimented legacies of colonial stratification, the contested architectures of postcolonial nation-building, and the homogenising pressures of contemporary globalisation. Pierre Bourdieu's analytic of *habitus* provides a generative lens for apprehending the manner in which educational institutions inculcate durable economic dispositions. Through the interplay of explicit curricular instruction and tacit socio-cultural cues, schools scaffold the embodied competencies necessary for navigating economic fields, whether through mastery of the hegemonic lingua franca of commerce, internalisation of dominant modes of corporeal and sartorial presentation, or assimilation of normative scripts regarding the moral economy of "*deserving*" success. In elite private schools in India, the pedagogical cultivation of English fluency, cosmopolitan etiquette, and transnational mobility is not reducible to exam preparation; it constitutes the intentional fabrication of a globalised economic persona primed for insertion into high-prestige corporate and bureaucratic transnational networks.

By contrast, public schools servicing structurally marginalised populations often generate alternative economic dispositions, shaped less by aspirational cosmopolitanism than by the constraints of resource deprivation. In South Africa's township schools, despite formal curricular parity with national standards, learners are routinely deprived of access to enrichment programmes, advanced technological infrastructures, and mentorship networks that serve as cultural signals of readiness for elite labour-market integration. The market's misrecognition of such structural inequities as deficits in individual merit exemplifies what Bourdieu theorises as *symbolic violence*, the naturalisation of socio-economic hierarchies through ostensibly neutral evaluative mechanisms. Embedded within these processes are culturally contingent constructions of "*success*" and "*failure*." In numerous Global South contexts, the archetypal figure of the successful graduate remains the urban-salaried professional embedded in globally integrated sectors such as finance, information technology, or engineering. Agricultural labour, artisanal craft production, and informal sector entrepreneurship, despite their macroeconomic indispensability, are discursively positioned as inferior or fallback occupational outcomes. A 2021 Philippine Institute for Development Studies survey revealed that 78 percent of urban secondary students aspired to overseas employment, predominantly in healthcare and engineering, highlighting the gravitational pull of remittance economies in structuring educational aspirations.

The matrix of caste, race, gender, and class further refracts these dynamics. In India, Dalit and Adivasi students face persistent obstacles in higher education access and high-wage labour market integration, notwithstanding formal credential equivalence with upper-caste peers. In Brazil, Afro-Brazilian students remain disproportionately underrepresented in high-status university programmes, even after the 2012 implementation of affirmative action statutes. Such disparities illustrate the mediating role of social identity markers in the translation of educational capital into economic capital, perpetuating intergenerational cycles of exclusion (Baerlocher et al., 2021; Ahmed et al., 2020). Educational institutions do not merely prepare entrants for pre-existing cultural markets; they actively participate in constructing these markets. University branding, school ranking systems, and alumni network capitalisation function as mechanisms for generating and circulating symbolic capital in labour markets, shaping hiring practices and remuneration trajectories (Kopnina, 2020; Oliinyk et al., 2021). Data from the University of Cape Town's 2020 Graduate Exit Survey indicate that graduates of historically advantaged institutions commanded average entry-level salaries 25 percent higher than those from newer institutions, independent of disciplinary field, highlighting the market premium attached to institutional prestige as a cultural asset.

Yet education also harbours the potential to disrupt these homogenising market narratives. Indigenous knowledge schools, cultural literacy programmes, and intercultural pedagogical initiatives are experimenting with alternative constructions of value and success. In Jharkhand, India, the Jan Sanskriti Shiksha Kendra network embeds Munda, Ho, and Santhali language instruction alongside community economic practices such as cooperative agriculture and forest-based livelihoods. This dual anchoring enables graduates to operate across both subsistence economies and regional markets without discarding Indigenous epistemologies. Similarly, Zambia's Cultural Literacy Programme, initiated by the Ministry of General Education in 2015, incorporates local historiography, artisanal craft pedagogy, and oral narrative traditions into primary curricula. A 2021 evaluation reported elevated parental engagement and improved retention rates, particularly for female students who identified schooling as an affirmational space for cultural identity. In Bolivia, the 2010 *Avelino Siñani–Elizardo Pérez* Law mandates intercultural and multilingual education, integrating Indigenous languages and locally grounded economic practices, thereby contesting the assimilationist assumption that national market integration necessitates epistemic and cultural erasure.

Culturally sustaining pedagogies, as conceptualised by Django Paris and H. Samy Alim, explicitly position education as a site for the maintenance and revitalisation of cultural practices under conditions of assimilationist duress. In Global South applications, this might entail teaching financial literacy via case studies of rotating community savings groups, or delivering environmental science curricula through the epistemic frames of traditional ecological knowledge. Feminist economic literacy interventions similarly expand the aperture of entrepreneurial subjectivity. BRAC's *Gender Quality Action Learning* initiative in rural Bangladesh trains adolescent girls in cooperative enterprise governance while providing education on gender rights and labour law. Alumni have established women-led cooperatives in poultry farming and handicrafts, directly contesting patriarchal restrictions on women's economic agency. In Kenya's Maasai Mara, the Olonana Women's Beadwork Cooperative integrates artisanal skill reproduction into tourism economies, supplementing production with training in pricing, marketing, and contractual negotiation, thus reframing cultural production as a dynamic economic resource rather than a static heritage artefact.

The capacity to cultivate symbolic capital outside elite institutional ecologies is contingent upon resisting the homogenisation imperatives embedded in global competitiveness discourses (Farooq et al., 2020; Jamaliah et al., 2022). Standardised testing regimes and uniform competency frameworks, when applied indiscriminately, risk erasing the cultural distinctiveness that could otherwise function as a comparative advantage in local and regional economic systems. Ultimately, the entanglement of cultural markets and educational practice in the Global South demonstrates that economic subject formation is irreducibly mediated by identity, aspiration, and belonging. Educational systems possess the structural capacity either to reproduce the symbolic hierarchies that underpin the global political economy or to engineer alternative circuits of value in which historically marginalised identities operate as reservoirs of economic strength (Duan et al., 2022; Zhao & Zhou, 2021). The following section advances this argument by examining how systemic design principles, in conjunction with decolonial pedagogies, can be mobilised to reconfigure educational systems as engines of economic justice rather than as conduits into inequitable market hierarchies.

7. SYSTEMIC DESIGN, DECOLONIAL PEDAGOGY, AND EDUCATION FOR ECONOMIC JUSTICE

Reconfiguring education in the Global South as a transformative driver of economic justice demands a decisive departure from the incrementalist logic of piecemeal reform toward the integrative praxis of systemic design. This paradigm conceptualises education not as a mechanistic input–output pipeline calibrated to labour-market absorption, but as an adaptive, interdependent socio-ecological assemblage in which curricula, pedagogical practices, governance structures, infrastructural arrangements, and community economies function as co-evolving subsystems (Doğanalp et al., 2021; Lin et al., 2021). When conjoined with the epistemic interventions of decolonial pedagogy, systemic design furnishes a composite framework for dismantling the colonial and neoliberal logics that still underpin educational priorities, while simultaneously cultivating the structural, cognitive, and affective capacities necessary for equitable, inclusive, and ecologically regenerative economic participation.

7.1 Decolonial Curricular Interventions

A systemic redesign grounded in decolonial pedagogy begins with curricular reconstitution. This entails more than tokenistic incorporation of local content; it requires reconfiguring curricular epistemologies to legitimise and embed pluriversal knowledge systems as central rather than peripheral (Amna Intisar et al., 2020; Mishra et al., 2020). Ecuador's 2016 Intercultural Bilingual Education Model exemplifies this shift by institutionalising Indigenous languages such as Kichwa and Shuar alongside Spanish, and aligning curricular content with community economic practices including agroforestry, artisanal fisheries, and ethnobotanical medicine. The result is a curriculum that not only affirms cultural identity but also fortifies community economic sovereignty. India's Swaraj University in Udaipur offers another radical departure from credentialist orthodoxy, operating entirely outside the degree-certification regime since 2010. Here, learners co-design their programmes around social and ecological projects, producing graduates who have initiated community seed banks, renewable energy cooperatives, and ethical fashion enterprises, thus contesting both the monopoly of centralised curriculum control and the fetishisation of formal degrees. South Africa's IkamvaYouth demonstrates a hybridised model wherein conventional subject tutoring is braided with project-based learning embedded in township economies, enabling students to meet national examination standards while simultaneously engaging in waste recycling cooperatives, solar lighting initiatives, and other community-focused ventures that embody economic justice principles.

7.2 Participatory Prototyping and Co-Creation

Participatory prototyping is a cornerstone of systemic design, privileging iterative, stakeholder-driven experimentation over prescriptive, top-down implementation (Zhang et al., 2021; Zhang et al., 2021). Kenya's Young Africa Works programme, launched in 2019 in partnership with the Mastercard Foundation, co-developed vocational training curricula through sustained dialogue with local artisans, agribusiness owners, and youth cooperatives. This process yielded training modules deeply attuned to market realities and cultural contexts, resulting in demonstrably higher rates of placement in both formal and informal sectors. Barefoot College in Tilonia, Rajasthan, founded by Bunker Roy in 1972, offers a globally recognised exemplar, training rural women, many of them illiterate grandmothers, as solar engineers through hands-on, non-textual workshops conducted in local languages. Its design emerged through collaborative engagement with village councils and has been replicated in over 90 countries, illustrating the scalability of co-created, context-specific innovation. In Brazil, the Rede de Escolas de Trabalho e Produção Coletiva integrates formal education with cooperative production, with students participating directly in governance, marketing strategy, and distribution, thereby transforming the school into a functioning prototype of an alternative, solidarity-based economic system.

7.3 Institutional Innovation and Governance

A systemic turn toward economic justice necessitates rethinking institutional governance to decentralise authority, embed local epistemologies, and legitimise non-hierarchical modes of decision-making (Zafar et al., 2020; Challoumis, 2024). Universidad de la Tierra in Oaxaca, Mexico, rejects conventional administrative hierarchies in favour of self-governed learning communities, where elders, activists, and practitioners mentor students in domains ranging from agroecology to community broadcasting. Without rigid timetables or degree requirements, learners engage at their own pace in projects directly aligned with community economic priorities. The African Rural University for Women in Uganda adopts a governance model that institutionalises intergenerational epistemic transfer through its Council of Grandmothers, ensuring that locally rooted wisdom informs all strategic decisions. By positioning women as epistemic agents capable of catalysing sustainable development in their home districts, this model fuses systemic design principles with feminist decolonial pedagogy.

7.4 Justice-Oriented Design Frameworks

The synthesis of systemic design and decolonial pedagogy yields a set of operational principles essential to embedding economic justice in educational ecosystems (Aiyar & Ebeke, 2020; Bertay et al., 2025). Inclusivity is realised when governance and design processes structurally integrate the voices of historically marginalised groups, ensuring that their priorities are not peripheral but constitutive of system design. Resilience emerges when curricular and organisational architectures are capable of adaptive reconfiguration in response to economic, ecological, and political disruptions. Intergenerational equity is embedded when ecological stewardship is treated as a transversal competency permeating all learning pathways. Local–global coherence is achieved when endogenous economic practices are articulated to broader markets without sacrificing cultural specificity or autonomy. Epistemic humility materialises when institutions recognise the partiality of their own knowledge systems and create structured spaces for reciprocal learning between formal and informal sectors. These principles move beyond rhetorical commitment when translated into concrete measures, such as pre-service teacher education that incorporates community facilitation and participatory planning, or donor funding models that reward long-term socio-economic resilience over short-term output metrics.

7.5 Scaling Without Erasure

One of the principal hazards in extending successful educational innovations is the erasure of the contextual particularities that underpin their efficacy. The Tanzanian experience of scaling the COSTECH science education model from a network of rural schools to the national system during the 1990s exemplifies this risk; under centralised control, much of the model's participatory ethos was diluted. Scaling within a systemic design framework must therefore prioritise the cultivation of enabling environments, comprising flexible policy architectures, decentralised resource flows, and peer-learning networks, that can sustain diverse, locally tailored models simultaneously. The African Union's Pan-African University network, established in 2011, offers an example of distributed scaling, with thematic institutes such as water and energy sciences in Algeria and governance and social sciences in Cameroon, each granted curricular and research autonomy to respond to regional priorities while remaining connected to a broader continental mission.

7.6 Linking Systemic Design to Policy

Embedding systemic design and decolonial pedagogy within educational policy frameworks requires more than programmatic rhetoric; it demands institutional mechanisms that facilitate continuous, bidirectional translation between practice and policymaking. South Africa's Community Education and Training Colleges system, formalised in 2015, incorporates local advisory boards composed of civil society actors, labour unions, and small business representatives, enabling real-time curricular recalibration to shifting local economic and environmental conditions. In Fiji, the Technical College Network collaborates with village councils to ensure that training in renewable energy and marine resource management aligns both with national sustainable development priorities and with the livelihood needs articulated by local communities. These governance arrangements exemplify how policy can be reconstituted as a dynamic interface rather than a static directive, enabling educational systems to function as living infrastructures for economic justice (Zhang et al., 2021; Chatterjee, 2020). By aligning the procedural intelligence of systemic design with the epistemic insurgency of decolonial pedagogy, education in the Global South can transition from a passive mechanism of socio-economic reproduction to an active laboratory for the co-production of just, plural, and sustainable economies (Bloom et al., 2024; Bulturbayevich & Jurayevich, 2020). Such transformation emerges not from isolated pilot projects or abstract policy commitments, but from the deliberate, structural redesign of institutional architectures, curricular regimes, and governance modalities to operationalise equity, resilience, and epistemic plurality as core economic competencies (Prasetyo & Kistanti, 2020; Zafar et al., 2021). The subsequent section examines how these principles can be embedded within broader institutional change strategies, translating design and pedagogical innovation into large-scale systemic reform across the diverse geographies of the Global South.

8. POLICY AND PRACTICE

Translating the composite principles of narrative economics, systemic design, and decolonial pedagogy into durable, system-wide reforms in the Global South necessitates more than the proliferation of innovative pilot projects; it requires deep-seated policy innovation, institutional restructuring, and sustained multi-scalar collaboration across state, market, and civil society domains. While there exist numerous microcosmic exemplarscommunity-administered rural schools, cooperative vocational centres, intercultural universitiesthe strategic challenge lies in institutionalising these approaches at national and regional levels without hollowing out the context-specific epistemic and socio-economic textures that render them effective.

8.1 Policy Blind Spots

Despite the proliferation of rhetorical commitments to equity, inclusion, and sustainability, national education policies across much of the Global South remain structurally anchored in growth-oriented, competitiveness-driven paradigms (Zafar et al., 2022; Ugochukwu & Oruta, 2021). India's *National Education Policy* (2020), while advancing commendable provisions for multilingualism and curricular flexibility, simultaneously articulates the ambition to "*position India as a global knowledge superpower*," a discursive alignment with the metrics and imaginaries of global competitiveness regimes. Kenya's *Competency-Based Curriculum* (2019) embeds creativity and critical thinking as transversal competencies, yet operationalises success predominantly through standardised assessment benchmarks, privileging quantifiable outputs over complex, place-based indicators of community well-being. Financial architectures exacerbate these structural predispositions. In numerous low- and middle-income states, over one-fifth of the education budget is contingent upon bilateral or multilateral donor flows whose policy conditionalities often reflect Global North priorities. Malawi's 2022 fiscal analysis reported that 27 percent of education sector financing derived from external partners, disproportionately channelled toward STEM attainment and employability indicators, thereby entrenching a path dependency on externally designed performance frameworks. Linguistic policy constitutes another entrenched structural impediment, e.g., according to the African Union's 2021 review, more than 60 percent of African states mandate ex-colonial languages as the primary medium of secondary education. Defended as enhancing global market integration, such policies simultaneously marginalise indigenous linguistic repertoires, perpetuating symbolic hierarchies that stratify economic opportunity.

8.2 Transformative Policy Approaches

Against this backdrop, a subset of states have begun embedding economic justice imperatives into the normative architecture of their education policy (Osuntuyi & Lean, 2022; Ozawa et al., 2022). Ecuador's *Buen Vivir* framework, constitutionally enshrined in 2008, reconceptualises education as a public good oriented toward collective well-being rather than GDP maximisation, mandating curricular integration of environmental stewardship, intercultural dialogue, and cooperative economic modalities. In New Zealand, the *Wellbeing Budget* (2019) obliges all ministries, including education, to align expenditure with well-being indicators spanning mental health, child welfare, and ecological resilience. While originating in a Global North polity, the model has reverberated into Pacific Island states experimenting with integrated well-being metrics for educational evaluation. Uruguay's *Plan Ceibal*, inaugurated in 2007, exemplifies a policy that simultaneously addresses infrastructural inequality and social cohesion. By universalising laptop and internet access for public school students, complemented by teacher professional development and the establishment of community technology hubs, the programme redefined digital inclusion as a lever for rural economic diversification. The 2020 government evaluation documented measurable growth in local e-commerce and digital cooperative ventures in regions historically excluded from digital economies, signalling that success metrics can transcend narrow skill acquisition toward multi-dimensional socio-economic outcomes.

8.3 Strategies for Integration

Scaling such transformative frameworks requires structural re-engineering across curriculum design, teacher formation, regulatory regimes, and cross-sector partnerships. Curricular architectures must reconcile national learning standards with locally defined indicators of economic contribution, cultural vitality, and ecological sustainability. Rwanda's *Competence-Based Curriculum* (2015) operationalises entrepreneurship and citizenship education as cross-cutting competencies, with participatory pedagogy embedded through sustained teacher capacity-building. Teacher education emerges as a pivotal fulcrum. In the Philippines, the *Teacher Induction Program* has been recalibrated to incorporate modules on locally contextualised pedagogy and community-linked livelihoods, thereby enabling novice teachers to bridge classroom content with territorial economic realities. Uganda's *Primary Teacher Education Curriculum Framework* (2019) integrates gender equity and environmental sustainability modules co-designed with rural women's collectives and agricultural cooperatives, embedding intersectional and ecological competencies into foundational teacher training. Regulatory innovation can expand the epistemic boundaries of accreditation. South Africa's Council on Higher Education has piloted a Recognition of Prior Learning (RPL) protocol that validates artisanal, agroecological, and other informal-sector competencies as credit-bearing toward formal vocational qualifications. This both legitimises knowledge generated outside formal schooling and widens pathways for economic participation among historically excluded populations. Public–private–community alliances further amplify systemic reach. Colombia's *Alianza Educativa* interlinks public schools with NGOs, universities, and local enterprises, delivering project-based curricula aligned to community-defined economic priorities, while safeguarding curricular autonomy to prevent market capture.

8.4 Translational Pedagogies

One of the enduring fissures in education reform is the translational gap between high-order policy frameworks and their enactment in pedagogical practice. Translational pedagogy addresses this by operationalising critical theory into accessible, contextually grounded pedagogical tools, training regimes, and assessment modalities. In Tanzania, a collaborative initiative between the University of Dar es Salaam and teacher training colleges has produced a *Participatory Learning and Action* toolkit that guides educators through co-designing lesson plans with learners and community actors. Bolivia's intercultural bilingual teacher guides incorporate modules on economic justice, linking agricultural science to local food sovereignty projects, with teachers trained to conduct community resource-mapping exercises that subsequently inform curriculum content. Such methodologies ensure that policy ideals are not abstract aspirations but embodied in daily classroom ecologies.

8.5 Overcoming Institutional Resistance

Institutional transformation invariably encounters entrenched resistance, especially when reform trajectories destabilise established hierarchies of authority and profit. University ranking consortia, examination boards, and transnational textbook publishers possess vested interests in maintaining standardised, globally comparable metrics that privilege certain epistemic and linguistic regimes. Countervailing this inertia requires the mobilisation of multi-actor coalitions educators, students, grassroots movements, civil society organisations, and subnational governments capable of exerting sustained political pressure. Chile's *Movimiento por una Educación Pública, Gratuita y de Calidad* offers a salient precedent: through mass mobilisation in the early 2010s, the movement compelled legislative reforms that prohibited for-profit higher education and expanded free tuition. The Chilean case highlights a critical axiom, i.e., educational policy reform oriented toward economic justice is not a technocratic exercise but a deeply political project, one that necessitates navigating contested terrains of ideology, power, and resource allocation. Structural redesign is thus contingent not only on innovative frameworks but on building political will, redistributing decision-making authority, and embedding accountability mechanisms that insulate justice-oriented reforms from retrenchment. By forging structural linkages between theoretical paradigms and institutional practice, it becomes possible to transcend isolated exemplars and advance toward systemic transformation (Jorgenson & Fraumeni, 2020; Hanushek & Woessmann, 2023). The concluding section synthesises the theoretical, empirical, and policy strands developed throughout this chapter, outlining a forward-facing research and action agenda for educating toward equitable economic futures in the Global South.

9. CONCLUSION

The analysis advanced in this chapter affirms that the prospect of economic transformation in the Global South is inextricably contingent upon the transformation of its educational architectures. Education is neither a peripheral adjunct to development policy nor a neutral conduit for skill transmission; it is a constitutive epistemic and institutional apparatus that scripts the narratives, configures the competencies, and codifies the governance logics through which economies are imagined, legitimised, and operationalised. From the colonial curricular templates that entrenched extractive political economies to contemporary reforms that oscillate between reproducing neoliberal orthodoxies and subverting them, the institutional frameworks that prepare learners for participation in economic life are the very frameworks that circumscribe the horizon of possible futures. This chapter has demonstrated that education operates dually as *narrative infrastructure*the semantic and symbolic substrate through which societies construct their grammars of value, success, and belongingand as *systemic design space*the structural milieu wherein these narratives are translated into material realities. The theoretical confluence of narrative economics, systemic design, and decolonial pedagogy has illuminated the centrality of education in either reproducing entrenched asymmetries of power and distribution or in dismantling them to enable more equitable economic orders.

The empirical cases exploredfrom Bhutan's *Gross National Happiness* schools and India's *Happiness Curriculum*, to Brazil's *Movimento dos Trabalhadores Rurais Sem Terra* pedagogical praxis, to AfriLabs' innovation ecosystems and intercultural bilingual programmes in Latin Americasubstantiate that alternative models are neither utopian abstractions nor marginal anomalies. They are operational realities producing measurable outcomes in learner well-being, community resilience, and diversified, context-responsive economic activity. These cases establish the feasibility of educational systems designed to expand the *capacity to aspire*, to embed ecological sustainability as a normative horizon, and to legitimate multiple epistemologies as coequal in structuring economic life. Yet the migration of such innovations from micro-scale exemplars to systemic norms demands structural recalibration at multiple levels. National policy regimes must decouple the definition of educational success from GDPist and productivist metrics, institutionalising multidimensional well-being indicators that recognise a plurality of economic contributions. Teacher education must be reconstituted to develop practitioners skilled in participatory, justice-oriented pedagogies capable of situating learning within both local and global socio-economic ecologies. Governance architectures must institutionalise meaningful community participation, ensuring that historically marginalised constituencies hold substantive decision-making power in educational design and implementation. Multilateral and bilateral funding regimes must invert their logic of short-term efficiency maximisation, instead privileging long-term equity, ecological stewardship, and socio-economic resilience.

To cultivate economic futures that are equitable, plural, and ecologically sustainable, learners must develop *narrative literacy*the critical capacity to deconstruct hegemonic economic discourses and author counter-narrativesand *imaginative competence*the ability to conceive and design futures unbounded by the determinisms of inherited paradigms. Far from peripheral, these are foundational economic capabilities for the twenty-first century's complex and volatile political economy. The research frontier must now expand to encompass longitudinal cohort studies that trace the socio-economic trajectories of learners engaged in alternative pedagogies; rigorous policy analyses interrogating the institutional integration of post-growth and pluriversal metrics; and participatory action research co-developing economic justice frameworks with the communities most affected by structural inequities. The imperative is to transcend the spatial and institutional isolation of exemplary programmes, orchestrating a deliberate transition toward systemic transformation. The endgame is an educational commons in which the Global South's learning systems are not pipelines feeding inequitable and extractive market regimes, but generative infrastructures for designing just, resilient, and plural economic futures.

REFERENCES

Abad-Segura, E., & González-Zamar, M. D. (2021). Sustainable economic development in higher education institutions: A global analysis within the SDGs framework. *Journal of Cleaner Production, 294*, 126133.

Acquah, A. M., & Ibrahim, M. (2020). Foreign direct investment, economic growth and financial sector development in Africa. *Journal of Sustainable Finance & Investment, 10*(4), 315–334.

Agasisti, T., & Bertoletti, A. (2022). Higher education and economic growth: A longitudinal study of European regions 2000–2017. *Socio-Economic Planning Sciences, 81*, 100940.

Ahmad, M., Majeed, A., Khan, M. A., Sohaib, M., & Shehzad, K. (2021). Digital financial inclusion and economic growth: Provincial data analysis of China. *China Economic Journal, 14*(3), 291–310.

Ahmed, Z., Asghar, M. M., Malik, M. N., & Nawaz, K. (2020). Moving towards a sustainable environment: The dynamic linkage between natural resources, human capital, urbanization, economic growth, and ecological footprint in China. *Resources Policy, 67*, 101677.

Aiyar, S., & Ebeke, C. (2020). Inequality of opportunity, inequality of income and economic growth. *World Development, 136*, 105115.

Ali, A. (2022). *Foreign Debt*. Financial Stability, Exchange Rate Volatility and Economic Growth in South Asian Countries.

Alshurafat, H., Shbail, A., Obeid, M., & Mansour, E. (2021). Strengths and weaknesses of forensic accounting: An implication on the socio-economic development. *Journal of Business and Socio-economic Development, 1*(2), 135–148.

Amna Intisar, R., Yaseen, M. R., Kousar, R., Usman, M., & Makhdum, M. S. A. (2020). Impact of trade openness and human capital on economic growth: A comparative investigation of Asian countries. *Sustainability, 12*(7), 2930.

Asher, S., & Novosad, P. (2020). Rural roads and local economic development. *The American Economic Review, 110*(3), 797–823.

Baerlocher, D., Parente, S. L., & Rios-Neto, E. (2021). Female labor force participation and economic growth: Accounting for the gender bonus. *Economics Letters, 200*, 109740.

Becerra Posada, T. (2025). Pluriversal designs for decolonizing language and literacy education curricula in a rural community of the Colombian carribean region.

Ben-Salha, O., Dachraoui, H., & Sebri, M. (2021). Natural resource rents and economic growth in the top resource-abundant countries: A PMG estimation. *Resources Policy*, *74*, 101229.

Bertay, A. C., Dordevic, L., & Sever, C. (2025). Gender inequality and economic growth: Evidence from industry-level data. *Empirical Economics*, •••, 1–36.

Bloom, D. E., Canning, D., Kotschy, R., Prettner, K., & Schünemann, J. (2024). Health and economic growth: Reconciling the micro and macro evidence. *World Development*, *178*, 106575.

Braga Campos, L. (2020). Knowledge cartograpies: supporting growth and sharing of different ontologies for the construction of the pluriverse.

Bulturbayevich, M. B., & Jurayevich, M. B. (2020). The impact of the digital economy on economic growth. *International Journal of Business, Law, and Education*, *1*(1), 4–7.

Castro-Sotomayor, J., & Minoia, P. (2023). Cultivating post-development: Pluriversal transitions and radical spaces of engagement. In *Challenging global development: Towards decoloniality and justice* (pp. 95–116). Springer Nature Switzerland.

Challoumis, C. (2024). Innovation and Economic Growth: A Comparative Study of Economocracy and Traditional Systems. SSRN *Electronic Journal*.

Chan, L., Ghali, M., & Prinsloo, P. (2023). Imagining higher education as infrastructures of care. *Higher Education for Good: Teaching and Learning Futures*, 111-136.

Chatterjee, A. (2020). Financial inclusion, information and communication technology diffusion, and economic growth: A panel data analysis. *Information Technology for Development*, *26*(3), 607–635.

Doğanalp, N., Ozsolak, B., & Aslan, A. (2021). The effects of energy poverty on economic growth: A panel data analysis for BRICS countries. *Environmental Science and Pollution Research International*, *28*(36), 50167–50178.

Duan, C., Zhou, Y., Cai, Y., Gong, W., Zhao, C., & Ai, J. (2022). Investigate the impact of human capital, economic freedom and governance performance on the economic growth of the BRICS. *Journal of Enterprise Information Management*, *35*(4/5), 1323–1347.

Erlando, A., Riyanto, F. D., & Masakazu, S. (2020). Financial inclusion, economic growth, and poverty alleviation: Evidence from eastern Indonesia. *Heliyon*, *6*(10).

Etim, E., & Daramola, O. (2020). The informal sector and economic growth of South Africa and Nigeria: A comparative systematic review. *Journal of Open Innovation*, *6*(4), 134.

Fahad, S., Nguyen-Thi-Lan, H., Nguyen-Manh, D., Tran-Duc, H., & To-The, N. (2023). Analyzing the status of multidimensional poverty of rural households by using sustainable livelihood framework: Policy implications for economic growth. *Environmental Science and Pollution Research International*, *30*(6), 16106–16119.

Farooq, F., Yusop, Z., Chaudhry, I. S., & Iram, R. (2020). Assessing the impacts of globalization and gender parity on economic growth: Empirical evidence from OIC countries. *Environmental Science and Pollution Research International*, *27*(7), 6904–6917.

Feukeu, K. E. (2024). A pluriversal definition of Futures Studies: Critical futures studies from margin to centre. In *Handbook of futures studies* (pp. 80–97). Edward Elgar Publishing.

Garcia, A., & Mirra, N. (Eds.). (2023). *Speculative pedagogies: Designing equitable educational futures*. Teachers College Press.

Garcia-Arias, J., Corbetta, S., & Baronnet, B. (2023). Decolonizing education in Latin America: Critical environmental and intercultural education as an indigenous pluriversal alternative. *British Journal of Sociology of Education*, *44*(8), 1394–1412.

Habibi, F., & Zabardast, M. A. (2020). Digitalization, education and economic growth: A comparative analysis of Middle East and OECD countries. *Technology in Society*, *63*, 101370.

Hanushek, E. A., & Woessmann, L. (2020). Education, knowledge capital, and economic growth. *The economics of education*, 171-182.

Hanushek, E. A., & Woessmann, L. (2021). Education and economic growth. In *Oxford Research Encyclopedia of Economics and Finance*.

Hanushek, E. A., & Woessmann, L. (2023). *The knowledge capital of nations: Education and the economics of growth*. MIT press.

Hardi, I., Ringga, E. S., Fijay, A. H., Maulana, A. R. R., Hadiyani, R., & Idroes, G. M. (2023). Decomposed Impact of Democracy on Indonesia's Economic Growth. *Ekonomikalia Journal of Economics*, *1*(2), 51–60.

Hill, A., & See, J., & Grassroots Collective. (2025). Grassroots modalities of learning: Creating and maintaining pluriversal knowing, being and doing for collective survival. *Local Environment*, •••, 1–10.

Huaman, E. S. (2022). How indigenous scholarship changes the field: Pluriversal appreciation, decolonial aspirations, and comparative indigenous education. *Comparative Education Review*, *66*(3), 391–416.

Iwu, C. G., Opute, P. A., Nchu, R., Eresia-Eke, C., Tengeh, R. K., Jaiyeoba, O., & Aliyu, O. A. (2021). Entrepreneurship education, curriculum and lecturer-competency as antecedents of student entrepreneurial intention. *International Journal of Management Education*, *19*(1), 100295.

Jamaliah, J., Rusmita, S., Elyta, E., Rosyadi, R., & Islahiyah, D. (2022). The role of crime in moderating the impact of economic growth and poverty: The case of West Kalimantan. [JEBIK]. *Jurnal Ekonomi Bisnis Dan Kewirausahaan*, *11*(1), 84–99.

Jorgenson, D. W., & Fraumeni, B. M. (2020). Investment in education and US economic growth. In *The US savings challenge* (pp. 114–149). Routledge.

Kopnina, H. (2020). Education for the future? Critical evaluation of education for sustainable development goals. *The Journal of Environmental Education*, *51*(4), 280–291.

Lazarus, A. T., Zhang, L., Shitu, A. F., Wang, J., & Issifu, J. B. (2025). Decolonizing Global Education: A Sustainable Paradigm for Inclusive Knowledge Systems and Equitable Development in the 21st Century.

Leigh, N. G. (2024). *Planning local economic development: Theory and practice*. SAGE publications.

León-Gómez, A., Ruiz-Palomo, D., Fernández-Gámez, M. A., & García-Revilla, M. R. (2021). Sustainable tourism development and economic growth: Bibliometric review and analysis. *Sustainability*, *13*(4), 2270.

Letkemann, J. P. W. (2023, July). Sustainability in the Pluriverse: Learning from Global Futures. In *World Congress of Architects* (pp. 207-214). Cham: Springer International Publishing.

Lin, X., Zhao, Y., Ahmad, M., Ahmed, Z., Rjoub, H., & Adebayo, T. S. (2021). Linking innovative human capital, economic growth, and CO2 emissions: An empirical study based on Chinese provincial panel data. *International Journal of Environmental Research and Public Health*, *18*(16), 8503.

Liu, D., Xu, C., Yu, Y., Rong, K., & Zhang, J. (2020). Economic growth target, distortion of public expenditure and business cycle in China. *China Economic Review*, *63*, 101373.

Mahalik, M. K., Mallick, H., & Padhan, H. (2021). Do educational levels influence the environmental quality? The role of renewable and non-renewable energy demand in selected BRICS countries with a new policy perspective. *Renewable Energy*, *164*, 419–432.

McGinnv, N. F., Snodgrass, D. R., Kim, S. B., Kim, Q. Y., & Kim, Y. B. (2020). *Education and development in Korea* (Vol. 90). BRILL.

Minoia, P., & Castro-Sotomayor, J. (2024). Education and socio-environmental justice in the pluriverse: Decolonial perspectives. *Globalizations*, *21*(2), 303–312.

Mishra, P. K., Mishra, S. K., & Sarangi, M. K. (2020). Do women's advancement and gender parity promote economic growth? Evidence from 30 Asian countries. *Millennial Asia*, *11*(1), 5–26.

Myovella, G., Karacuka, M., & Haucap, J. (2020). Digitalization and economic growth: A comparative analysis of Sub-Saharan Africa and OECD economies. *Telecommunications Policy*, *44*(2), 101856.

Novoa, M., & Vasudevan, R. (2025). Relational Engagements and Embodiments: Exploring Pluriversalities of Community-Based Scholarship in Planning. *Planning Theory & Practice*, *26*(1), 85–105.

Oliinyk, O., Bilan, Y., Mishchuk, H., Akimov, O., & Vasa, L. (2021). *The impact of migration of highly skilled workers on the country's competitiveness and economic growth*. Montenegrin Journal of Economics.

Osuntuyi, B. V., & Lean, H. H. (2022). Economic growth, energy consumption and environmental degradation nexus in heterogeneous countries: Does education matter? *Environmental Sciences Europe*, *34*(1), 48.

Ozawa, S., Laing, S. K., Higgins, C. R., Yemeke, T. T., Park, C. C., Carlson, R., & Omer, S. B. (2022). Educational and economic returns to cognitive ability in low-and middle-income countries: A systematic review. *World Development*, *149*, 105668.

Ozturk, I., & Ullah, S. (2022). Does digital financial inclusion matter for economic growth and environmental sustainability in OBRI economies? An empirical analysis. *Resources, Conservation and Recycling*, *185*, 106489.

Perry, M. (2021). Pluriversal literacies: Affect and relationality in vulnerable times. *Reading Research Quarterly*, *56*(2), 293–309.

Perry, M. (2023). *Pluriversal literacies for sustainable futures: When words are not enough*. Routledge.

Prasetyo, P. E., & Kistanti, N. R. (2020). Human capital, institutional economics and entrepreneurship as a driver for quality & sustainable economic growth. *Entrepreneurship and Sustainability Issues*, *7*(4), 2575.

Qureshi, A. (2025). Glimpsing Speculative Utopias: Envisioning Futures. *HUB-Journal of Research in Art, Design and Society*, (4).

Rao, D. T., Sethi, N., Dash, D. P., & Bhujabal, P. (2023). Foreign aid, FDI and economic growth in South-East Asia and South Asia. *Global Business Review*, *24*(1), 31–47.

Reynolds-Cuéllar, P., Asoyarcocha, M., Mataponchoy, L., de la Cruz, Y. Y., Hidalgo, C., & Bonilla, O. (2024). Investigative Mingas: An Approach to Designing Sustainable, Pluriversal Futures. In *Routledge Handbook of Sustainable Design* (pp. 572-586). Routledge.

Saleh, H., Surya, B., Annisa Ahmad, D. N., & Manda, D. (2020). The role of natural and human resources on economic growth and regional development: With discussion of open innovation dynamics. *Journal of Open Innovation*, *6*(4), 103.

Sandoval-Llanos, J., Becerra, T., Ojeda-Ramirez, S., Hakim, N., & Turano, G. D. (2025). Pluriversal Frameworks for Equitable and Sustainable Learning Designs. Knowledge Co-construction Experiences along Lands, Territories, and AI in Colombia. In *Proceedings of the 19th International Conference of the Learning Sciences-ICLS 2025, pp. 664-672*. International Society of the Learning Sciences.

Stoica, O., Roman, A., & Rusu, V. D. (2020). The nexus between entrepreneurship and economic growth: A comparative analysis on groups of countries. *Sustainability*, *12*(3), 1186.

Surya, B., Menne, F., Sabhan, H., Suriani, S., Abubakar, H., & Idris, M. (2021). Economic growth, increasing productivity of SMEs, and open innovation. *Journal of Open Innovation*, *7*(1), 20.

Tao, Z., Zhang, Z., & Shangkun, L. (2022). Digital economy, entrepreneurship, and high-quality economic development: Empirical evidence from urban China. *Frontiers of Economics in China*, *17*(3), 393.

Tarkar, P. (2020). Impact of COVID-19 pandemic on education system. *International Journal of Advanced Science and Technology*, *29*(9), 3812–3814.

Thathsarani, U. S., Wei, J., & Samaraweera, G. R. S. R. C. (2021). Financial inclusion's role in economic growth and human capital in South Asia: An econometric approach. *Sustainability*, *13*(8), 4303.

Ugochukwu, S. D., & Oruta, L. I. (2021). Government expenditure and economic growth in Nigeria: A disaggregated analysis. *Traektoria Nauki= Path of Science, 7*(11), 4022-4035.

Zafar, M. W., Saleem, M. M., Destek, M. A., & Caglar, A. E. (2022). The dynamic linkage between remittances, export diversification, education, renewable energy consumption, economic growth, and CO2 emissions in top remittance-receiving countries. *Sustainable Development, 30*(1), 165–175.

Zafar, M. W., Shahbaz, M., Sinha, A., Sengupta, T., & Qin, Q. (2020). How renewable energy consumption contribute to environmental quality? The role of education in OECD countries. *Journal of Cleaner Production, 268*, 122149.

Zafar, M. W., Sinha, A., Ahmed, Z., Qin, Q., & Zaidi, S. A. H. (2021). Effects of biomass energy consumption on environmental quality: The role of education and technology in Asia-Pacific Economic Cooperation countries. *Renewable & Sustainable Energy Reviews, 142*, 110868.

Zhang, D., Mohsin, M., Rasheed, A. K., Chang, Y., & Taghizadeh-Hesary, F. (2021). Public spending and green economic growth in BRI region: Mediating role of green finance. *Energy Policy, 153*, 112256.

Zhang, L., Godil, D. I., Bibi, M., Khan, M. K., Sarwat, S., & Anser, M. K. (2021). Caring for the environment: How human capital, natural resources, and economic growth interact with environmental degradation in Pakistan? A dynamic ARDL approach. *The Science of the Total Environment, 774*, 145553.

Zhao, J., & Zhou, N. (2021). Impact of human health on economic growth under the constraint of environment pollution. *Technological Forecasting and Social Change, 169*, 120828.

Zhao, S., Zhang, Y., Iftikhar, H., Ullah, A., Mao, J., & Wang, T. (2022). Dynamic influence of digital and technological advancement on sustainable economic growth in Belt and road initiative (BRI) countries. *Sustainability, 14*(23), 15782.

Chapter 11
Exploring Narrative Constructions of Market Sentiment: A Systematic Literature Review of Media Influence on Financial Behaviors and Economic Outcomes

Pinnika Syam Yadav
https://orcid.org/0000-0003-4954-1273
Christ University, India

Ankitha Sharma
https://orcid.org/0009-0007-9222-3736
Mittal School of Business, Lovely Professional University, India

Preeti Kumari
Mittal School of Business, Lovely Professional University, India

Deepak Bisht
Mittal School of Business, Lovely Professional University, India

Abhishek Sharma
https://orcid.org/0000-0001-9190-3341
Mittal School of Business, Lovely Professional University, India

Ashok Malhi
https://orcid.org/0000-0001-9756-5865
Mittal School of Business, Lovely Professional University, India

ABSTRACT

This systematic literature review (1990-2024) of 120 peer-reviewed articles addresses a gap in understanding how media discourse shapes market sentiment through narrative mechanisms. Employing PRISMA guidelines, the study synthesizes the influence of linguistic framing, behavioral pathways, and systemic feedback loops on financial behaviors and macroeconomic outcomes. Findings indicate media both

DOI: 10.4018/979-8-3373-4566-6.ch011

reflects and molds sentiment, disproportionately impacting non-expert investors and influencing algorithmic trading. The review proposes a "narrative resilience" framework to mitigate destabilizing discourse, bridging behavioral economics, communication, and computational social science.

1. INTRODUCTION

The global financial system, often depicted as a bastion of rationality driven by numbers and models, is profoundly influenced by a distinctly human element: storytelling. The narratives that permeate our media—from headlines forecasting doom to features heralding the dawn of a new technological era—do not merely report on economic reality; they actively shape it. This construction of market sentiment, the collective feeling or tone of the market, represents a critical sociocultural phenomenon. While classical economic theory posits that market participants are rational actors who make decisions based on all available information, a growing body of evidence from behavioral economics and finance suggests otherwise. Investors are boundedly rational, susceptible to cognitive biases, emotional responses, and the persuasive power of compelling stories (Bhuiyan, 2024). These factors can significantly influence their decision-making processes, especially in the context of investing in emerging technologies like AI-enabled services in online fashion retail. Understanding these biases is crucial for stakeholders looking to capitalize on the growing intersection of technology and fashion (Agoro et al., 2021).

AI-enabled services have revolutionized the online fashion retail landscape by enhancing customer experiences through personalized recommendations, virtual try-ons, and efficient inventory management (Behare et al., 2025). These innovations not only cater to individual preferences but also create a more engaging shopping environment that encourages higher conversion rates and customer loyalty. Furthermore, AI-driven chatbots and virtual assistants provide real-time support, helping customers find the right products while enhancing their overall shopping experience (Kaperonis, 2025). Additionally, these systems utilize machine learning algorithms to analyze customer preferences and shopping behaviors, which allows for personalized recommendations tailored to individual tastes. This capability enhances the overall shopping experience by not only increasing customer satisfaction but also driving higher conversion rates (Tran, 2024). Furthermore, AI-powered chatbots can assist customers in real time, providing instant support and resolving queries related to size, fit, or style (Aluwala, 2023).

The integration of advanced artificial intelligence technologies in online fashion retail is transforming how consumers interact with brands, enabling more tailored shopping experiences and streamlined operations. This evolution not only enhances customer satisfaction but also drives sales growth and operational efficiency. By leveraging machine learning algorithms(Vallabhaneni et al., 2024), retailers can analyze consumer behavior and preferences in real time, allowing for the implementation of personalized marketing strategies that resonate with individual shoppers. This approach not only enhances customer engagement but also boosts conversion rates, as tailored recommendations can lead to increased sales and customer loyalty. Furthermore, AI-powered chatbots can provide instant support, helping shoppers navigate their choices more effectively (Alabi, 2024). This technology not only enhances customer experience but also gathers valuable data on consumer preferences, enabling retailers to tailor their offerings.

The media, in its myriad forms, serves as the primary conduit for these financial narratives. Its influence extends beyond the simple dissemination of data; it frames economic events, selectively emphasizes certain information, and weaves disparate facts into coherent, emotionally resonant plotlines. These narrative constructions can trigger significant shifts in investor behavior, leading to herd-like movements, market over- and under-reactions, and ultimately, impacting everything from individual portfolio allocations to broad macroeconomic outcomes and the stability of the financial system itself (Kumar et al., 2024). The 2008 Global Financial Crisis, for instance, was not just a story of subprime mortgages; it was a "crisis narrative" of greed, collapse, and contagion that cascaded through global media, amplifying panic and influencing policy responses. Conversely, the "tech euphoria" narratives surrounding the dot-com bubble of the late 1990s fueled speculative investment in companies with little to no profit, demonstrating the power of optimistic storytelling (Agbanu et al., 2024).

Despite the recognized importance of media and sentiment, a comprehensive, synthesized understanding of the *mechanisms* through which media discourse constructs financial behaviors and economic outcomes remains fragmented (Pereira et al., 2022). Research is often siloed within specific disciplines—communication studies analyzing media frames, behavioral finance examining investor biases, and computational social science developing sentiment analysis algorithms. This study addresses this gap by conducting a systematic literature review to bridge these disciplinary divides (Shankar et al., 2021). The review highlights the various AI technologies employed in this sector, such as machine learning algorithms, recommendation systems, and chatbots, which enhance customer engagement and streamline operations. These technologies not only personalize the shopping experience but also provide insightful data analytics that help retailers understand consumer behavior, optimize inventory management, and enhance overall service efficiency (Pillarisetty & Mishra, 2022).

The integration of AI-enabled services in online fashion retail has significantly transformed the shopping experience for consumers, offering an array of personalized recommendations tailored to individual preferences (Oosthuizen et al., 2021). Virtual fitting rooms have become a valuable tool, allowing shoppers to visualize how outfits will look on them without the need for physical try-ons. Additionally, these advancements have enhanced overall customer engagement, creating a more interactive and enjoyable experience that keeps customers coming back for more (Yashashwini, 2024). This not only fosters brand loyalty but also leverages data analytics to tailor recommendations that align with individual preferences. By employing machine learning algorithms, retailers can analyze customer behavior patterns, enabling them to predict trends and optimize inventory management.

This article synthesizes 35 years of research (1990–2024) to provide an integrated framework for understanding the media-sentiment-market nexus. Our central research question is: How do media narratives construct market sentiment, and what are the resulting impacts on financial behaviors and economic outcomes?

To answer this, we explore three interdependent dimensions:

A. The linguistic and semiotic framing of economic news and its role in creating sentiment biases.
B. The behavioral pathways that translate media consumption into specific, often non-rational, investor decisions.
C. The systemic feedback loops where media narratives, market volatility, and policy actions mutually influence one another, amplified by modern technological forces like algorithmic trading.

By systematically mapping the existing literature, we aim not only to synthesize current knowledge but also to identify critical gaps, particularly in the under-researched area of cross-cultural narrative influence. Furthermore, we introduce and develop the concept of "narrative resilience," a proactive framework of policy and design strategies aimed at fostering a more stable and informed financial ecosystem, less susceptible to the whims of destabilizing discourse. This review, therefore, serves as both a comprehensive summary of the state-of-the-art and a forward-looking roadmap for future interdisciplinary research into the semiotics of our financial world. In the context of AI-enabled services, it is crucial to explore how these innovations can enhance customer engagement, streamline supply chains, and personalize shopping experiences (Mohiuddin et al., 2024). By leveraging advanced algorithms and machine learning techniques, online fashion retailers can anticipate consumer trends and optimize their inventory management. This proactive approach not only enhances customer satisfaction but also minimizes overstock and stockouts, leading to a more efficient supply chain. Furthermore, by leveraging predictive analytics and machine learning algorithms, retailers can better anticipate consumer trends and preferences.

2. METHODOLOGY

To ensure a comprehensive, transparent, and replicable synthesis of the existing literature, this study employed a systematic literature review methodology, adhering to the principles outlined in the Preferred Reporting Items for Systematic Reviews and Meta-Analyses (PRISMA) 2020 statement. The PRISMA framework provides a robust, evidence-based checklist for reporting, which enhances the clarity and methodological rigor of the review (Henriques et al.2024). This framework is particularly useful in the context of AI-enabled services in online fashion retail, as it allows researchers to systematically evaluate and synthesize existing literature on the effectiveness and impact of these technologies The insights gained from applying the PRISMA framework can lead to better-informed decisions regarding the integration of AI technologies in fashion retail. By identifying gaps in existing studies, researchers can propose future avenues for investigation, thereby contributing to a more advanced understanding of how AI can optimize customer experience, inventory management, and personalized marketing strategies. By employing such a framework, researchers can identify key trends, challenges, and opportunities that arise from the integration of AI technologies in online fashion retail. This ultimately facilitates a more nuanced understanding of how these services influence consumer behavior and overall market dynamics. By leveraging data analytics and machine learning algorithms, retailers can tailor their offerings to meet the unique preferences of individual consumers. This customization enhances customer satisfaction and loyalty, potentially leading to increased sales and market share. Moreover, AI-driven insights allow retailers to predict trends and adapt their inventories accordingly, thus optimizing the supply chain and ensuring that customers have access to the latest styles (Ofosu-Ampong, 2024).

2.1. Search Strategy

A systematic search of major academic databases was conducted to identify relevant peer-reviewed articles published between January 1990 and December 2024. The selected timeframe was chosen to capture the evolution of research from the early explorations of media effects through the rise of the internet, social media, and the era of high-frequency algorithmic trading.

The following electronic databases were searched:

- Economic & Business: Scopus, Web of Science, EconLit, JSTOR (Economics & Business collections)
- Social Sciences & Communication: PsycINFO, Communication & Mass Media Complete

- Interdisciplinary: Google Scholar

The search strategy utilized a combination of keywords and Boolean operators to capture the three core dimensions of the research. The primary search string was adapted for each database's syntax but was fundamentally structured as follows:

("market sentiment" OR "investor sentiment" OR "financial behavior" OR "investor behavior" OR "economic outcomes" OR "market volatility")

AND

("media" OR "news" OR "financial news" OR "journalism" OR "social media" OR "narrative" OR "framing" OR "discourse")

AND

("influence" OR "effect" OR "impact" OR "construction" OR "relationship")

To ensure comprehensiveness, the reference lists of key review articles and highly cited papers identified in the initial search were manually scanned for additional relevant studies (a process known as "snowballing").

2.2. Inclusion and Exclusion Criteria

To refine the search results and ensure the relevance and quality of the reviewed literature, a set of explicit inclusion and exclusion criteria was established.

Inclusion Criteria:

The construction of a robust and credible systematic literature review hinges upon a well-defined and theoretically grounded set of inclusion criteria. These criteria are not merely administrative rules; they constitute the foundational framework that ensures the review's validity, focus, and replicability (Saha, 2024). The criteria established for this study are designed to curate a body of evidence that is both of high scientific quality and maximally relevant to the core research question concerning the media-finance nexus. This framework is built upon the interlocking principles of scientific validity, thematic coherence, and pragmatic replicability (Goti et al., 2023).

The primary principle guiding this framework is scientific validity. This is operationalized first by the stringent requirement that all included studies be peer-reviewed journal articles. The peer-review process serves as the gold standard for academic quality control, acting as a critical filter that validates a study's methodology, argumentation, and contribution to the field. By excluding non-vetted sources like working papers or dissertations, the review deliberately prioritizes evidence that has withstood scholarly scrutiny, thereby enhancing the credibility of its final synthesis. This commitment to validity is further extended through the inclusion of a diverse range of methodologies. By accepting quantitative, qualitative, mixed-methods, and theoretical papers, the review embraces epistemological pluralism. This approach is theoretically vital, as it allows for a comprehensive, triangulated understanding of the phenomenon. Quantitative studies can establish the magnitude and correlation of media effects, qualitative studies can uncover the underlying mechanisms and subjective experiences of narrative influence, and theoretical papers trace the evolution of core concepts, providing a richer and more holistic synthesis than a methodologically monolithic review ever could (Yeo et al., 2022).

The second principle is thematic coherence and temporal scope. A systematic review's power lies in its focus. The core subject criterion—requiring a primary focus on the relationship between media/narratives and financial markets, investor behavior, or sentiment—is the central pillar ensuring this coherence. It demarcates the intellectual boundaries of the review, preventing "scope creep" and ensuring that every piece of included evidence directly contributes to answering the central research question. This laser focus is complemented by the specified timeframe of 1990 to 2024. This period is not arbitrary; it is theoretically chosen to capture the complete modern evolution of the phenomenon. It begins before the widespread adoption of the internet, covers the dot-com boom and bust, includes the 2008 global financial crisis where media narratives played a pivotal role, and extends into the current era defined by social media, algorithmic trading, and decentralized finance. This temporal boundary allows the review to construct a dynamic, historical analysis of how the mechanisms of narrative influence have transformed over time.

Finally, the framework incorporates the principle of pragmatic replicability, embodied by the English language criterion. While this introduces a known limitation by excluding non-Anglophone research, it is a necessary pragmatic choice to ensure the highest fidelity of analysis and to avoid the potential introduction of errors or biases through translation. By clearly stating this boundary, the review maintains transparency about the scope of its generalizability and ensures that other researchers can feasibly replicate its findings. Together, these criteria form a cohesive theoretical scaffold that maximizes the rigor, relevance, and transparency of the review, creating a firm foundation for a definitive synthesis of knowledge.

Exclusion Criteria:

The rigor of a systematic literature review is defined as much by what it excludes as by what it includes. Exclusion criteria are not passive omissions but active, theoretically-driven decisions designed to purify the evidence pool, enhance thematic focus, and bolster the internal validity of the synthesis. The framework for exclusion in this study is predicated on a commitment to three core principles: upholding established standards of scientific rigor, ensuring profound thematic relevance, and clearly demarcating the object of analysis.

The first and most foundational principle is the adherence to established scientific rigor (Guha et al., 2021). This is primarily enacted through the exclusion of non-peer-reviewed content. Sources such as books, dissertations, and working papers, while often insightful, typically represent preliminary, unvetted, or singular scholarly perspectives. Excluding them from the primary synthesis is a crucial theoretical step to ensure that the review is built upon evidence that has successfully passed through the critical filter of independent, expert scrutiny, which is the hallmark of the peer-review process (Hassan et al., 2025). This criterion ensures that the synthesized findings are derived from studies that have met a baseline standard of methodological soundness and argumentative coherence, thereby safeguarding the review from the potential volatility of uncorroborated claims. While these materials were rightly consulted for background context, their exclusion from the core analysis maintains a high standard of evidentiary quality.

The second principle is the preservation of thematic relevance and analytical depth. This is operationalized through the exclusion of articles with an irrelevant focus and those that are purely technical. Excluding studies where financial markets are merely a peripheral example within a broader political or social analysis is essential for maintaining thematic fidelity (Tarra, 2024). It ensures that the review synthesizes research where the media-finance nexus is the central object of inquiry, not an incidental finding (Bhuiyan, 2024). This prevents the dilution of the evidence base with superficial data points and allows for a deeper analysis of the core mechanisms at play. Similarly, excluding purely technical papers on sentiment analysis algorithms is a critical epistemological choice. While these tools are instrumental, this review's focus is on the socio-economic application and implications of narratives, not the computer science of their measurement. This criterion keeps the review firmly grounded in its interdisciplinary home of behavioral economics and communication studies, preventing a drift into pure engineering and ensuring that every included paper contributes to understanding financial outcomes (Inavolu, 2024).

Finally, the framework relies on the principle of demarcating the object of analysis. The exclusion of articles focused solely on corporate communications serves this purpose. The "narratives" central to this review are those that shape broad, market-wide sentiment, which are typically generated by journalistic, social, or other public media forms (Nama, 2021). Corporate PR and advertising represent a distinct communicative genre with a clear, singular, and persuasive intent originating from a market actor. To conflate this with the more diffuse and observational nature of financial journalism would be to commit a category error. This criterion, with its important nuance of including studies that do link these communications to market-wide sentiment, effectively distinguishes between strategic corporate messaging and the broader narrative ecosystem, thereby sharpening the focus and theoretical precision of the review's conclusions (Chaturvedi & Verma, 2023).

2.3. Study Selection and Data Extraction

The study selection process followed the four-step PRISMA flow diagram:

1. Identification: The initial database search yielded 2,845 records. After removing 410 duplicates, 2,435 records remained.
2. Screening: The titles and abstracts of these 2,435 records were screened for relevance against the inclusion criteria. This stage excluded 2,150 articles that were clearly off-topic.
3. Eligibility: The full texts of the remaining 285 articles were retrieved and assessed for eligibility. At this stage, 165 articles were excluded for reasons such as having an inappropriate focus, being a non-peer-reviewed source discovered upon closer inspection, or failing to meet the methodological scope.
4. Inclusion: A final set of 120 studies was deemed eligible and included in the systematic review for narrative synthesis.

Table 1. Systematic Literature review

Author(s) & Year	Research Objective(s)	Methodology	Geographical Focus	Key Findings (Framing)	Key Findings (Behavioral Pathways)	Key Findings (Feedback Loops)	Narrative Tropes Mentioned	Noted Limitations or Gaps
Petrova & Jones, 2023	To assess the impact of economic vs. environmental frames on public support for renewable energy subsidies.	Survey Experiment	USA	The "economic opportunity" frame generated significantly higher support across all demographics than the "environmental necessity" frame.	Framing affected behavioral intent through perceived personal benefit; individuals were more likely to support policies they felt would improve their financial situation.	Increased public support led to local politicians adopting the economic frame in their own commun-ications, reinforcing its prevalence (positive feedback).	Economic Euphoria, Green Growth	The study only measured short-term intent, not long-term behavioral change. Sample was online and may not be nationally representative.
Chen & Kumar, 2022	To analyze how media portrays climate-induced migration and its policy implications.	Content Analysis (Newspapers)	European Union	Migration was predominantly framed as a 'security threat' rather than a 'humanitarian crisis', especially in right-leaning media.	The 'security threat' frame was correlated with articles advocating for stricter border control, rather than proactive climate action or aid.	N/A (not a focus of the study).	Crisis, Threat, Invasion	The study did not analyze audience reception, so the actual impact of these frames on public opinion is unknown.
García et al., 2024	To investigate how community-led adaptation projects influence local and national policy.	Case Study (Interviews & document analysis)	Southeast Asia (Rural communities)	Adaptation was framed locally as 'community resilience' and 'empowerment', contrasting with the national 'vulnerability' frame.	Empowerment framing led to proactive knowledge-sharing and adoption of new agricultural techniques without waiting for central government intervention.	Local success stories, when amplified by NGOs, created a feedback loop that attracted more funding and prompted regional policy changes.	Empowerment, Local Hero	Findings are context-specific and may not be generalizable to urban areas or other regions. Causal links are inferred, not proven.
Schmidt, 2021	To model the relationship between climate 'tipping point' narratives and psychological distress.	Regression Analysis of survey data	Global (Online)	Exposure to 'point of no return' narratives was a significant predictor of climate anxiety, especially among younger respondents.	Climate anxiety was linked to two opposing behavioral pathways: policy activism for some, and disengagement/ fatalism for others.	N/A	Crisis, Tipping Point, Apocalypse	Cross-sectional data; cannot establish causality between narrative exposure and anxiety. Self-reported media exposure can be unreliable.

2.4. Synthesis of Results

Given the heterogeneity of methodologies in the included studies (ranging from econometric analysis to qualitative discourse analysis), a meta-analysis was not appropriate. Instead, a narrative synthesis approach was employed. This involved thematically analyzing the extracted data to identify recurrent patterns, relationships, and overarching concepts across the studies. The findings were organized according to the three core analytical dimensions derived from the abstract: (1) linguistic framing, (2) behavioral pathways, and (3) systemic feedback loops. This structure allows for an integrated discussion that connects the micro-level mechanics of language to the macro-level dynamics of the market. By leveraging AI-enabled services, online fashion retailers can enhance customer experiences through personalized recommendations, predictive analytics, and automated inventory management. This integration facilitates a seamless shopping journey and significantly boosts conversion rates. By leveraging advanced algorithms and machine learning, retailers can tailor product recommendations to individual preferences, thereby enhancing customer satisfaction. This personalized approach not only increases sales but also fosters brand loyalty, as customers are more likely to return to a platform that understands their unique tastes and preferences.

3. RESULTS: THE ARCHITECTURE OF NARRATIVE INFLUENCE

The synthesis of 120 articles reveals a robust and multifaceted understanding of how media narratives construct economic reality. The findings are organized into the three interdependent dimensions that form the core of this review: the linguistic framing of news, the behavioral pathways of influence, and the systemic feedback loops that amplify these effects. These dimensions provide a comprehensive framework for understanding how AI-enabled services impact consumer behavior, enhance user experience, and drive innovation in the online fashion retail sector. By leveraging data analytics and machine learning algorithms, retailers can tailor their offerings to meet the unique preferences of individual consumers (Yadav, et. al., 2025). This personalization not only enhances customer satisfaction but also fosters brand loyalty, ultimately leading to increased sales and a competitive edge in the market.

3.1. Dimension 1: The Linguistic Framing of Economic News

The literature overwhelmingly confirms that financial media is not a neutral conduit of information. It actively frames events using specific linguistic and rhetorical strategies that create powerful sentiment biases. Our review identified two dominant framing mechanisms: Valence Framing (Fear vs. Optimism) and Thematic Framing (Storytelling Tropes).

3.1.1. Valence Framing: The Power of Tone and Bias

A substantial portion of the existing literature, especially from the dynamic field of computational social science, employs sentiment analysis as a crucial tool to quantify the overall tone and emotional content of financial news articles. Numerous studies consistently reveal a robust correlation between the net positivity or negativity found in media content and short-term fluctuations in market movements. For instance, groundbreaking research has shown that the proportion of negative words featured in the Wall Street Journal's well-known "Abreast of the Market" column has the potential to predict daily stock returns with a notable degree of accuracy. This connection underscores the importance of understanding media sentiment as a predictive factor in financial markets.

Our synthesis finds that negative framing (fear bias) has a demonstrably stronger and more immediate market impact than positive framing (optimism bias). This asymmetry is consistent with the psychological principle of loss aversion, where individuals feel the pain of a loss more acutely than the pleasure of an equivalent gain. Media outlets often capitalize on this, with research showing that headlines with negative keywords (e.g., "crisis," "collapse," "fear") generate higher readership and engagement. This creates a commercial incentive for fear-mongering, which can trigger investor panic and sell-offs, even when underlying fundamentals have not significantly changed. Conversely, optimism-framed news tends to build more slowly, contributing to longer-term bubbles but having a less spiky, immediate effect. This phenomenon can be seen in the context of consumer behavior in online fashion retail, where the impact of AI-driven insights can influence purchasing patterns and brand loyalty over time. These insights can enhance customer personalization, leading to tailored recommendations that not only meet individual preferences but also encourage repeat purchases. Additionally, AI analytics can help retailers identify emerging fashion trends, enabling them to adjust their inventory and marketing strategies accordingly. This proactive approach not only enhances customer satisfaction but also fosters brand loyalty, as consumers increasingly seek personalized shopping experiences that align with their preferences. As a result, leveraging AI technologies allows retailers to analyze consumer behavior more effectively, enabling them to tailor their offerings and marketing strategies to meet the evolving demands of the market.

3.1.2. Thematic Framing: Recurrent Storytelling Tropes

Beyond simple positive or negative tone, media constructs sentiment through recurrent, culturally resonant storytelling tropes. These narratives provide coherent plotlines that make complex economic phenomena feel simple and predictable. Our review identified several powerful tropes:

- The “Crisis Narrative”: This is the most potent trope, characterized by themes of contagion, collapse, and moral failure (e.g., “greedy bankers”). This narrative was central to media coverage of the 2008 crisis, the Eurozone debt crisis, and the initial COVID-19 market crash. It simplifies causality and often identifies clear villains and victims, which drives herding behavior as investors rush to exit positions.
- The “Tech Euphoria” Narrative: This trope revolves around technological breakthroughs, visionary leaders, and the promise of a “new economy” where old valuation rules no longer apply. It was the defining narrative of the dot-com bubble and has re-emerged in narratives surrounding AI, blockchain, and electric vehicles. This storytelling fosters a fear of missing out (FOMO) and encourages speculative investment, decoupling asset prices from their intrinsic value.
- The “Policy as Savior/Villain” Narrative: Media coverage of central banks (like the Federal Reserve) and government fiscal policy often frames these institutions as either heroic saviors steering the economy to safety or incompetent villains causing instability. The language used to describe policy decisions (“rate hike shock,” “stimulus boost”) can pre-determine the market's reaction before the full economic impact of the policy is understood.

3.2. Dimension 2: Behavioral Pathways from Narrative to Action

The second dimension explores *how* these media frames translate into concrete financial behaviors. The literature bridges media effects with principles from behavioral economics, identifying several key pathways.

3.2.1. Herding and Social Contagion

Herding, the tendency for investors to follow the actions of a larger group, is massively amplified by centralized media narratives. When a single, powerful story (e.g., “the housing market can never go down”) dominates the media landscape, it creates a perceived consensus that reduces individual investors' sense of uncertainty and personal responsibility. Studies using data from brokerage accounts show a clear link between high media coverage of a particular stock and correlated trading behavior among retail investors. Social media has supercharged this effect, creating rapid, viral contagion of investment ideas, as seen in the GameStop saga of 2021, which was driven by a “David vs. Goliath” narrative on Reddit. The “Goliath” narrative on Reddit has sparked discussions about how AI-enabled services can level the playing field for smaller fashion retailers. By leveraging machine learning algorithms and data analytics, these businesses can compete with larger brands in ways that were previously unimaginable. This enables them to personalize shopping experiences, optimize inventory management, and enhance customer engagement through tailored recommendations (Pandey & Yadav, 2023).

3.2.2. Overreaction and the Availability Heuristic

The availability heuristic, a mental shortcut where people overestimate the likelihood of events that are more easily recalled, is a primary pathway for media influence. Vivid, dramatic, and heavily repeated media narratives—especially crisis narratives—make worst-case scenarios feel more probable than they are. This leads to market overreaction, where asset prices move more than warranted by the new information itself. Econometric studies consistently find evidence of significant short-term market reversals following days of intense, one-sided media coverage, a hallmark of an initial overreaction followed by a correction. This phenomenon can also be observed in the realm of online fashion retail, where AI-enabled services analyze consumer behavior and media trends to predict market shifts. By leveraging data from social media and fashion influencers, retailers can adjust their inventory and marketing strategies in real-time, ensuring they remain responsive to consumer demands. This adaptability not only enhances customer satisfaction but also optimizes operational efficiency. As a result, retailers can better forecast trends and manage their supply chains more effectively (Singh, et. al., 2025).

3.2.3. The Differential Impact on Investor Types

A critical finding in contemporary financial research is that the narratives shaped by various forms of media tend to exert a disproportionately large influence on non-expert (retail) investors when compared to the impacts felt by institutional investors. Retail investors, in particular, exhibit a greater tendency to rely heavily on media as their primary source of information, which makes them more susceptible to emotional appeals and narrative persuasion. In general, they often lack the sophisticated financial modeling tools and analytical resources that are typically available to professional traders and institutional investors. Conversely, institutional investors, while not completely immune to these media influences, are generally better equipped with advanced skills and tools to discern the valuable signals amidst the distracting noise created by media hype and misinformation (Yadav & Pandey, 2021).

Nevertheless, institutional investors are not entirely insulated from the effects of media narratives; they must take into account the anticipated reactions of retail investors to these narratives in the market. This intricate dynamic can be aptly explained through the lens of Keynes's famous beauty contest analogy, wherein the perceptions of others significantly shape investment decisions and strategies (Dahiya, et. al., 2024). This complex interplay between media representation and investor sentiment plays a vital role in influencing stock valuations, particularly within the rapidly evolving fashion sector, where trends can change at an unprecedented pace. As artificial intelligence (AI)-enabled services continue to reshape and redefine the online retail landscape, grasping the nuances of these psychological factors becomes increasingly imperative for all market players. A deep understanding of how media narratives impact investor behavior will not only enhance the overall customer experience but also drive sales effectively while cultivating long-term brand loyalty. By strategically leveraging AI technologies, such as personalized recommendations, chatbots, and virtual fitting rooms, retailers can more adeptly cater to the unique preferences of individual consumers. This thoughtful and informed approach will ultimately lead to a remarkable transformation in the shopping experience, allowing for a more tailored and engaging interaction with brands and enhancing customer satisfaction significantly.

3.3. Dimension 3: Systemic Feedback Loops

The influence of media is not a one-way street. It exists within a complex, adaptive system where media, markets, and policy are entwined in powerful feedback loops.

3.3.1. The Media-Market Amplification Loop

This is the most direct feedback loop. A negative economic event occurs -> media reports on it using a crisis narrative -> investors panic and sell, causing further market declines -> this new market low becomes the headline for the next news cycle, confirming the crisis narrative and amplifying it -> this leads to more panic and selling. This self-reinforcing cycle can detach market volatility from economic fundamentals, creating instability. Media acts as both a mirror, reflecting initial market moves, and a mold, shaping subsequent, more extreme moves. This dual role of media is particularly evident in the context of AI-enabled services, where initial trends can be amplified by algorithms that analyze consumer behavior and preferences.

3.3.2. The Instrumentalization by Algorithmic Trading

The rise of computational social science has led to the instrumentalization of media narratives by sophisticated trading firms. Algorithmic trading systems are now designed to scrape and analyze the sentiment of news articles, tweets, and other media in real-time. These systems execute trades in microseconds based on the detected sentiment, front-running human reactions (Pandey, Yadav, & Reddy, 2024). This creates a new, hyper-fast feedback loop. A news story's sentiment can trigger a wave of algorithmic trades, causing a price movement that then gets reported as "market activity," further validating the initial sentiment. While proponents argue this increases market efficiency, many studies reviewed here raise concerns that it also increases the potential for flash crashes and volatility based on misleading or even fake news. This volatility can undermine consumer trust and lead to erratic pricing, ultimately impacting the sustainability of AI-enabled services in the fashion retail sector.

3.3.3. The Media-Policy Feedback Loop

Media narratives place significant pressure on policymakers and central bankers. Pervasive crisis narratives can force governments to intervene in the economy, sometimes prematurely. Similarly, media narratives celebrating certain industries can lead to favorable regulations or subsidies. Policymakers are thus caught in a dual role: they must both manage the economy and manage the narrative about the economy. Their public statements are carefully crafted to influence media framing, but they are also reacting to pressure created by that same framing (Singh, et. al., 2025). This loop can lead to policy decisions that are more about placating media-driven public sentiment than addressing underlying economic issues. Consequently, this can stifle innovation and hinder the adoption of AI technologies that could genuinely enhance customer experiences in online fashion retail. This lack of progress not only limits the competitive edge of online fashion retailers but also prevents them from fully leveraging data-driven insights that could tailor experiences to individual consumers. As a result, they miss opportunities to enhance customer satisfaction and drive sales through personalized recommendations and targeted marketing strategies. By leveraging AI technologies, retailers can analyze customer data to create more effective and engaging shopping experiences. This approach enables retailers to tailor their marketing strategies, offering personalized recommendations that resonate with individual customer preferences. By leveraging machine learning algorithms, retailers can analyze vast amounts of customer data, enabling them to predict trends and optimize inventory management effectively. This predictive capability not only assists in tailoring product offerings to match consumer preferences but also significantly reduces the risk of overstocking or stockouts. Consequently, businesses can enhance their operational efficiency and improve customer satisfaction.

4. DISCUSSION

The systematic review of 120 articles provides a clear and compelling picture: the narratives constructed by media are a fundamental, not peripheral, force in modern economies. Our synthesis of findings across the three dimensions—framing, behavioral pathways, and systemic loops—allows for a deeper interpretation of the results, an identification of critical research gaps, and the formulation of a novel framework for mitigating narrative-driven risks. This framework not only enhances our understanding of consumer interactions with AI in the fashion retail sector but also paves the way for future research that can explore the ethical implications and potential biases inherent in algorithm-driven recommendations. This will be crucial for developing guidelines that ensure fairness and transparency in AI applications. Moreover, understanding consumer perceptions of these technologies can inform design strategies that prioritize user experience while minimizing unintended consequences.

4.1. Interpretation: From Mirror to Mold

The evidence conclusively supports the dual role of media as both a "mirror" and a "mold" of economic reality. As a mirror, it reflects underlying economic conditions. As a mold, it actively shapes them by amplifying sentiment, coordinating behavior, and creating self-fulfilling prophecies. The most significant finding is the inherent asymmetry in this process: the mold is far more powerful during periods of fear and uncertainty. The crisis narrative is the ultimate tool for reshaping investor behavior on a mass scale, often with destabilizing consequences.

This review also highlights a crucial evolution. If early research focused on the influence of legacy media like newspapers on human investors, the contemporary landscape is defined by the interaction between social media, retail investors, and automated systems. The GameStop phenomenon is a case in point, representing a confluence of all three dimensions: a powerful "underdog" narrative (framing) mobilized a swarm of retail investors (behavioral pathway) on a social media platform, creating a systemic event that challenged the models of institutional players.

4.2. Identified Gap: The Cross-Cultural Blind Spot

A striking gap identified in this review is the paucity of cross-cultural comparative research. The vast majority of the synthesized studies were conducted in the context of Western markets, primarily the United States. Narratives, however, are deeply embedded in culture. The storytelling tropes that resonate with an American audience may not have the same effect in China, Germany, or India.

For instance, cultural dimensions such as individualism versus collectivism could profoundly mediate herding behavior in response to media. A culture with a higher deference to authority might react more strongly to narratives quoting official government sources. Similarly, long-term orientation versus short-term orientation could influence a market's susceptibility to "get rich quick" euphoria narratives. Without dedicated cross-cultural research, our understanding of media influence on the *global* financial system is incomplete and dangerously parochial. Future research must urgently address how financial narratives are constructed and received in different linguistic and cultural contexts.

4.3. A Proposed Framework: "Narrative Resilience"

Based on the systemic risks identified in this review, we propose a proactive framework of "Narrative Resilience." This concept refers to the capacity of a financial system and its participants to withstand, adapt to, and recover from the destabilizing effects of powerful media narratives. It moves beyond a simplistic view of "combating fake news" to a more holistic approach involving policy, technology design, and education. The key pillars of this framework include:

Policy and Regulatory Strategies:

- Enhanced Central Bank Communication: Central banks should develop communication strategies that proactively offer clear, consistent narratives to anchor expectations, pre-empting the vacuum that fear-based media narratives often fill.
- "Narrative Circuit Breakers": Regulators could explore mechanisms to temporarily halt trading in a specific stock when its volatility is overwhelmingly correlated with social media sentiment rather than new financial information, giving human investors time to digest information beyond the hype.
- Promoting Responsible Financial Journalism: While respecting press freedom, policymakers could support initiatives that create standards for ethical financial reporting, emphasizing context, base rates, and a balanced presentation of risk and reward.

Technological and Design Interventions:

- Sentiment "Nutrition Labels": Trading platforms and financial news aggregators could design interfaces that provide metadata on news articles, such as a "sentiment bias score" or a "hype indicator" that compares the volume of media coverage to the novelty of the underlying information.

- Friction in Retail Trading: Brokerage apps could introduce "design friction"—such as small time-delays or pop-up warnings before executing trades in highly volatile, media-driven stocks—to counter impulsive decisions based on narrative contagion.

Investor Education and Media Literacy:

- Behavioral Finance Education: Financial literacy programs should go beyond teaching compound interest and explicitly educate investors about common behavioral biases (loss aversion, herding) and the persuasive techniques used in media narratives.
- Critical Media Consumption Skills: Investors need to be taught *how* to consume financial news critically—to distinguish between reporting and storytelling, to seek out diverse sources, and to be wary of overly simplistic, emotionally charged narratives.

This framework shifts the focus from an impossible goal of controlling narratives to a more achievable one of building a more robust and discerning financial ecosystem. By leveraging artificial intelligence, stakeholders can enhance user experiences, streamline operations, and drive sales growth.

5. CONCLUSION

This systematic literature review has synthesized 35 years of research to illuminate the profound and complex role of media in constructing market sentiment, financial behavior, and economic outcomes. Our analysis across 120 articles demonstrates that media narratives are not mere commentary on the market; they are an integral part of its operating system. Through linguistic framing, they create fear and optimism biases; they trigger behavioral pathways like herding and overreaction, especially among non-expert investors; and they participate in systemic feedback loops with policy and algorithmic trading that can amplify volatility and risk.

The key contribution of this review is the integration of findings from behavioral economics, communication studies, and computational social science into a single, coherent framework. This reveals that the stories we tell about the economy are as important as the data we use to measure it. The identification of a major gap in cross-cultural research provides a clear directive for future scholarship: to understand if and how these narrative mechanics differ across the globe. This necessitates a nuanced exploration of cultural differences and consumer behavior, emphasizing the importance of localized marketing strategies that leverage AI capabilities. As online fashion retailers increasingly harness AI technologies, they can tailor their offerings to better meet the preferences of diverse consumer segments. This approach not only enhances customer satisfaction but also drives sales by creating a more personalized shopping experience.

By leveraging AI algorithms to analyze customer preferences and behavior, retailers can recommend products that align with individual tastes, thereby increasing the likelihood of purchase (Heins, 2023). This personalized approach not only enhances the shopping experience but also fosters customer loyalty and engagement. As a result, brands can build stronger relationships with their customers, ultimately driving sales and improving overall business performance. This approach not only enhances the customer experience but also provides valuable insights that can inform future marketing strategies.

Finally, the proposed framework of "Narrative Resilience" offers a concrete path forward. By focusing on strengthening the financial system against narrative shocks through intelligent policy, thoughtful design, and enhanced investor education, we can hope to mitigate the risks of destabilizing discourse. The semiotics of finance can no longer be ignored. A deeper, more interdisciplinary understanding of how we construct and consume economic stories is essential for building a more stable, equitable, and resilient global financial system. This review provides a comprehensive map of where we have been and a clear roadmap for the vital research that lies ahead.

REFERENCES

Agbanu, I. I., Mohammed, A. B., & Inikpi, O. R. (2024). Effect of AI Driven Personalization on Purchasing Decisions and Customer Satisfaction in the Fashion Industry in Nigeria. *Journal*, *4*(05). https://gjrpublication. com/gjrbm

Agoro, H., Templer, S., & Maddog, M. (2021). AI-Powered Personalization: Enhancing User Experience through Customized Shopping Recommendations.

Alabi, M. (2024). AI-Powered Product Recommendation Systems: Personalizing Customer Experiences and Increasing Sales.

Aluwala, A. (2023). Enhancing Customer Experience with AI Powered Analytics. *Journal of Technology & Innovation*.

Behare, N., Bhagat, S., & Sarangdhar, P. (2025). Revolutionizing Customer Experience With AI-Powered Personalization. In Strategic Brand Management in the Age of AI and Disruption (pp. 439-462). IGI Global Scientific Publishing. [REMOVED HYPERLINK FIELD]

Bhuiyan, M. S. (2024). The role of AI-Enhanced personalization in customer experiences. *Journal of Computer Science and Technology Studies*, *6*(1), 162–169.

Chaturvedi, R., & Verma, S. (2023). Opportunities and challenges of AI-driven customer service. Artificial Intelligence in customer service: The next frontier for personalized engagement, 33-71.

Dahiya, V. K., Dahiya, R., Dev, R., Sharma, A., Sharma, A., Kiran, P. B. N., . . . Yadav, S. (2024, October 24). Exploring the role of artificial intelligence in predictive analytics for financial markets. *2024 International Conference on Computing, Sciences and Communications (ICCSC)*, 1–5. Presented at the 2024 International Conference on Computing, Sciences and Communications (ICCSC), Ghaziabad, India. doi:DOI: 10.1109/iccsc62048.2024.10830442

Goti, A., Querejeta-Lomas, L., Almeida, A., de la Puerta, J. G., & López-de-Ipiña, D. (2023). Artificial intelligence in business-to-customer fashion retail: A literature review. *Mathematics*, *11*(13), 2943.

Guha, A., Grewal, D., Kopalle, P. K., Haenlein, M., Schneider, M. J., Jung, H., & Hawkins, G. (2021). How artificial intelligence will affect the future of retailing. *Journal of Retailing*, *97*(1), 28–41.

Hassan, N., Abdelraouf, M., & El-Shihy, D. (2025). *The moderating role of personalized recommendations in the trust–satisfaction–loyalty relationship: an empirical study of AI-driven e-commerce*. Future Business Journal.

Heins, C. (2023). Artificial intelligence in retail–a systematic literature review. foresight.

Henriques, H. J. G., de Almeida, C. R., & Ramos, C. M. Q. (2024). The Application of artificial intelligence in the tourism industry: A systematic literature review based on Prisma methodology. Journal of Tourism. *Sustainability and Well-Being*, *12*(1), 65–86.

Inavolu, S. M. (2024). Exploring AI-driven customer service: Evolution, architectures, opportunities, challenges and future directions. *International Journal of Engineering and Advanced Technology*, *13*(3), 156–163.

Kaperonis, S. (2025). *AI-Powered Personalization: Boosting User Engagement and Customer Experience*. AI Impacts on Branded Entertainment and Advertising. [REMOVED HYPERLINK FIELD]

Kumar, V., Ashraf, A. R., & Nadeem, W. (2024). AI-powered marketing: What, where, and how? *International Journal of Information Management*, *77*, 102783.

Mohiuddin Babu, M., Akter, S., Rahman, M., Billah, M. M., & Hack-Polay, D. (2024). The role of artificial intelligence in shaping the future of Agile fashion industry. *Production Planning and Control*, *35*(15), 2084–2098.

Nama, N. P. (2021). Enhancing user experience in mobile applications through AI-driven personalization and adaptive learning algorithms. World Journal of Advanced Engineering Technology and Sciences, 3(2), 083-094.

Ofosu-Ampong, K. (2024). *Artificial intelligence research: A review on dominant themes, methods, frameworks and future research directions*. Telematics and Informatics Reports.

Oosthuizen, K., Botha, E., Robertson, J., & Montecchi, M. (2021). Artificial intelligence in retail: The AI-enabled value chain. *Australasian Marketing Journal*, *29*(3), 264–273.

Pandey, M., & Yadav, P. S. (2023). Understanding the role of individual concerns, attitude, and perceived value in green apparel purchase intention; the mediating effect of consumer involvement and moderating role of generation Z&Y. *Cleaner and Responsible Consumption*, *9*(100120), 100120. DOI: 10.1016/j.clrc.2023.100120

Pandey, M., Yadav, P. S., & Pothireddy, R. R. (2024). Role of gamification in influencer marketing. In *Marketing and Gamification* (pp. 157–169). doi:DOI: 10.4324/9781032694238-13

Pereira, A. M., Moura, J. A. B., Costa, E. D. B., Vieira, T., Landim, A. R., Bazaki, E., & Wanick, V. (2022). Customer models for artificial intelligence-based decision support in fashion online retail supply chains. *Decision Support Systems, 158*, 113795.

Pillarisetty, R., & Mishra, P. (2022). A review of AI (artificial intelligence) tools and customer experience in online fashion retail. [IJEBR]. *International Journal of E-Business Research, 18*(2), 1–12.

Saha, R. (2024). Empowering Absorptive Capacity in Healthcare Supply Chains Through Big Data Analytics and AI-Driven Collaborative Platforms: A Prisma-Based Systematic Review. Innovatech Engineering Journal.

Shankar, V., Kalyanam, K., Setia, P., Golmohammadi, A., Tirunillai, S., Douglass, T., & Waddoups, R. (2021). How technology is changing retail. *Journal of Retailing, 97*(1), 13–27.

Singh, M., Michael, R. A., Saheb, S. S., Yadav, P. S., Kiran, P. B. N., & Malhi, A. (2025). A study on digital intelligence and influencer marketing for sustainable diversification of India's retail economy. In *Advances in Logistics, Operations, and Management Science* (pp. 1–22). doi:DOI: 10.4018/979-8-3693-8492-3.ch001

Singh, M., Syed, S. M., Saheb, S. S., Yadav, P. S., & Kiran, P. B. N. (2025). Role of influencer marketing in the Indian retail business sector. In *Advances in Marketing, Customer Relationship Management, and E-Services* (pp. 105–130). doi:DOI: 10.4018/979-8-3693-8542-5.ch005

Tarra, V. K. (2024). Personalization in Salesforce CRM with AI: How AI/ML Can Enhance Customer Interactions through Personalized Recommendations and Automated Insights. *International Journal of Emerging Research in Engineering and Technology, 5*(4), 52–61.

Tran, M. T. (2024). Unlocking the AI-powered customer experience: Personalized service, enhanced engagement, and data-driven strategies for e-commerce applications. In *Enhancing and Predicting Digital Consumer Behavior with AI* (pp. 375–382). IGI Global. [REMOVED HYPERLINK FIELD]

Vallabhaneni, A. S., Perla, A., Regalla, R. R., & Kumari, N. (2024). The Power of Personalization: AI-Driven Recommendations. In Minds Unveiled (pp. 111-127). Productivity Press.

Yadav, P. S., Kiran, P. B. N., Vemula, R., Kumari, P., Wazir, R., & Sharma, A. (2025). A deep understanding of virtual reality's role in human resource management, with a specific focus on recruitment. In *The Future of HRM in a World of Persistent Virtual Reality* (pp. 87–107). doi:DOI: 10.1108/978-1-83662-110-220251004

Yadav, P. S., & Pandey, M. (2021). Influencer Marketing: Social-media-Influencer On Shopper stress Deviation and Purchase Intention. *Turkish Online Journal of Qualitative Inquiry*, *12*(8), 6926–6947.

Yashashwini, A. (2024). *Unveiling the Role of Artificial Intelligence and its Impact on Consumer Buying Behaviour in Online Fashion Retail: A Comprehensive Study*. RVIM Journal of Management Research.

Yeo, S. F., Tan, C. L., Kumar, A., Tan, K. H., & Wong, J. K. (2022). Investigating the impact of AI-powered technologies on Instagrammers' purchase decisions in digitalization era–A study of the fashion and apparel industry. *Technological Forecasting and Social Change*, *177*, 121551.

Chapter 12
Global R&D Strategies of German MNEs in the Changing World:
Exploring Co-Invention Trends Between 2019 and 2023

Irina Ervits
https://orcid.org/0000-0002-4729-9498
CBS International Business School, Germany

ABSTRACT

This paper traces the changes in the co-invention patterns of the largest German companies from 2019 to 2023 using patent statistics. The assumption about the withdrawal of R&D activities from emerging economies is tested. Based on descriptive statistics and binary logistic regression, we observe three trends. The first trend concerns the increase of sole patenting activities in the subsidiaries in emerging economies, especially in China. We also observe the decrease in the number of patent filings generated by the headquarters and subsidiaries in advanced economies. There is an increase in co-invention with inventors from external organizations, including by subsidiaries in emerging economies. There is a plethora of possible explanations. One of them is the benefits of exercising a recombinant advantage implying embeddedness in local innovation systems and collaborations with external partners. Another explanation is the nature of corporate decision-making, including loss-aversion that allows for high levels of risk toleration.

DOI: 10.4018/979-8-3373-4566-6.ch012

INTRODUCTION

Motivation: Rising Techno-nationalism

Against the backdrop of the large-scale Russia's invasion of Ukraine in early 2022 that ushered the new era of military conflict right at the border with the European Union and NATO, the technological competition between the North and the South intensifies. The West is becoming increasingly weary of autocratic tendencies[1] in the so-called emerging economies like India, China and, of course, Russia. This growing divide must influence the global networks of innovation that multinational companies have so laboriously spung for the past three decades, riding the wave of the alleged triumph of liberal democracies, market-based economic systems and technological globalization (Fukuyama, 1992). These geographically spread innovation networks included subsidiaries that received R&D mandates and their vertical (between headquarters and subsidiaries), as well as horizontal (inter-subsidiary) knowledge-sharing links. Open innovation was also utilized not only as a way to engage with consumers, but also learn from suppliers, competitors, universities and other organizations in host economies where subsidiaries operated. Different patterns of innovation management implying exchanges of knowledge, ideas, people and technical solutions across countries hailed praise and acknowledgement as prerequisites of technological competitiveness. Now more and more analysts warn of the risks involved in continuing the path of global R&D strategies and denying the signs of increasing techno nationalism, ideological and physical clashes and increasing concerns over national security everywhere. There is a paucity of research on MNEs´ global international innovation strategies in the times of techno nationalism.

This paper traces the changes in the co-invention patterns of the largest German companies from 2019 to 2023. Co-invention measured as co-patenting in this paper serves as an indicator of cross-border R&D collaborations. The manifestations of cross-border R&D collaboration include inter-subsidiary co-patenting, as well as co-patenting between subsidiaries and the headquarters and with external organizations at home and abroad. Based on patent data, we test the assumption about the withdrawal of R&D activities from emerging economies like China as a tribute to recent geopolitical developments. We also challenge the narrative of re-concentration of R&D activities at the headquarters and less collaboration with external organizations in emerging economies. This paper contributes to the understanding of the MNEs´ R&D internationalization patterns by scrutinizing their internal and external networks of knowledge generation and sharing. Even though, as Papanastassiou et al. (2020) caution, the times of the taxonomies of R&D activities have passed, we believe there is room for deep diving into the co-invention mechanisms. Our approach offers a more comprehensive view of the "double network structure" (Zanfei, 2000) of the R&D effort at the intersection of external and internal knowledge networks of MNEs. Plus, we concretize the geographical patterns of co-invention based on a sample of German MNEs inspired by a broader literature exploring the domestic and international influences on R&D activities (Cantwell & Mudambi, 2011; Cantwell & Piscitello, 2014).

Most importantly, this paper addresses the gap in the literature on the MNEs´ response to current techno nationalism (Godley et al., 2025). And, more specifically, we zoom into the effects of techno-nationalism on the ability of MNEs to engage in geographically distributed strategies of sourcing and integrating knowledge. These effects have not been yet explored. The recent study by Li et al. (2024) looked at the cross-border acquisitions of technology assets at the backdrop of the increasing tension between the US and China. This paper contributes by addressing the empirical gap by focusing on an important aspect of MNEs´ global R&D strategy - the co-invention patterns measured by patent statistics. Our empirical contribution is in exploring a very concrete and pragmatic evidence of R&D activity – co-patenting. Patent data, of course, only provides insight into the invention output, it does not capture the entire innovation cycle. But in the atmosphere of political instability, ideological clashes, propaganda and other forms of information manipulation, we look at patent filings as a trustworthy, fact-based data source, which is hard to misinterpret or distort.[2] We use German MNEs as a case study.

Papanastassiou et al. (2020) describe a fundamental shift in literature in the past three decades from observing MNEs from advanced economies mostly pursuing their R&D activities at home, towards a more complex and geographically dispersed approach to innovation. This recently documented mode of innovation implies cross-border generation and diffusion of knowledge (Criscuolo et al., 2005; Meyer et al., 2011; Dachs, 2017). It also involves vertical and horizontal knowledge sharing within co-invention networks across advanced and developing economies (Ervits, 2018; Ervits 2021), as well as sourcing from and co-creation with external partners from the host economies (Castellani & Zanfei, 2002; Le Bas & Sierra, 2002; De Beule & Van Beveren, 2019; Ervits, 2023a). These patterns of international innovation are consistent, however, with the internalization theory (Buckley & Casson,1976), which has been inclusive enough to allow for a possibility of subsidiaries engaging in R&D independently from headquarters. The innovation effort does not have to be centralized following this global innovation strategy. The existence of such complex patterns of R&D is a result of economic globalization (Dunning & Lundan, 2008), emergence of new technologies (Alcacer et al., 2016) and the growing role of emerging markets in the generation of knowledge (Demirbag & Glaister, 2010).

The war in Ukraine that though started in 2014 escalated to a full-scale Russian invasion in February 2022. Wars are notorious for disrupting trade, capital, and technology flows. Western economies imposed economic sanctions on Russia.

Their relationship with India and China is deteriorating over the continuing trade of the two economic powers with Russia (Ribakova, 2024). And overall, the US and the EU became increasingly suspicious of Chinese political and economic ambitions. For example, the US embarked on the policy of restricting access of Chinese companies to latest semiconductor technology alluding to national security concerns (Liu, 2024). Luo and Van Assche (2023) view the US CHIPS (Creating Helpful Incentives to Produce Semiconductors) and Science Act (signed into law in 2022) as a demonstration of a paradigm shift from market-oriented liberalism to intervention-oriented techno-nationalism. The Act involves a combination of subsidies and export controls mainly aimed at counterposing China. Tung et al. (2023) characterize the contention between the US and China as the "Tech Cold War", a state of antagonistic geopolitical rivalry between the superpowers for achieving supremacy over technologies of importance to national security.

The Case of German Multinationals

And techno-nationalism is on the rise beyond China and the US. Tung et al. (2023) warn about a deeper division between the Global North and Global South as governments everywhere become increasingly competitive and guarding of their technological advantage. Casella and Bolwijn (2024) note that even before Russia´s invasion of Ukraine in 2022, investments between geopolitically distant countries decreased by 10 percent between 2013 and 2022 (p. 29). Germany is the third-largest economy in the world after the United States and China. It is Europe´s largest economy with a GDP of 4,186 billion Euros in 2023 (KPMG, 2024). According to Belitz (2017), Germany together with France, Great Britain and the US has intermediate levels of R&D globalization. This group of countries is ahead of South Korea and Japan but behind Switzerland and Sweden (Belitz, 2017). Thus, we use the DAX index companies as a showcase of the "middle-of-the-road" R&D globalization strategies. But there are also specific reasons why German multinationals serve as an important case for understanding global R&D strategies now. One of the reasons is their close ties with China.

China is Germany´s biggest trading partner and there is a long history of German interests in the region. Volkswagen, for example, has been in China since 1978 (Volkswagen Group China, 2025). Siemens is active in China since 1872 (Siemens, 2025). Secondly, after the Russia´s invasion of Ukraine in 2022, many German companies present in China were confronted with an existential dilemma. On the one hand, because of national security concerns and politization of technology, they found themselves pressured to leave or at least decrease their exposure to the Chinese market, on the other hand, the economic logic is in favor of staying.

In 2022, German politicians like Minister for Economic Affairs Robert Habeck became increasingly concerned about the dependence of German companies on the Chinese market (Rinke et al., 2022). In 2023, German government warned indigenous companies not to depend on imports from China especially for primary products, which are difficult to replace: "the government would not pick up the tab if they fell victim to mounting geopolitical risks" (Arnold and Chazan, 2023). In describing Germany´s "Future Research and Innovation Strategy" German Federal Ministry of Education and Research clearly stated: "Germany and the European Union must be able to understand, develop and produce key technologies on their own" (BMBF, 2024). At the beginning of 2024, based on company surveys, the number of German companies considering withdrawing from the Chinese market increased compared to 2022. Reportedly companies were concerned about increased competition from local Chinese companies, restrictions on market access and overall geopolitical risks (ChinaScope, 2024).

The former German government led by the Green Party with Robert Habeck as Vice-Chancellor and Annalena Baerbock as foreign minister had been pushing German companies to de-risk and minimize their exposure to the Chinese market. The expressed concerns included industrial espionage, unfair competition or human rights violations and potential aggression against Taiwan (Rinke et al., 2022). On the other hand, despite warnings by politicians back in 2022 (Hage et al., 2022) or recently at the beginning of 2025 (Milne, 2025), for companies like Volkswagen withdrawing from China remains a difficult decision. In 2022, Chinese imports to Germany grew by 34%, and Germany's three largest automakers — Mercedes-Benz, BMW and Volkswagen — continued to export over a third of all their cars to China (Schmitz, 2024). According to Milne (2025), withdrawing from China is a potentially lose-lose situation, it implies losing market share in China and facing an increasing competition of Chinese companies at home. Plus, according to a Siemens top executive, it would take decades to reduce dependence on the Chinese market (Neil Alim and Jones, 2024). In the first weeks of his second term in office, the US President Trump waged a global trade war. The US "beggar thy neighbor" policies seem to push German companies even more in the direction of Beijing, as Xinhua, the Chinese news agency, enthusiastically reports (Xinhua, 2025).

The recent statements of German politicians reflect a transition from multilateralism towards geopolitical realism and economic nationalism (Luo, 2024). As per Luo (2024), the features of this economic nationalism are national security and techno-economic security concerns. They stand in contrast to the long presence of the largest German companies in China and the practical difficulties of exiting a market they have been operating in for decades. Thus, the main research inquiry is if we would see disruptions in the established knowledge networks as a result of the geopolitical polarization discussed above.

Research Questions and Structure

The main research question is whether German MNEs are descaling their R&D activities in emerging economies as a result of political instability and polarization. This decoupling might manifest in fewer Patent Cooperation Treaty (PCT) applications produced by the largest German companies in countries like China. This might be a consequence of multiple factors, including the full-scale invasion of Ukraine by Russian in 2022 and more incremental hostility in economic and political dealings between the EU and China. We choose 2019 and 2023 as markers of the "before" the Russia´s second invasion in 2023 and "after." The COVID-pandemic that started in 2019 did not immediately have an effect on patent production reflected in filings made in 2019, its negative effects on the global R&D networks materialized in the years after. The second question is whether we observe the re-concentration of R&D activities at the headquarters in this period. Headquarters of Western MNEs have always been more prolific than their subsidiaries (Belderbos et al., 2013; Belitz and Lejpras, 2021). However, we assume that current political polarization might involve even more R&D activity being pursued at home than before due to emerging techno-nationalism. Our third research question is whether there is less collaboration with external organizations in emerging economies between 2019 and 2023.

In this paper we utilize patent statistics to trace the effects of recent geopolitical developments on the co-invention patterns in the largest German companies from 2019 to 2023. We pursue an inductive approach as the argument that growing techno-nationalism (noted by many observers) should lead to technological decoupling by European multinationals from the global South does not require intensive theoretical framing. It is based on common sense. Nationalism, especially if it escalates to military aggression, is not conducive to R&D internationalization or MNEs´ global business activities in general. Nationalism disrupts global supply chains, trade flows and financial flows. Theoretical explanations would be in order to clarify the patterns we identify. For example, we see, based on the data analysis, that the withdrawal from emerging markets and curtailing of R&D activities there, especially with external partners, is not taking place. We make an attempt to explain this outcome with relevant theoretical arguments.

The following sections discuss fitting theoretical concepts like "recombinant advantage" that explain why multinationals engage in cross-border R&D collaborations leading to co-invention or co-patenting. Recombinant advantage serves as a prerequisite of their technological competitiveness globally and involves integrating and recombining knowledge sourced from different locations and entities. Then, we discuss our methodological approach, present and reflect on the results. Finally, we draw conclusions and discuss practical implications.

THEORETICAL FRAMEWORK

Recombinant Advantage

When a patent application is filed by an MNE in a foreign country we assume that this act implies considerable financial involvement and legal control over a subsidiary or affiliate in this host economy. Thus, it is a case of "home-base augmenting" R&D activities as opposed to "home-base exploiting" (Kuemmerle, 1999). Another term frequently used to indicate R&D by foreign subsidiaries is a "competence-creating mandate", which implies their embeddedness into the local, regional or national systems of innovation (Cantwell & Mudambi, 2000). These competence-creating subsidiaries engage in the innovation process by sourcing knowledge, technology and creative inspiration from indigenous companies, universities, or research institutes. Narula (2014) refers to "recombinant advantage" as the ability to source knowledge from internal (within a company) and external networks (R&D collaboration partners like universities, suppliers, industry competitors, NGOs and research institutes) and integrate knowledge. MNEs source ideas from their own knowledge network of subsidiaries scattered globally and integrate them with knowledge or assets sourced locally in host economies. Indeed, the ability to recombine knowledge is one of the prerequisites of technological leadership and market competitiveness (Zanfei, 2000; Rugman and Verbeke; 2001, Yamin and Otto, 2004, Verbeke, 2009).

Similar to the concept of "recombinant advantage," Kramer et al. (2010) discuss the concept of "network capital" as the capacity of MNEs to exchange information and knowledge and to promote "collective learning" with external organizations, such as other firms, universities and research institutes. This "network capital" is an important prerequisite of building a successful global R&D strategy. Co-invention is at the nexus of augmenting home competencies and combining them with internally generated ideas and technologies and, thus, it is one of the forms of obtaining and exercising "recombinant advantage" through collective learning and sharing.

Even though R&D activities mostly take place at home (Belderbos et al., 2013), in the past decades emerging markets began to serve as the locations of competence creating activities by MNEs from advanced economies (Shimizutani & Todo, 2008; Demirbag & Glaister, 2010). The dominant role of headquarters in producing patents of high technological value has been questioned (Ervits, 2023a). Even back in 1990, Kotabe suggested that the US firms' competitiveness depended as much on overseas R&D as on domestic R&D and that international collaboration was important. Indeed, MNEs´ competence building in emerging economies has been driven by sourcing ideas locally. Technological solutions in developing countries frequently address local social and environmental problems (Prabhu & Jain, 2015). But this locally generated innovation can target global needs as well through exercising "reverse innovation" (Zhao et al., 2020). One reason for the emerging markets like China and India becoming an attractive destination for R&D activities is the emergence of regional technological hubs (Chen & Vang, 2008; Cuervo-Cazurra & Genc, 2011). Emerging markets also offer abundant highly skilled labor and a large consumer base.

One way to capture how MNEs leverage their recombinant advantage is to look at co-patenting patterns, how subsidiaries co-invent with the headquarters and other subsidiaries, as well as external partners. Different forms of co-patenting vis-à-vis sole patenting by headquarters, especially with external partners, have produced more patent citations (Ervits, 2023a). Empirical evidence shows that inter-subsidiary or horizontal co-patenting serves as a mechanism of new knowledge creation by increasing the technological diversification of patents produced by such companies as Siemens (Ervits, 2023b). Based on the assumption that companies want to maximize their recombinant advantage, it is strategically important for Western MNEs to enter and entrench themselves in powerful emerging markets like China or India. These are dynamic innovation environments because of the support of the government, including investments in technological infrastructure and universities, and, of course, large markets and skilled and motivated labor force.

There are also sunk costs involved, they cannot be recovered if MNEs decides to decouple. Plus, psychologically the parting can be traumatizing since there is also a social aspect reflected in such concepts as "network capital." Everything we know about the power of networking, creating connections, institutions and routines (putting aside the physical infrastructure), exchanging and applying knowledge to co-create patented technology speaks to the difficulty of disrupting R&D engagement in emerging markets by European companies. Recombinant advantage is hard-wired into the MNEs´ DNA as part of the strategies to reach higher levels of competitiveness. On the other hand, geopolitical realities might force companies to decrease the intensity of R&D activities in emerging markets. The question is if the geopolitical developments between 2019 and 2023 were taken seriously by German MNEs as warnings to reevaluate their global R&D strategies.

Techno-Nationalism and Recombinant Advantage

Researchers have tackled the MNEs´s strategies to navigate diverse political contexts (Lubinski and Wadhwani, 2020; Buckley, 2022; Bhaumik et al., 2024). Luo (2022), focusing on the ramifications of the current form of techno nationalism, outlined some MNEs´ coping mechanisms and Lynn and Salzman (2023) offer advice to MNEs in China of how to tap into the Chinese science and technology innovation capabilities. But there is still a gap in our understanding of what MNEs´ response to current techno nationalism is (Godley et al., 2025). And, more specifically, it is important to explore the effects of techno-nationalism on the ability to exercise recombinant advantage. There is a paucity of research zooming into R&D strategies even though it is exactly what techno nationalism attempts to control – flow of knowledge and technology. There are some exceptions, for example, Li et al. (2024) look at how nationalistic rivalry between the US and China affected the acquisition of technology assets in the US by Chinese companies. This paper contributes to the debate by focusing on a different aspect of R&D strategies: R&D collaboration patterns measured by co-patenting in the era of techno nationalism.

Luo (2022) points out that the current manifestations of techno nationalism involve an active role played by the host government to curtail economic ties with rivals if they believe that these ties benefit their rivals more. It is a zero-sum, *realpolitik* mentality exemplified by the recent policies in the US and EU to minimize technology transfer to and cooperation with China as such collaborations became to be perceived as national security threats (Luo, 2022). Luo (2022) describes how the Chinese leadership retaliated with policies aimed at increasing technological autonomy like ordering public institutions and government agencies to stop using foreign-made computers and software. Techno-nationalist policies increase the costs for MNEs everywhere (Godley et al., 2025). Trade barriers like subsidies or tariffs can become deterrents of geographical expansion and entrenchment in foreign markets.

Godley et al. (2025) summarize the main features of techno nationalism described in Luo (2022) as decoupling, countries categorize some technologies as strategically significant and there is perceived threat to national interests. Decoupling implies that the economic returns to co-operation diminish; it is about zero-sum thinking (Godley et al., 2025). We believe that slowing down of co-inventive activity in host countries can serve as a concrete indicator of decoupling. This feature directly relates to our first and second research questions about the descaling of inventive activities in subsidiaries and the re-concentration of R&D at the headquarters. The designation of some technologies as strategically significant and the perceived threat to national interests should lead to more government monitoring in the host economy. This close monitoring and possessive tendencies should lead, we assume, to the decrease of the patenting collaboration of MNEs with external organizations, which is the focus of our third research question.

METHODS

Co-patenting as an Indicator of Co-invention

We use Patent Cooperation Treaty (PCT) statistics from PatentScope, a patent statistics database administered by the World Intellectual Property Organization

(WIPO). We use the classification of co-patenting patterns developed by Ervits (2023a) and presented in Table 1 below. These different forms of co-patenting, which we classify as forms of co-invention, range from sole patenting by headquarters to inter-subsidiary co-invention.

Patent statistics have been traditionally used as a measure of invention or inventive activity even though not all intentions are patentable (Griliches, 1990). Still, "a patent represents a minimal quantum of invention" (Griliches, 1990, p. 1669). The use of patents as a measure of invention, not innovation, goes back to the interpretation of Schumpeter´s insights on the process of innovation. Conceptually Schumpeter drew a demarcation line between the functions of an inventor and entrepreneur: an inventor invents a new product or process, an entrepreneur – commercializes it (Schumpeter, 1942). Thus, the first stage of the innovation process involving the generation of ideas is distinct from the second stage when ideas are transformed into marketable products and processes (Stoneman, 1995). Patents that include a technological description of an invention are closer in spirit to the first stage of the innovation process, namely invention, as they serve as a legal claim on an idea that usually precedes its materialization, especially on a mass-produced scale and commercialization. A patent is an expression of intent to commercialize an invention. Many patents do not get commercialized because they are hoarded by assignees who are either unwilling or unable to apply them (Cayton, 2020). Thus, patents serve as an output measure, though imperfect, of invention, not the entire innovation cycle.

Broadly defined, co-invention happens when different inventors participate in an act of invention. They may belong to the same or different organizations. Geographically co-inventors can come from different regions (Scherngell et al., 2023) or countries (Acosta et al., 2023). International co-invention is about inventors from different countries who combine their knowledge to create a new idea (Tsukada and Nagaoka, 2015). This idea may or may not materialize in the form of a patent application. This is a drawback of using patent statistics to trace co-invention. Furthermore, Palomeras and Wehrheim (2021) argue that staffing or inventor allocation decisions in the context of a R&D alliance are strategic. Strategically picking who gets to work on a joint research project is a key consideration to prevent unwanted knowledge leakage, which is as important as the IP protection mechanisms such as patents. Bergek and Bruzelius (2010) also argue that the application process, including who is listed as an inventor, can be politicized and might not reflect a meaningful contribution. Thus, it can be argued that patent statistics cannot properly reveal co-invention patterns not only because not all collaborated ideas end up being patented, but also because there are alternative ways to protect knowledge, and the list of inventors might not be accurate. This all might be true and the limitations of patent data to reflect inventive output have been widely acknowledged (Griliches, 1990). Still, we argue that co-patenting is a valid indicator of co-invention.

First, broadly speaking, if co-invention implies knowledge sharing, the fact that multiple inventors can be listed on a patent application serves as a testament to their collaborative effort. They are co-inventing a new technical solution described in the patent application. Co-invention of patents is admittedly one of the forms of collaboration seen in R&D (Hazir et al., 2018). It does not automatically imply that co-invention is an act of collaboration involving only a tandem of internal and external inventors, an R&D strategic alliance. More generally, every co-invented patent involving more than one inventor is an act of collaboration. Patents can be co-invented within a company through horizonal or vertical knowledge flows (Michailova and Zaidah, 2012), as well as across organizations (Belderbos et al., 2014; Agostini and Caviggioli, 2015; Caviggioli et al., 2023). And co-invention patterns using patent statistics have been investigated both in the context of developed economies (Berry, 2013), as well as involving inventors from emerging markets (Branstetter et al., 2013). Secondly, like the ethical standards of co-authorship in academic circles, there is an expectation that the listed inventors somehow contributed to the invention being patented. It is not a random list, and each individual enters into a contractual relationship with the applicant, be it a private company or a university, about the form of individual reward in case a patent is licensed or otherwise commercialized. Legally outlining the terms of engagement be it between organizational partners or between organizations and individuals is a prerequisite of R&D collaboration and a resultant patent application. Thus, co-patenting reflects co-invention and the two concepts, if not synonymous, are closely linked. Plus, there is a practical issue of patent data to be publicly available, which makes it a valuable data source for tracing the patterns of co-invention, including the geographical location of co-inventors. The inner workings of the invention process remain largely impenetrable (at least patent statistics cannot reveal this information).

Data Sources and Collection

We use Patent Cooperation Treaty (PCT) statistics from PatentScope by the World Intellectual Property Organization (WIPO) for a number of reasons. First, PCT patent data in the form of weekly gazettes (the list of filings per week) is available for the past six years on its website (PatentScope, 2025). The gazettes contain the identification numbers and technology field codes of patent applications, as well as names of the applicants be it a company or a university or an unaffiliated inventor. Secondly, PCT applications can be filed for patent protection in multiple jurisdictions simultaneously. Thus, PCT applications are "international" as opposed to the data from the United States Patent and Trademark Office (USPTO), which is US-centric. This international orientation is a testament of international commitment (Quinn (2011). And this commitment, involving financial implications, may reflect indirectly the quality of patent application (Zhao, 2022). Thirdly, the international orientation of PCT filings makes them a perfect data source for exploring the international co-invention patterns.

We scrutinize PCT applications filed by the forty major companies in the DAX index. The DAX, the Deutscher Aktien Index, or the GER40 is a stock index that represents forty largest German companies that trade on the Frankfurt Stock Exchange (Chen, 2022). We filtered close to a million of PCT applications published on PatentScope in 2019 and 2023 using a search algorithm designed in Perl to produce the list of PCT applications filed by the DAX companies. After excluding companies that had fewer than 50 patent[3] applications per year, the list of 9,597 applications filed by 15 DAX companies[4] was finalized. Each PCT application is treated as a unit of analysis and classified based on the type of invention or co-invention as per Table 1. Then every patent application was explored using Google Patents. Google Patents´ interface provides a link to the PDF of the patent application and other documents revealing information on inventors, including their mailing addresses and their patenting history. A patent application usually features an applicant, an organization or an individual, who will retain IP rights if the application is granted. A PCT application also contains a list of inventors, people who contributed to generating the technological innovation described in the patent application. If an application was filed with the help of an agent, frequently a law firm with specific legal and technological expertise, this agent is also mentioned in a PCT application. The mailing addresses of the inventors were cross-referenced with the location of the applicant (headquarters or subsidiaries that filed for patent protection) and a respective code based on the classification in Table 1 was assigned. The co-invention categories in Table 1 represent moving along three vectors: 1) vertical co-invention patterns: from sole patenting by the headquarters to co-invented filings between headquarters and subsidiaries in developed and emerging markets; 2) horizontal co-invention patterns: from sole patenting by subsidiaries to inter-subsidiary co-invention in developed markets to inter-subsidiary co-invention in emerging markets; 3) co-invention with external inventors: filings co-invented between the headquarters and external inventors and between foreign subsidiaries (developed and emerging markets) and external inventors. Unfortunately, PatentScope database does not have automation tools to download and process data efficiently. Cross-referencing with Google Patents is a manual exercise, but it is important to capture international patterns of co-invention as revealed by PCT patent statistics.

Table 1. Categories (types of co-invention) of PCT patent applications[5]

Category of PCT patent applications	Code
HQ	1
Subsidiary/affiliate/joint venture advanced	2
Subsidiary/affiliate/joint venture emerging	3
HQ+ from advanced	4
HQ + from emerging	5
HQ + from advanced and emerging	6
Subsidiary/affiliate/joint venture advanced + from advanced	7
Subsidiary/affiliate/joint venture advanced + from emerging	8
Subsidiary/affiliate/joint venture advanced + from advanced and emerging	9
Subsidiary/affiliate/joint venture emerging + from advanced	10
Subsidiary/affiliate/joint venture emerging + from emerging	11
Subsidiary/affiliate/joint venture emerging + advanced and from emerging	12
HQ + external partner	13
Subsidiary/affiliate/joint venture + external partner advanced	14
Subsidiary/affiliate/joint venture + external partner emerging	15

Source: Adopted from Ervits (2023a).

The ability to source knowledge from internal (within a company) and external networks (R&D collaboration partners like universities, suppliers, industry competitors, NGOs and research institutes) and integrate knowledge creates to the so-called recombinant advantage, one of the prerequisites of technological leadership and market competitiveness (Zanfei, 2000; Rugman & Verbeke, 2001, Yamin and Otto, 2004, Verbeke, 2009; Narula, 2014). That is why it is important to identify these external influences and account for their effect on co-invention. Looking at each patent application filed by the sample DAX companies, we identified internal and external members of the inventor team. Inventors working for a company´s subsidiaries or joint ventures were treated as belonging to the inner knowledge network and hence, classified as internal. We developed a system for identifying "external" inventors by screening the output history of each inventor on Google Patents. If an inventor was involved in at least two (or one, if it is the only application filed by the inventor) patent applications filed by our focal company then this inventor was classified as an "internal" inventor. This means that the inventor has an established patenting history with the focal firm. If the inventor mostly filed applications with other firms, then we classify this inventor as "external."

We look at each co-invented application as a unit of analysis and make assumptions about the "affiliation status" of each inventor in the team based on their patenting history. Of course, the limitation of this approach is the inability to trace job mobility effects. Indeed, researchers change jobs, and their professional life might revolve around a local industrial cluster consisting of the same pool of companies. We took two random samples of 100 patent applications in order to estimate the amount of human error since the classification of patents is conducted manually and is vulnerable to the bias of labeling "internal" inventors as "external" representing outside organizations. We cross-referenced the patenting history of individuals with the open-source information about their professional affiliations. The human error was estimated at 2 percent.

Descriptive Statistics and Regression

Using descriptive statistics, we compare the patterns of co-invention in the two years: 2019 and 2023. We run Chi-Square tests to ascertain if the differences between the two years are statistically significant. We use Siemens as an example to map the trends of co-patenting in both years. Siemens by far is the most prolific patent applicant among the DAX companies. Then, we perform binary logistic regression based on patent data from all 15 DAX companies. The goal is to ascertain the likelihood that a randomly selected patent application in each co-patenting category (Table 1) will be filed in 2019 or 2023. Thus, the two years (2019 and 2023) serve as categories of the binary dependent variable. We treat the type of co-invention (Table 1) as an ordinal scale variable with scores ranging between 1 and 15 reflecting a continuum of different levels of co-invention internationalization from low (sole patenting by headquarters at 1) to high (subsidiary/affiliate/joint venture and an external partner in an emerging economy at 15).

The equation for the logistic regression is Log-odds = A + B(X), where

a) A is the constant;
b) B is the logistic regression coefficient measured on the log scale, and
c) X is value of the independent variable (Type of co-invention patent application in Table 1).

RESULTS

General Trends

Table 2 compares the percentages of each category in 2019 and 2023 in all sample companies. On average the headquarters are responsible for 50 percent of PCT applications in 2019, in 2023 – for 45 percent. Foreign subsidiaries in advanced economies filed on average 18 percent of applications, in 2023 – 15 percent, which is a decrease. However, filings by headquarters with external inventors increased, from 12 percent to 15 percent. Filings by subsidiaries with external inventors have increased from 7 percent to 12 percent. External inventors seem to play a more important role, which corroborates the thesis of "home-base augmenting" R&D activities (Kuemmerle, 1999) or the "competence-creating mandate" (Cantwell & Mudambi, 2000). Headquarters and subsidiaries from advanced economies seem to have intensified their collaborations with local companies, universities, or research institutes. Co-invention is at the nexus of augmenting home competencies and combining them with internal knowledge networks´ output. There is a slight increase in the involvement of subsidiaries from emerging economies – from 1 percent to 2 percent – in PCT filings. Emerging subsidiaries also filed more with external inventors in 2023 compared to 2019: there is an increase from 2 percent to 3 percent.

Table 2. PCT patenting activity in 2019 and 2023 based on the type of co-invention and company

Code	Categories of PCT patent applications	Airbus		Basf		Bayer		BMW		Conti-nental		Covestro		Daimler		Fresenius Med. Care		Fresenius		Henkel		Linde		Merck		Sartorius		Siemens		Volks-wagen		Average 2019	Average 2023
		2019	2023	2019	2023	2019	2023	2019	2023	2019	2023	2019	2023	2019	2023	2019	2023	2019	2023	2019	2023	2019	2023	2019	2023	2019	2023	2019	2023	2019	2023		
1	HQ	31%	39%	43%	49%	53%	46%	74%	71%	48%	45%	43%	41%	58%	42%	40%	36%	35%	36%	59%	44%	45%	70%	32%	30%	68%	22%	43%	35%	82%	70%	50%	45%
2	Subsidiary/affiliate/joint venture advanced	38%	27%	8%	9%	11%	10%	2%	4%	29%	22%	9%	4%	17%	25%	28%	19%	30%	26%	12%	15%	16%	3%	31%	28%	18%	17%	26%	18%	0%	0%	18%	15%
3	Subsidiary/affiliate/joint venture emerging	0%	0%	2%	4%	1%	1%	3%	0%	0%	0%	2%	0%	0%	0%	2%	2%	3%	2%	2%	3%	1%	0%	1%	2%	0%	0%	2%	10%	1%	0%	1%	2%
4	HQ+ from advanced	5%	1%	13%	13%	13%	15%	1%	3%	3%	1%	11%	14%	1%	0%	5%	3%	4%	3%	9%	4%	3%	0%	5%	4%	4%	3%	2%	6%	2%	3%	5%	5%
5	HQ + from emerging	0%	0%	5%	3%	0%	1%	1%	1%	0%	0%	3%	1%	0%	0%	0%	5%	1%	3%	0%	2%	1%	0%	2%	1%	0%	0%	0%	3%	0%	0%	1%	1%
6	HQ + from advanced and emerging	0%	0%	2%	2%	0%	1%	0%	0%	0%	0%	1%	0%	0%	0%	0%	0%	0%	0%	1%	0%	1%	0%	0%	0%	0%	0%	0%	1%	0%	0%	0%	0%
7	Subsidiary/affiliate/joint venture advanced + from advanced	0%	3%	1%	2%	2%	2%	0%	0%	1%	0%	0%	0%	0%	0%	1%	2%	5%	1%	3%	1%	3%	0%	3%	5%	0%	7%	5%	4%	0%	0%	2%	2%
8	Subsidiary/affiliate/joint venture advanced + from emerging	0%	0%	0%	1%	0%	0%	0%	0%	0%	1%	0%	0%	0%	0%	2%	0%	0%	0%	0%	0%	0%	0%	1%	0%	0%	0%	0%	2%	0%	0%	0%	0%
9	Subsidiary/affiliate/joint venture advanced + from advanced and emerging	0%	0%	0%	0%	0%	0%	0%	0%	0%	0%	0%	0%	0%	0%	0%	0%	0%	0%	0%	0%	0%	0%	0%	0%	0%	0%	0%	0%	0%	0%	0%	0%
10	Subsidiary/affiliate/joint venture emerging + from advanced	0%	1%	0%	0%	0%	0%	0%	0%	0%	0%	0%	0%	0%	2%	0%	0%	1%	0%	1%	1%	0%	0%	0%	0%	0%	0%	0%	0%	0%	0%	0%	0%
11	Subsidiary/affiliate/joint venture emerging + from emerging	0%	0%	0%	0%	0%	0%	0%	0%	0%	0%	0%	0%	0%	0%	0%	0%	0%	0%	0%	0%	0%	0%	0%	0%	0%	0%	0%	0%	0%	0%	0%	0%
12	Subsidiary/affiliate/joint venture emerging + advanced and from emerging	0%	0%	0%	0%	0%	0%	0%	0%	0%	0%	0%	0%	0%	0%	0%	0%	0%	0%	0%	0%	0%	0%	0%	0%	0%	0%	0%	0%	0%	0%	0%	0%
13	HQ + external partner	7%	18%	17%	9%	12%	14%	16%	21%	13%	25%	15%	11%	21%	13%	16%	16%	14%	17%	2%	8%	18%	20%	6%	5%	4%	15%	10%	11%	13%	27%	12%	15%
14	Subsidiary/affiliate/joint venture + external partner advanced	18%	10%	7%	3%	7%	10%	2%	1%	5%	5%	4%	18%	2%	17%	5%	17%	6%	9%	7%	13%	11%	5%	17%	22%	2%	37%	8%	6%	2%	1%	7%	12%
15	Subsidiary/affiliate/joint venture + external partner emerging	0%	0%	1%	6%	1%	2%	1%	1%	0%	0%	11%	11%	1%	2%	1%	2%	1%	1%	5%	8%	2%	3%	2%	2%	4%	0%	3%	4%	1%	0%	2%	3%
	TOTAL (PCT filings)	55	67	682	693	284	132	515	534	628	402	233	135	149	48	104	64	139	88	256	202	101	80	311	250	50	60	1630	1277	254	174	9597	

Source: PatentScope (2024) (accessed November 2020-April 2024) and Google Patents (2024) (accessed November 2020-October 2024).

Figure 1 below visualizes the results in Table 2. The patterns are clear: co-invention with external partners is on the rise (Categories 13, 14 and 15). Subsidiaries in emerging economies have produced more sole applications in 2023 than in 2019 (Category 3). Headquarters and subsidiaries in advanced economies (Categories 1 and 2 respectively) produced fewer PCT applications in 2023 than in 2019.

Figure 1. PCT patenting activity in 2019 and 2023 based on the type of co-invention

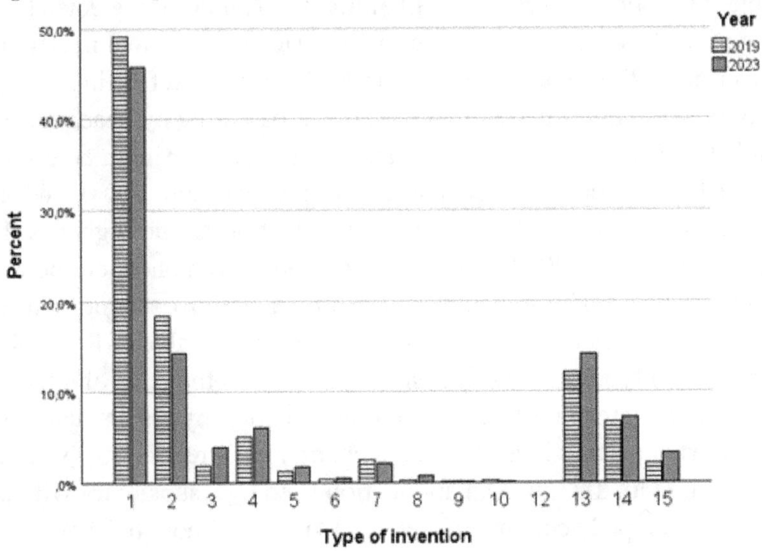

Source: PatentScope (2024) (accessed November 2020-April 2024) and Google Patents (2024) (accessed November 2020-October 2024).

Comparing the PCT filings in 2019 and 2023, we ascertained a statistically significant difference between the two years as per the Pearson Chi-Square (2-sided) test. We see more patenting activity involving subsidiaries in emerging economies: increase in sole patenting in emerging economies (Category 3, which mostly includes Chinese subsidiaries); co-patenting between headquarters and subsidiaries in emerging economies (Category 5); co-patenting between subsidiaries in advanced countries and in emerging economies (Category 8, 44 applications in total); and co-patenting of subsidiaries (both from advanced and emerging economies) with external partners from emerging economies (Category 15, which mostly includes external partners in China). Thus, despite concerns about the geopolitical tension between the West and China, we observe more R&D activity involving subsidiaries in China and other emerging economies. More sole patenting by subsidiaries in emerging economies and their co-patenting efforts with local external partners might be a sign of more independence from their German headquarters. Subsidiaries in China are building up their own R&D capabilities. Of course, German MNEs maintain legal rights over the inventions conceived at their subsidiaries in emerging economies. However, at the backdrop of the erosion of trust between developed and emerging economies and growing national security concerns, more attention must be paid to managing the co-patenting networks.

Table 3 below presents the results of the logistic regression involving two variables: Type of co-invention (Table 1) and the two years (2019 and 2023) as a binary dependent variable. The model ascertains the likelihood of a randomly selected PCT application (categorized based on the co-invention type) to be filed either in 2019 or in 2023. We treat co-invention type as an ordinary scale variable reflecting a scale from low levels of co-invention internationalization (so patenting by headquarters at 1) to high levels (subsidiary patenting with an external partner in an emerging economy at 15). The logistic regression coefficient is a positive value at 0.02, which indicates that the more internationalized a patent application is, the higher is the likelihood that it will be filed in 2023. For every movement from one increment to the next (toward higher levels of co-invention internationalization), the log-odds of being filed in 2023 increase by 0.02. Or, based on the exponent value of B (ExpB), each one-unit increase on type of co-invention scale increases the odds of a patent being filed in 2023 by a factor of 1.021. A patent application by a subsidiary co-patented with an external partner in an emerging economy (Category 15) is 1.021 times as likely to be filed in 2023 than an application filed by a subsidiary with an external partner in a developed economy (Category 14). The Although Cox& Snell and Nagelkerke R square measures can be seen as similar to the R^2 value in linear regression in that they provide an indication of the explanatory power of the model. The values are low. The reason might be a lean structure of the model – it does not contain control variables as they are hard to identify with a single PCT application being the unit of analysis. Another equivalent of R^2 statistics is the Chi-square tests that indicate if the model as a whole predicts occurrence better than chance. In this case, the model can be considered as statistically significant – predicts occurrences in the population.

Table 3. Results of binary logistic regression (comparing 2019 and 2023)

Predictor	**B**	**S.E.**	**Wald**	**df**	**Sig.**	**Exp(B)**
Constant	-0.341	0.028	152.893	1	<0.001	0.711
Type of invention	0.02	0.004	25.521	1	<0.001	1.021
Cox & Snell R^2	0.003					

continued on following page

Table 3. Continued

Predictor	B	S.E.	Wald	df	Sig.	Exp(B)
Nagelkerke R^2	0.004					
Chi-square	25.482			1	<0.001	

Source: PatentScope (2024) (accessed November 2020-April 2024) and Google Patents (2024) (accessed November 2020-October 2024).

Siemens dominates our dataset by a sheer number of patents and drives the trends discussed above. We explore the case of Siemens below to understand the company´s strategic priorities for building a global R&D network. Indeed, we observe that Siemens technological priorities in China include new digital technologies, as acknowledged in Dachs et al. (2023).

Case of Siemens

Siemens is the most prolific patent applicant in our sample. Siemens increased, not contracted, its R&D presence in China between 2019 and 2023. Table 2 presents the summation of the Siemens co-patenting patterns from 2019 based on the 1630 PCT applications filed by Siemens[6]. It also presents the categories of Siemens´ PCT applications (1277 in total) in 2023.[7]

Based on Table 2, most research activity and the resultant patent generation happens at the headquarters in Germany. This pattern is observable in both 2019 and 2023. Subsidiaries of Siemens in advanced countries like the US or Austria produce patent applications "solo" and in collaboration with the headquarters. Most inter-subsidiary collaborations happen among advanced economies. However, subsidiaries in emerging economies like China are also active. Siemens´ R&D activity in China has intensified between 2019 and 2023. Chinese subsidiaries are frequently self-sufficient and generate patent filings on their own involving local inventors. Despite the Western geopolitical distancing from China (or maybe exactly because of this distancing), the number of sole patent applications generated in China increased in 2023. The sole patenting by subsidiaries in emerging economies is an indicator of their autonomy. Another trend is that more external inventors are involved in patent creation at the headquarters´, as well at subsidiary levels. This implies more embeddedness in the local, regional or national-level networks of innovation building connections with various partners like universities, suppliers, private and public research organizations, start-ups and even competitors. Kramer et al. (2010) particularly stress the benefits of regional knowledge networks. And Siemens subsidiaries´ R&D activities are geographically clustered (Ervits, 2023a). The practical challenge is being able to coordinate and leverage the R&D output produced in China or other emerging economies as the political distance between the Global North and Global South increases.

Beyond the analysis of our dataset of patent statistics we tried to understand the company´s motivations based on press releases and news publications. Siemens seems to be unaffected by the growing political tensions and technological competition between China and the West. On the contrary, the company is set to pursue market opportunities in digital technologies and AI solutions in China (Siemens, 2024a, Siemens 2024b). In 2019, the company opened its first AI Lab outside Germany. This R&D facility is part of a global innovation network that covers Europe, North America and Asia (Siemens, 2024a). The company partners with local governments, companies, start-ups, as well as universities and scientific research institutions to jointly develop innovations (Siemens, 2024a). On this occasion a Siemens top executive reassured that Siemens was "well prepared to support China into the digital age" (Siemens, 2019). Thus, Siemens is committed to serve the Chinese market by generating digital solutions for China in China, as stressed by another Siemens top executive (Siemens, 2019).

Siemens has a long history of the "open innovation" strategy relying on geographically bound relationships with a variety of partners in host economies like government-funded research institutes, universities, startups leading to R&D collaborations (Ervits, 2023a). And Siemens strategy of co-patenting with external partners paid off. Ervits (2023a) demonstrates that the company´s PCT applications with inventors from external organizations attracted more forward citations than sole applications filed by the headquarters.

In 2021, Siemens signed an investment agreement with the Hi-Tech Industrial Development Zone in Suzhou to establish the Siemens Electrical Products China headquarters. The new headquarters were planned to cover the complete value chain including business development, product management, digitalization center, R&D and manufacturing. Andreas Matthé, Global CEO of Electrical Products, reassured that it was an important strategic step "to strengthen our set-up in China and strongly support our future development" (Siemens, 2021). Siemens Electrical Products business in China has four factories and four R&D centers. The unit is dedicated to the development and manufacturing of smart electrical products, including digital application services, software and hardware applied in smart infrastructure, new energy, smart automation and IoT applications (Siemens, 2021).

In 2022, Siemens opened an R&D facility, Digital Native Factory in Nanjing, to create a wide range of digital solutions to its customers in China (Siemens, 2024b). In 2023, Siemens announced further investments in boosting its digitalization and high-tech manufacturing R&D capacity in China. Specifically, Siemens pledged to invest into a new digital research center in Shenzhen in South China's Guangdong province. The center focuses on motion control systems technology. Roland Busch, president and chief executive officer of Siemens AG, was quoted stating that the Chinese market was important for Siemens and would continue to grow and that the company did not want "to miss out on these opportunities" (Xinhua, 2023).

Above are a few examples of Siemens´ strategic investments in China. Siemens stays in China for market-seeking reasons. There are also asset-seeking (technology, knowledge and skills) reasons involved. The fact that Simens´ subsidiaries in China are active producers of PCT applications, which, by definition, target international markets, implies their commitment to pursing knowledge augmenting strategies. Siemens in China generates inventions in partnerships with indigenous inventors from external organizations. And Siemens does not seem to respond to the German government´s expressions of political and economic concerns about China´s authoritarian tendencies.

Siemens´ attitude toward China is a showcase of the overall lack of unity between German government and German MNEs. In 2023, German politicians signaled a new strategy toward China that the government called "de-risking (Schmitz, 2024). Even though the EU in general and German authorities in particular wanted to "de-risk" their connection to China, as Kraemer reports, German companies continued to invest in China (2023). Despite China´s repressive policies inside and aggressive posturing outside, there is no significant diversification away from China. According to experts, German companies seek out suppliers and source knowledge (Kraemer, 2023). German companies also seek out the Chinese market. Basf and Volkswagen have been heavily betting on China pledging billions of Euros in hopes of solidifying and expanding their market share in the country (Hage et al., 2022). In fact, a survey published by the German Chamber of Commerce in China found that more than half of the 566 firms surveyed in 2023 said they planned to increase investments in China to stay competitive (Martin, 2024).

DISCUSSION OF RESULTS

Inertia and Slow Adjustments to R&D Global Strategy

Overall, we do not observe the retreatment of the DAX companies´ R&D activities from emerging economies. On the contrary, their patenting efforts in emerging economies between 2019 and 2023 have intensified, including collaborations with local external inventors. Why is this the case? One explanation is the relatively recent nature of paradigmatic changes in the industrial and national security priorities of the West. Reorganizing global R&D activities takes time and the economic globalization locomotive is still in motion even though it might be slowing down. We might be observing the inertia effect – the contraction from emerging economies might happen in the near future as a testament to the growing divide between the Global North and Global South. The changing political context means that Western multinationals must reevaluate their strategies.

Tung et al. (2023) note that current geopolitical challenges imply both short-term and long-term consequences for MNEs' ability to coordinate their internationally dispersed R&D activities depending on when "normality resumes." If current developments are of a temporary nature, then MNEs will make small adjustments to the R&D strategy. Altman et al. (2024) also point out that the announcements about the demise of economic globalization might be premature as the intensity of international trade was at a record high in 2022, despite an unprecedented use of trade sanctions. The authors suggest that we observe a reconfiguration of global flows, not deglobalization. Companies and governments adjust their strategies, for example, by switching from direct to indirect trade as a response to trade barriers. Such configuration illustrates the resilience of economic globalization as it still offers advantages to Western MNEs.

We observed the de-centralization of R&D activities in the last decade of the 20th century and the first two decades of the 21st century (Castellani & Zanfei, 2002; Le Bas & Sierra, 2002; Criscuolo et al., 2005; Meyer et al., 2011; Dachs, 2017; De Beule & Van Beveren, 2019; Papanastassiou et al. 2020; Ervits, 2023a). Even though the headquarters still played a dominant role in R&D (Belderbos, Leten & Suzuki, 2013), Western multinationals flocked to China, Russia, India in search for new markets, human capital, fresh perspectives and frugal approaches to innovation. They set up their R&D facilities there, taking advantage of local expertise and knowledge networks. We still see evidence of these activities in our data. Our results are consistent with Belitz and Lejpras (2021) who ascertained the growing levels of R&D internationalization by MNEs based on global patent data from the 2012-2014 period. In their sample the level of technological internationalization of MNEs was higher than that in the studies for earlier periods. However, one fifth of MNEs still kept R&D close to home (Belitz and Lejpras, 2021). Dachs et al. (2023) are optimistic about the future of R&D internationalization. They believe that the current trend of "techno-nationalism" will not prevent MNEs from pursuing investment opportunities abroad. The authors allude to the imperative of combatting climate change, as well as the rise of digital technologies, that will facilitate international research collaborations and knowledge exchange.

Growing Autonomy of Subsidiaries

If signs of deglobalization persist then long-term political instability implies re-integration and re-centralization, which will be hard to implement as successful and independent subsidiaries abroad would be clinging to their autonomy (Tung et al., 2023). Plus, a more nationalistic and protectionist attitude in host economies might lead to increased independence of some subsidiaries and more integration of others located in still more accessible regions (Luo, 2024). This, indeed, implies abandoning aspirations of a balanced network linking subsidiaries in both developed and emerging economies. Empirically this trend is manifested in the propensity of subsidiaries in China to patent their inventions "solo," as our data demonstrates. And another manifestation of this trend is their entrenchment with the local or regional systems of innovation exemplified by more co-patenting with inventors from external organizations. From the competence creating point of view, it is a desirable outcome. Of course, the question is how to leverage this embeddedness against the backdrop of techno nationalism.

Loss-aversion

The explanations above take into account a more general, macro look at the reactions of MNEs. If we look more closely at their decision-making process and motivation, further insights appear. Jallat and Coronati (2022) argue that business leaders tend to ignore political risk for two reasons: they allow their heuristics and biases to intervene into the decision-making process and because of the elusive nature of political risk as a remote possibility. This argument is particularly relevant to understand the responses of MNEs to geopolitical risks that might affect their R&D activities abroad, especially in emerging markets. Jallat and Coronati (2022) refer to the concept of loss-aversion elaborated in Tversky and Kahneman (1974) that leads to high levels of risk acceptance to regain previous losses or revert to an earlier position. In the context of FDI, considering abandoning a competence-creating subsidiary is an uncomfortable thought. The decision-making process in this case is naturally hindered by the loss-aversion mechanism, clinging to sunk costs. In this scenario of postponing taking action in an attempt to maintain status quo, risks are tolerated until it is too late and then "real" losses are incurred. Simon (1984) notes that until MNEs had encountered considerable losses due to the Iranian and Nicaraguan revolutions in late 1970s, business leaders did not consider studies of national interest or identity or political risks as relevant. Simon says that "MNEs were interested only in the applied side of political risk" (p. 124).

Competence Creating Networks

One of the convincing arguments explaining why German MNEs are in China and other emerging markets (despite intensifying political tensions) is the intensity of their "home-base augmenting" or competence creating R&D activities. These activities feed into maintaining a recombinant advantage. Competence-creating subsidiaries engage in the innovation process by sourcing knowledge, technology and creative inspiration from local companies, universities, or research institutes. MNEs combine ideas from their own knowledge network of subsidiaries scattered globally with knowledge or assets sourced locally in host economies. This ability to source and recombine knowledge is one of the prerequisites of their technological leadership and market competitiveness (Zanfei, 2000; Rugman and Verbeke, 2001, Yamin and Otto, 2004, Verbeke, 2009; Narula, 2014). Recently scholars investigated how the global competence networks are linked to the systems of value chain organization.

Turkina et al. (2025) emphasize the practical underpinnings of managing an effective global value chain as opposed to a narrower focus on sourcing and re-combining local knowledge. The two perspectives are not mutually exclusive, the only difference is in acknowledging that the highly complex organization of global value chains must affect the decisions of MNEs regarding where and how to engage in R&D. The authors borrow from the systems integration literature focusing on the "technical understanding of how different components are linked together in a coherent organizational system" (Turkina et al., 2025). The systems integration argument serves as an explanation of why decoupling from markets where processes and physical infrastructure, as well as human resources have become invaluable and entrenched parts of the value chain systems is such a mammoth task. No wonder that managers resist acknowledging political risks and acting on them. Path dependence might also play a role when we compare organizational systems with natural organisms that grow in an unpredictable way, one event leading to another. The challenge now is how to leverage existing global knowledge infrastructure in the hostile environment of political competition, maintain links with subsidiaries and continue the knowledge augmenting process safely.

CONCLUSIONS

We traced the changes in the co-invention patterns of the largest German companies from 2019 to 2023. With PCT data we tested the assumption about the withdrawal of R&D activities from emerging economies like China as a tribute to the recent geopolitical developments such as the Russia´s full scale invasion of Ukraine in 2022. Following an inductive approach we addressed a number of research questions. They concerned the re-concentration of R&D activities at the headquarters and less collaboration with external organizations in emerging economies. None of these assumptions were confirmed by comparing the number of PCT applications produced by the DAX index companies in 2019 and 2023. Based on descriptive statistics and binary logistic regression, we observe three trends. The first trend concerns the increase of sole patenting activities in the subsidiaries in emerging economies, especially in China. We also observe the decrease in the number of PCT filings generated by the headquarters and subsidiaries in advanced economies. There is, however, an increase in co-invention with inventors from external organizations, including by subsidiaries in emerging economies. And, as the fourth observation, inter-subsidiary co-invention remains marginal.

There is a plethora of possible explanations. One of them is the benefits of exercising a recombinant advantage – sourcing and integrating knowledge gained in different host economies, developed and developing, for decades. The recombination processes explain why subsidiaries become more embedded into local innovation systems and collaborate with external partners. Sourcing knowledge from local universities and companies is a logical consequence of exercising a recombinant advantage. Turkina et al. (2025) build on this argument to suggest that the architecture of global supply chains underpin global R&D strategies, which implies the difficulties of recoupling and retrenching from established positions in markets like China. Another explanation is the nature of corporate decision-making, the psychological pains of acknowledging threats and cutting loose considering massive prior investments into physical infrastructure and social networks. Current R&D activities of companies like Siemens in China either indicate a slow reaction to political risks or their denial. Global political tensions might also lead to subsidiaries becoming more autonomous in their R&D mandate, generating patent applications independently from the headquarters or other subsidiaries.

This paper provides insights into the landscape and intensity of global R&D activities by large German MNEs and reflects on their co-invention strategies. It contributes empirically by focusing on a specific aspect of MNEs´ global R&D strategy - the co-invention patterns measured by patent statistics. The obvious limitation of this paper is in a very short time frame. At the moment, the implications of techno-nationalism are difficult to evaluate. Techno nationalistic attitudes can take different forms and include new policies being announced and then taken back, hostile rhetoric against trading partners, formal warnings issued by governments toward indigenous MNEs, and informal signaling. The complexity and speed of these developments are mudding up the analytical lens. It is still important to try to capture their effects and analyze them as objectively as possible. We believe that patent data offers a solid ground for objectivity as it is a fact-based, hard data. Taking the rapidly changing nature of techno nationalism, we try to provide explanations – some grounded in theory, some – in common sense. Avenues for future research might involve interviews with top executives responsible for international R&D strategies of German multinational companies. These interviews should provide more insight into their motivation and reasoning, as well as their political risk perception and understanding of legal implications.

What we can say for certain, based on our data, is that empirically German MNEs are entrenched in emerging markets and their reactions to warnings emitted by their home government are either slow or they do not take them into account. At least in the period between 2019 and 2023, when truly paradigmatic changes happened in the global economic and political landscape. The changing political context, however, means that European multinationals, and German MNEs in particular, sooner or later must reevaluate their strategies. They must acknowledge the trade-off between the advantages of market-seeking and asset-seeking activities in large emerging economies like China and the higher risks of engagement due to political distancing between the Global North and Global South.

The COVID-pandemic (2019-2022), Russia´s treacherous full-scale attack on Ukraine in early 2022, the Hamas attack on Israel in October 2023 and the brutal retaliation by the Netanyahu government that followed, Trump administration aggressive policies against the US trading partners and threats to political allies at the start of 2025 and other geopolitical developments of similar magnitude and human costs teach us an important lesson. In this increasingly volatile global environment political risks matter. Our data show that German MNEs continue the path of R&D globalization. This indicates a clash of economic and political realities.

REFERENCES

Acosta, M., Coronado, D., & Medina, J. (2023). Effects of co-patenting across national boundaries on patent quality. An exploration in pharmaceuticals. *Economics of Innovation and New Technology*. Advance online publication. DOI: 10.1080/10438599.2023.2167201

Agostini, L., & Caviggioli, F. (2015). R & D collaboration in the automotive innovation environment: An analysis of co-patenting activities. *Management Decision*, *53*(6), 1224–1246.

Alcacer, J., Cantwell, J., & Piscitello, L. (2016). Internationalization in the information age: A new era for places, firms, and international business networks? *Journal of International Business Studies*, *47*(5), 499–512.

Altman, S. A., Bastian, C. R., & Fattedad, D. (2024). Challenging the deglobalization narrative: Global flows have remained resilient through successive shocks. *Journal of International Business Policy*, *7*, 416–439.

Arnold, M., & Chazan, G. (2023). German companies must cut exposure to China, warns Bundesbank. *Financial Times*, September 18, 2023, available at: https://www.ft.com/content/4b7ad09a-b05c-4651-b6ff-43a448a9bd76

Belderbos, R., Cassiman, B., Faems, D., Leten, B., & Van Looy, B. (2014). Co-ownership of intellectual property: Explaining the value-appropriation and value-creation of the copatenting with different partners. *Research Policy*, *43*(5), 841–852.

Belderbos, R., Leten, B., & Suzuki, S. (2013). How global is R&D? Firm-level determinants of home-country bias in R&D. *Journal of International Business Studies*, *44*(8), 765–786.

Belitz, H. (2017). Companies with R&D abroad make Germany a strong research location. *DIW Economic Bulletin*, *7*, 477–487.

Belitz, H., & Lejpras, A. (2021). *R&D internationalization strategies of the world's top corporate R&D investors. DIW Discussion Papers, No. 1942, Deutsches Institut für Wirtschaftsforschung*. DIW.

Bergek, A., & Bruzelius, M. (2010). Are patents with multiple inventors from different countries a good indicator of international R&D collaboration? The case of ABB. *Research Policy*, *39*(10), 1321–1334.

Berry, H. (2013). Global integration and innovation: Multi-country knowledge generation within MNCs. *Strategic Management Journal*, *35*(69), 869–890.

Bhaumik, S. K., Estrin, S., & Narula, R. (2024). Integrating host-country political heterogeneity into MNE–state bargaining: Insights from international political economy. *Journal of International Business Studies*, *55*, 157–171.

BMBF. (2024). Federal Ministry of Education and Research. The Future Research and Innovation Strategy, available at: https://www.bmbf.de/bmbf/en/research/hightech-and-innovation/future-research-and-innovation-strategy/future-research-and-innovation-strategy_node.html (accessed 28 October 2024).

Branstetter, L., Li, G., & Veloso, F. (2013). The polyglot patent boom. *Scientific American*, *309*(4), 62–63. PMID: 24137857

Buckley, P. J. (2022). Navigating three vectors of power: Global strategy in a world of intense competition, aggressive nation states, and antagonistic civil society. *Global Strategy Journal*, *12*(3), 543–554.

Buckley, P. J., & Casson, M. C. (1976). *The Future of the Multinational Enterprise*. Macmillan.

Cantwell, J., & Mudambi, R. (2000). The location of MNE R&D activity: The role of investment incentives. *Management International Review*, *40*(1), 127–148.

Cantwell, J., & Piscitello, L. (2014). Historical changes in the determinants of competence creation in MNC subunits: The increasing role of international knowledge. *Industrial and Corporate Change*, *23*(3), 633–660.

Cantwell, J. A., & Mudambi, R. (2011). Physical attraction and the geography of knowledge sourcing in multinational enterprises. *Global Strategy Journal*, *1*(3–4), 206–232.

Casella, B., & Bolwijn, R. (2024). Global economic fracturing and shifting investment patterns - A diagnostic of 10 FDI trends and their development implications (UNCTAD/DIAE/2024/1), 23 April, available at: https://unctad.org/publication/global-economic-fracturing-and-shifting-investment-patterns

Castellani, D., & Zanfei, A. (2002). Multinational experience and linkages with local firms. Evidence from the electronics industry. *Cambridge Journal of Economics*, *26*(1), 1–25.

Caviggioli, F., Colombelli, A., De Marco, A., Scellato, G., & Ughetto, E. (2023). Co-evolution patterns of university patenting and technological specialization in European regions. *The Journal of Technology Transfer*, *48*, 216–239.

Cayton, S. (2020). The "Green Patent Paradox" and fair use: The intellectual property solution to fight climate change. *Seattle Journal of Technology, Environmental & Innovation Law*, *11*(1), 8. https://digitalcommons.law.seattleu.edu/sjteil/vol11/iss1/8

Chen, J. (2022). DAX Stock Index: Definition and member companies. *Investopedia,* April 5, available at https://www.investopedia.com/terms/d/dax.asp#:~:text=As%20of%20Jan.%2025%2C%202022%2C%20four%20months%20after,AG%20%28BMW%3AGR%29%208%20Brenntag%20SE%20%28BNR%3AGR%29%2A%20Weitere%20Elemente (accessed 1 March 2024).

Chen, Y. C., & Vang, J. (2008). MNCs, global innovation networks and developing countries: Insights from Motorola in China. *International Journal of Business and Management Science*, *1*(1), 11–30.

ChinaScope. (2024). Lianhe Zaobao: More German companies considering withdrawal from Chinese market, January 31, available at: https://chinascope.org/archives/34223#:~:text=Top%20concerns%20cited%20by%20German%20companies%20include%20increased,with%20a%20total%20of%20566%20member%20companies%20polled (accessed 2 April 2025).

Criscuolo, P., Narula, R., & Verspagen, B. (2005). Role of home and host country innovation systems in R&D internationalisation: A patent citation analysis. *Economics of Innovation and New Technology*, *14*(5), 417–433.

Cuervo-Cazurra, A., & Genc, M. (2011). Obligating, pressuring, and supporting dimensions of the environment and the non-market advantages of developing-country multinational companies. *Journal of Management Studies*, *48*(2), 441–455.

Dachs, B. (2017). Internationalisation of R&D: A review of drivers, impacts and new lines of research. *Asian Research Policy*, *8*(2), 17–39.

Dachs, B., Amoroso, S., Castellani, D., Papanastassiou, M., & von Zedtwitz, M. (2023). The internationalisation of R&D: Past, present and future. *International Business Review*, *33*(1), 102191.

De Beule, F., & Van Beveren, I. (2019). Sources of open innovation in foreign subsidiaries: An enriched typology. *International Business Review*, *28*(1), 135–147.

Demirbag, M., & Glaister, K. W. (2010). Factors determining offshore location choice for R&D projects: A comparative study of developed and emerging regions. *Journal of Management Studies*, *47*(8), 1534–1560.

Dunning, J. H., & Lundan, S. M. (2008). *Multinational Enterprises and the Global Economy*. Edward Elgar.

Ervits, I. (2018). Geography of corporate innovation: Internationalization of innovative activities by MNEs from developed and emerging markets. *Multinational Business Review, 26*(1), 25–49.

Ervits, I. (2021). R&D by US multinationals in China: Geography and patterns of co-invention. *International Journal of Emerging Markets, 16*(8), 1525–1555.

Ervits, I. (2023a). Co-patenting and technological value of patent applications: The case of Siemens. *International Journal of Innovation and Technology Management, 21*(2), 2450009.

Ervits, I. (2023b). The effect of co-patenting as a form of knowledge meta-integration on technological differentiation at Siemens. *European Journal of Innovation Management, 27*(8), 2575–2596.

Fukuyama, F. (1992). *The End of History and the Last Man.* The Free Press.

Godley, A., Buckley, T. & Joseph, M. (2025). Techno-nationalism and capability development in the global pharmaceuticals industry, 1918–1970. *Journal of International Business Policy,* doi: .DOI: 10.1057/s42214-025-00210-0

Google Patents. (2024), available at: https://patents.google.com/ (accessed November 2020–October 2024).

Griliches, Z. (1990). Patent statistics as economic indicators: A survey. *NBER Working Papers,* Vol. 3301, National Bureau of Economic Research, Inc.

Hage, S., Hesse, M., Sauga, M., Müller-Arnold, B., & Giesen, C. (2022). German Companies Ignore major risks in China. *Spiegel International,* January 11, available at: https://www.spiegel.de/international/business/business-as-usual-german-companies-ignore-major-risks-in-china-a-27b66d20-f228-47d0-b2bf-95f3cc976e6c (accessed 2 April 2025).

Hage, S., Hesse, M., Sauga, M., Müller-Arnold, B., & Giesen, C. (2022). German companies ignore major risks in China. *Spiegel International,* November 1, available at: https://www.spiegel.de/international/business/business-as-usual-german-companies-ignore-major-risks-in-china-a-27b66d20-f228-47d0-b2bf-95f3cc976e6c

Hazir, C. S., LeSage, J., & Autant, B. C. (2018). The role of R&D collaboration networks on regional knowledge creation: Evidence from information and communication technologies. *Papers in Regional Science, 97*(3), 549–567.

Jallat, F., & Coronati, L. (2022). Why are leaders not taking political risks seriously enough despite a turbulent international business environment and geopolitical instability? *ESCP Impact Paper,* No. 2022-09-EN, available at: https://academ.escpeurope.eu/pub/IP%20N%C2%B02022-09-EN.pdf

Kotabe, M. (1990). The relationship between offshore sourcing and innovativeness of US multinational firms: An empirical investigation. *Journal of International Business Studies, 21*(4), 623–638.

KPMG. (2024). Economic Key Facts Germany, available at: https://kpmg.com/de/en/home/insights/overview/economic-key-facts-germany.html (accessed 1 November 2024).

Kraemer, C. (2023). German firms keep investing in China despite moves to 'de-risk´. *Reuters,* May 10, available at: https://www.reuters.com/business/german-firms-keep-investing-china-despite-moves-de-risk-2023-05-10/ (accessed April 3, 2024).

Kramer, J. P., Revilla Diez, J., Marinelli, E., & Iammarino, S. (2010). Intangible assets and MNEs' locational strategies for innovation—or: Why the *regional* matters. [Review of Regional Research]. *Jahrbuch für Regionalwissenschaft, 30*, 129–157.

Kuemmerle, W. (1999). The drivers of foreign direct investment into research and development: And empirical investigation. *Journal of International Business Studies, 30*(1), 1–24.

Le Bas, C., & Sierra, C. (2002). Location versus country advantages' in R&D activities: Some further results on multinationals' locational strategies. *Research Policy, 31*(4), 589–609.

Li, J., Shapiro, D. M., Ufimtseva, A., & Zhang, P. (2024). Techno-nationalism and cross-border acquisitions in an age of geopolitical rivalry. *Journal of International Business Studies, 55*, 1190–1203.

Liu, Q. (2024). China on cusp of next-generation chip production despite US curbs. *The Financial Times,* February 6, available at: https://www.ft.com/content/b5e0dba3-689f-4d0e-88f6-673ff4452977 (accessed 16 March 2024).

Lubinski, C., & Wadhwani, R. D. (2020). Geopolitical jockeying: Economic nationalism and multinational strategy in historical perspective. *Strategic Management Journal, 41*, 400–421.

Luo, Y. (2022). Illusions of techno-nationalism. *Journal of International Business Studies, 53*, 550–567. PMID: 34511654

Luo, Y. (2024). Paradigm shift and theoretical implications for the era of global disorder. *Journal of International Business Studies*, *55*, 127–135.

Luo, Y., & Van Assche, A. (2023). The rise of techno-geopolitical uncertainty: Implications of the United States CHIPS and Science Act. *Journal of International Business Studies*, *54*, 1423–1440. PMID: 37359753

Lynn, L., & Salzman, H. (2023). Techno-nationalism or building a global science and technology commons? (but what about China?). *Global Policy*, *14*, 832–846.

Martin, N. (2024). German firms ignore calls to shrink China investments. *Deutsche Welle*. August 26, available at: https://www.dw.com/en/germanys-investments-in-china-increase-despite-warnings/a-69957810 (accessed 2 March 2025).

Meyer, K. E., Mudambi, R., & Narula, R. (2011). Multinational enterprises and local contexts: The opportunities and challenges of multiple embeddedness. *Journal of Management Studies*, *48*(2), 235–252.

Michailova, S., & Zaidah, M. (2012). Subsidiary knowledge flows in multinational corporations: Research accomplishments, gaps, and opportunities. *Journal of World Business*, *47*(3), 383–396.

Milne, R. (2025). German companies are risking a 'lose-lose' decoupling with China. *Financial Times*, January 30, available at https://www.ft.com/content/f61d6b7b-0250-4edf-aac1-ba0ca18545cd (accessed 1 April 2025).

Narula, R. (2014). Exploring the paradox of competence-creating subsidiaries: Balancing bandwidth and dispersion in MNEs. *International Journal of Strategic Management*, *47*(1–2), 4–15.

Neil Alim, A., & Jones, S. (2024). German companies' dependence on China will last decades, warns Siemens. *Financial Times*, April 14, available at: https://www.ft.com/content/e8634f5c-5a28-4741-a13d-024665456f43?sharetype=blocked (accessed 2 April 2025).

Palomeras, N., & Wehrheim, D. (2021). The strategic allocation of inventors to R&D collaborations. *Strategic Management Journal*, *42*, 144–169.

Papanastassiou, M., Pearce, R., & Zanfei, A. (2020). Changing perspectives on the internationalization of R&D and innovation by multinational enterprises: A review of the literature. *Journal of International Business Studies*, *51*, 623–664.

PatentScope. (2024). WIPO. Simple search, available at: https://patentscope.wipo.int/search/en/search.jsf (accessed November 2020-April 2024).

PatentScope. (2025). Browse by week, available at: https://patentscope.wipo.int/search/en/ resultWeeklyBrowse.jsf (accessed 29 March 2025).

Prabhu, J., & Jain, S. (2015). Innovation and entrepreneurship in India: Understanding jugaad. *Asia Pacific Journal of Management*, *32*(4), 843–868.

Quinn, G. (2011). PCT Basics: Understanding the international filing process, available at: https://www.ipwatchdog.com/2011/11/03/pct-basics-understanding-the-international-¯ling-process/id=19960/ (accessed 3 November 2021).

Ribakova, E. (2024). Our experience with Russia holds lessons for future sanctions. *The Financial Times*, February 27, available at: https://www.ft.com/content/784eae1d-1fb8-455b-b3d3-2ddc4715499d (accessed 16 March 2024).

Rinke, A., Waldersee, V., & Marsh, S. (2022). German business chiefs clash with Berlin over China policies. *Reuters*, October 14, available at: https://www.euronews.com/next/2022/10/14/germany-china-insight (accessed 15 November 2024).

Rugman, A. M., & Verbeke, A. (2001). Subsidiary-specific advantages in multinational enterprises. *Strategic Management Journal*, *22*(3), 237–250.

Scherngell, T., Schwegmann, K., and Zahradnik, G. (2023). The geographical dynamics of global R&D collaboration networks in robotics: Evidence from co-patenting activities across urban areas worldwide. *PLoS One*, Vol. 18 No. 4, doi: . pone.0281353.DOI: 10.1371/journal

Schmitz, R. (2024). German government wants companies to 'de-risk' from China, but business is reluctant. National Public Radio, January 19, available at: https://www.kpbs.org/news/international/2024/01/19/german-government-wants-companies-to-de-risk-from-china-but-business-is-reluctant (accessed 1 November 2024).

Schumpeter, J. A. (1942). *Capitalism, Socialism and Democracy*. Harper & Row.

Shimizutani, S., & Todo, Y. (2008). What determines overseas R&D activities? The case of Japanese multinational firms. *Research Policy*, *37*(3), 530–544.

Siemens (2019). Siemens catalyzes digital transformation of China with innovation ecosystem, May 15, available at: https://w1.siemens.com.cn/press/NewsDetail_en.aspx?ColumnId=9&ArticleId=8014 (accessed 1 November, 2024).

Siemens (2021). Press Releases. Siemens to establish electrical products China headquarters in Suzhou, China, June 21, available at: https://w1.siemens.com.cn/news_en/news_articles_en/17472.aspx (accessed 20 October 2024).

Siemens (2024a). Siemens R&D in China. Innovation for China's development, available at: https://www.siemens.com/cn/en/company/about/research.html (accessed 30 October 2024).

Siemens (2024b). Across borders. Siemens and China – Milestones in a successful partnership, available at: https://www.siemens.com/global/en/company/about/history/stories/siemens-in-china.html (accessed 1 November 2024).

Siemens (2025). Über Grenzen hinweg. Siemens und China – Meilensteine einer erfolgreichen Partnerschaft (Across borders. Siemens and China – Milestones of a Successful Partnership), available at: https://www.siemens.com/de/de/unternehmen/konzern/geschichte/stories/siemens-in-china.html (accessed 6 April 2025).

Simon, J. A. (1984). Theoretical perspective on political risk. *Journal of International Business Studies*, *15*, 123–143.

Stoneman, P. (1995). *Handbook of the Economics of Innovation and Technological Change*. Wiley-Blackwell.

Tsukada, N., & Nagaoka, S. (2015). International Research Collaboration. *Asian Economic Policy Review*, *10*, 96–119.

Tung, R. L., Zander, I., & Fang, T. (2023). The Tech Cold War, the multipolarization of the world economy, and IB research. *International Business Review*, *32*(6), 102195.

Turkina, E., Van Assche, A., & Mudambi, R. (2025). Cross-border innovation for global value chain orchestration. *International Business Review*, *102450*. Advance online publication. DOI: 10.1016/j.ibusrev.2025.102450

Tversky, A., & Kahneman, D. (1974). Judgment under uncertainty: Heuristics and biases. *Science*, *185*(4157), 1124–1131. PMID: 17835457

Verbeke, A. (2009). *International Business Strategy*. Cambridge University Press.

Volkswagen Group China. (2025). Chronology", available at: https://volkswagengroupchina.com.cn/en/volkswagengroupchina/chronology (accessed 6 April 2025).

Xinhua (2023). Siemens expands China business. June 16. 2023, available at: https://global.chinadaily.com.cn/a/202306/16/WS648bee18a31033ad3f7bcac3.html (accessed 29 October 2024).

Xinhua (2025). German enterprises double down on China investment amid growing global uncertainties. *China Daily*, March 28, available at: https://www.chinadaily.com.cn/a/202503/28/WS67e6102ea3101d4e4dc2b69e.html (accessed 2 April 2025).

Yamin, M., & Otto, J. (2004). Patterns of knowledge flows and MNE innovative performance. *Journal of International Management*, *10*(2), 239–258.

Zanfei, A. (2000). Transnational firms and the changing organisation of innovative activities. *Cambridge Journal of Economics*, *24*(5), 515–542.

Zhao, L. (2022). On the grant rate of Patent Cooperation Treaty applications: Theory and evidence. *Economic Modelling*, *117*, 106051.

Zhao, S., Tan, H., & Papanastassiou, M.. (2020). The internationalization of innovation towards the South: A historical case study of a global pharmaceutical corporation in China (1993–2017). *Asia Pacific Journal of Management*, *37*, 553–585.

ENDNOTES

1 Of course, lately the Trump administration in the US has been also criticized for displaying authoritarian tendencies.

2 Nevertheless, we reflect on the limitations of patent data in the methods section. It is not entirely devoid of potential to be misinterpreted. For example, the definition of "internal" versus "external" inventor might be open to debate. We explain what we mean by both and apply accordingly.

3 With the exception of Daimler that filed 48 PCT applications in 2023.

4 The list of includes the following companies: Airbus, Basf, Bayer, BMW, Continental, Covestro, Daimler, Fresenius Medical Care, Fresenius, Henkel, Linde, Merck, Sartorius, Siemens and Volkswagen. Airbus operational headquarters are in France. France is also home to Airbus Defense and Space SAS, which a military division of Airbus. Airbus Operations GmbH is a subsidiary of Airbus in Germany.

5 Apart from international subsidiaries and joint ventures, Category 3 also contains joint ventures based in Germany such as BSH Bosch und Siemens Hausgeräte GmbH and Valeo Siemens Automotive Germany GmbH. These joint ventures were dissolved in 2015 and 2022 respectively (Ervits, 2023a).

6 The data combines patent application filed by Siemens AG, Siemens Energy AG, and Siemens Healthineers AG. Category "

7 The data combines applications filed by Siemens Healthcare Diagnostics, Inc., Siemens Energy Global GMBH and Siemens Gamesa Renewable Energy Gmbh & Co. Kg.

Compilation of References

Abaci, N. (2022). Relationship between entrepreneurship perception and communication skill: A structural equation model. *International Journal of Management Education*. Advance online publication. DOI: 10.1016/j.ijme.2022.100725

Abad-Segura, E., & González-Zamar, M. D. (2021). Sustainable economic development in higher education institutions: A global analysis within the SDGs framework. *Journal of Cleaner Production*, *294*, 126133.

Abbott, H. P. (2008). *The Cambridge introduction to narrative*. Cambridge University Press.

Aburous, D., & Kamla, R. (2022). Linguistic tensions in a professional accounting field: English linguistic capital, hierarchy, prestige, and distinction among accountants. *Contemporary Accounting Research*, *39*(2), 1120–1149.

Acemoglu, D., & Robinson, J. A. (2012). *Why Nations Fail: The Origins of Power, Prosperity, and Poverty*. Crown Business.

Ackoff, R. L. (1979). The future of operational research is past. *The Journal of the Operational Research Society*, *30*(2), 93–104.

Acosta, M., Coronado, D., & Medina, J. (2023). Effects of co-patenting across national boundaries on patent quality. An exploration in pharmaceuticals. *Economics of Innovation and New Technology*. Advance online publication. DOI: 10.1080/10438599.2023.2167201

Acquah, A. M., & Ibrahim, M. (2020). Foreign direct investment, economic growth and financial sector development in Africa. *Journal of Sustainable Finance & Investment*, *10*(4), 315–334.

Adams, C. A. (2015). The International Integrated Reporting Council, a call to action. *Critical Perspectives on Accounting*, *27*, 23–28. DOI: 10.1016/j.cpa.2014.07.001

Agasisti, T., & Bertoletti, A. (2022). Higher education and economic growth: A longitudinal study of European regions 2000–2017. *Socio-Economic Planning Sciences, 81*, 100940.

Agbanu, I. I., Mohammed, A. B., & Inikpi, O. R. (2024). Effect of AI Driven Personalization on Purchasing Decisions and Customer Satisfaction in the Fashion Industry in Nigeria. *Journal, 4*(05). https://gjrpublication. com/gjrbm

Agoro, H., Templer, S., & Maddog, M. (2021). AI-Powered Personalization: Enhancing User Experience through Customized Shopping Recommendations.

Agostini, L., & Caviggioli, F. (2015). R & D collaboration in the automotive innovation environment: An analysis of co-patenting activities. *Management Decision, 53*(6), 1224–1246.

Agustina, T., Isnaini, S., & Said, I. (2025). Examining the Importance of Developing Entrepreneurial Communication for Students. Teumulong: Journal of Community Service. https://doi.org/.DOI: 10.62568/jocs.v3i3.35

Ahl, H. (2006). Why research on women entrepreneurs needs new directions. *Entrepreneurship Theory and Practice, 30*(5), 595–621.

Ahl, H., & Marlow, S. (2012). Exploring the dynamics of gender, feminism and entrepreneurship: Advancing debate to escape a dead end? *Organization, 19*(5), 543–562.

Ahluwalia, M. S. (2002). Economic Reforms in India Since 1991: Has Gradualism Worked? *The Journal of Economic Perspectives, 16*(3), 67–88. DOI: 10.1257/089533002760278721

Ahmad, M., Majeed, A., Khan, M. A., Sohaib, M., & Shehzad, K. (2021). Digital financial inclusion and economic growth: Provincial data analysis of China. *China Economic Journal, 14*(3), 291–310.

Ahmed, Z., Asghar, M. M., Malik, M. N., & Nawaz, K. (2020). Moving towards a sustainable environment: The dynamic linkage between natural resources, human capital, urbanization, economic growth, and ecological footprint in China. *Resources Policy, 67*, 101677.

Aiyar, S., & Ebeke, C. (2020). Inequality of opportunity, inequality of income and economic growth. *World Development, 136*, 105115.

Akerlof, G. A., & Shiller, R. J. (2009). *Animal spirits: How human psychology drives the economy, and why it matters for global capitalism*. Princeton University Press.

Alabi, M. (2024). AI-Powered Product Recommendation Systems: Personalizing Customer Experiences and Increasing Sales.

Alcacer, J., Cantwell, J., & Piscitello, L. (2016). Internationalization in the information age: A new era for places, firms, and international business networks? *Journal of International Business Studies*, *47*(5), 499–512.

Aldrich, H. E., & Fiol, C. M. (1994). Fools rush in? The institutional context of industry creation. *Academy of Management Review*, *19*(4), 645–670.

Aldrich, H. E., & Ruef, M. (2006). *Organizations evolving* (2nd ed.). Sage Publications.

Ali, A. (2022). *Foreign Debt*. Financial Stability, Exchange Rate Volatility and Economic Growth in South Asian Countries.

Allison, T. H., Davis, B. C., Short, J. C., & Webb, J. W. (2015). Crowdfunding in a prosocial microlending environment: Examining the role of intrinsic versus extrinsic cues. *Entrepreneurship Theory and Practice*, *39*(1), 53–73.

Alshurafat, H., Shbail, A., Obeid, M., & Mansour, E. (2021). Strengths and weaknesses of forensic accounting: An implication on the socio-economic development. *Journal of Business and Socio-economic Development*, *1*(2), 135–148.

Altman, S. A., Bastian, C. R., & Fattedad, D. (2024). Challenging the deglobalization narrative: Global flows have remained resilient through successive shocks. *Journal of International Business Policy*, *7*, 416–439.

Aluwala, A. (2023). Enhancing Customer Experience with AI Powered Analytics. *Journal of Technology & Innovation*.

Alvarez, S. A., Young, S. L., & Woolley, J. L. (2015). Opportunities and institutions: A co-creation story of the king crab industry. *Journal of Business Venturing*, *30*(1), 95–112.

Amirapu, A., & Subramanian, A. (2015). *Manufacturing or services? An Indian illustration of a development dilemma* (Center for Global Development Working Paper No. 409). Center for Global Development. https://doi.org/DOI: 10.2139/ssrn.2623158

Amna Intisar, R., Yaseen, M. R., Kousar, R., Usman, M., & Makhdum, M. S. A. (2020). Impact of trade openness and human capital on economic growth: A comparative investigation of Asian countries. *Sustainability*, *12*(7), 2930.

Anand, S., & Ravallion, M. (1993). Human Development in Poor Countries: On the Role of Private Incomes and Public Services. *The Journal of Economic Perspectives*, *7*(1), 133–150. DOI: 10.1257/jep.7.1.133

Anderson, E. (1993). *Value in Ethics and Economics*. Harvard University Press.

Arnold, M., & Chazan, G. (2023). German companies must cut exposure to China, warns Bundesbank. *Financial Times*, September 18, 2023, available at: https://www.ft.com/content/4b7ad09a-b05c-4651-b6ff-43a448a9bd76

Arquilla, J., & Ronfeldt, D. (2001). *Networks and netwars: The future of terror, crime, and militancy*. Rand Corporation.

Arthur, W. B. (2021). Foundations of complexity economics. *Nature Reviews. Physics*, *3*(2), 136–145. DOI: 10.1038/s42254-020-00258-7

Asheim, B. T., Boschma, R., & Cooke, P. (2011). Constructing regional advantage: Platform policies based on related variety and differentiated knowledge bases. *Regional Studies*, *45*(7), 893–904.

Asher, S., & Novosad, P. (2020). Rural roads and local economic development. *The American Economic Review*, *110*(3), 797–823.

Atkins, J., & Maroun, W. (2015). Integrated reporting in South Africa in 2012, perspectives from preparers and early adopters. *Meditari Accountancy Research*, *23*(2), 197–221. DOI: 10.1108/MEDAR-07-2014-0047

Audretsch, D. B., & Belitski, M. (2017). Entrepreneurial ecosystems in cities: Establishing the framework conditions. *The Journal of Technology Transfer*, *42*(5), 1030–1051.

Audretsch, D. B., & Lehmann, E. E. (2025). A knowledge spillover narrative theory of entrepreneurship. *The Journal of Technology Transfer*, •••, 1–17.

Autio, E., Nambisan, S., Thomas, L. D. W., & Wright, M. (2018). Digital affordances, spatial affordances, and the genesis of entrepreneurial ecosystems. *Strategic Entrepreneurship Journal*, *12*(1), 72–95.

Bachmann, N., Rose, R., Maul, V., & Hölzle, K. (2024). What makes for future entrepreneurs? The role of digital competencies for entrepreneurial intention. *Journal of Business Research*. Advance online publication. DOI: 10.1016/j.jbusres.2023.114481

Baerlocher, D., Parente, S. L., & Rios-Neto, E. (2021). Female labor force participation and economic growth: Accounting for the gender bonus. *Economics Letters*, *200*, 109740.

Baig, U., Hussain, B., Meidutė-Kavaliauskienė, I., & Davidavičius, S. (2022). Digital Entrepreneurship: Future Research Directions and Opportunities for New Business Model. *Sustainability*. Advance online publication. DOI: 10.3390/su14095004

Bailey, K., Gliba, J., & O'Toole, J. (2018). Bridging the policy-practice gap: A narrative approach. [Publisher or source details unavailable; format as book chapter or article if full source confirmed.]

Bakan, J. (2020). *The New Corporation: How "Good" Corporations Are Bad for Democracy*. Vintage Canada.

Baker, S. R., Bloom, N., & Davis, S. J. (2016). Measuring economic policy uncertainty. *The Quarterly Journal of Economics, 131*(4), 1593–1636. DOI: 10.1093/qje/qjw024

Bakshi, A., & Kaur, K. (2020). Dragon versus Elephant: A Comparison of Pattern of Agriculture in Two Asian Economies - China and India. *PRAGATI : Journal of Indian Economy, 7*(1), 48. DOI: 10.17492/pragati.v7i1.195701

Baldassarre, B., Calabretta, G., Bocken, N. M. P., & Jaskiewicz, T. (2017). Bridging sustainable business model innovation and user-driven innovation: A process for sustainable value proposition design. *Journal of Cleaner Production, 147*, 175–186.

Baldwin, R., Cave, M., & Lodge, M. (2012). *Understanding regulation, theory, strategy, and practice* (2nd ed.). Oxford University Press.

Bamberg, M., & Georgakopoulou, A. (2008). Small stories as a new perspective in narrative and identity analysis. *Text & Talk, 28*(3), 377–396.

Bamman, D., Eisenstein, J., & Schnoebelen, T. (2014). Gender identity and lexical variation in social media. *Journal of Sociolinguistics, 18*(2), 135–160.

Banerjee, S. B. (2008). Corporate social responsibility: The good, the bad and the ugly. *Critical Sociology, 34*(1), 51–79.

Banerjee, S. B. (2008). Necrocapitalism. *Organization Studies, 29*(12), 1541–1563.

Banerjee, S. B., & Linstead, S. (2001). Globalization, multiculturalism and other fictions: Colonialism for the new millennium? *Organization, 8*(4), 683–722.

Banet-Weiser, S. (2018). *Empowered: Popular Feminism and Popular Misogyny*. Duke University Press.

Bank of England. (2019). *Brexit and the Bank of England's preparations*. https://www.bankofengland.co.uk

Bardhan, P. (1970). On the long-run stability of the Indian economy. *Journal of Political Economy, 78*(3), 456–468.

Barry, D., & Elmes, M. (1997). On paradigms and narratives: Barry and Elmes' response. *Academy of Management Review*, •••, 847–849.

Basole, A. (2022). Structural transformation and employment generation in India: Past performance and the way forward. *The Indian Journal of Labour Economics : the Quarterly Journal of the Indian Society of Labour Economics*, *65*(2), 295–320. DOI: 10.1007/s41027-022-00380-y PMID: 35874181

Battilana, J., Leca, B., & Boxenbaum, E. (2009). How actors change institutions: Towards a theory of institutional entrepreneurship. *The Academy of Management Annals*, *3*(1), 65–107.

Bebbington, J., & Unerman, J. (2018). Achieving the United Nations Sustainable Development Goals, an enabling role for accounting research. *Accounting, Auditing & Accountability Journal*, *31*(1), 2–24. DOI: 10.1108/AAAJ-05-2017-2929

Bebchuk, L., & Tallarita, R. (2021). The Illusory Promise of Stakeholder Governance. *Cornell Law Review*, *106*(1), 91–178.

Becerra Posada, T. (2025). Pluriversal designs for decolonizing language and literacy education curricula in a rural community of the Colombian carribean region.

Beckert, J. (2016). *Imagined futures: Fictional expectations and capitalist dynamics*. Harvard University Press.

Behare, N., Bhagat, S., & Sarangdhar, P. (2025). Revolutionizing Customer Experience With AI-Powered Personalization. In Strategic Brand Management in the Age of AI and Disruption (pp. 439-462). IGI Global Scientific Publishing. [REMOVED HYPERLINK FIELD]

Bejjani, M., Göcke, L., & Menter, M. (2023). Digital entrepreneurial ecosystems: A systematic literature review. *Technological Forecasting and Social Change*. Advance online publication. DOI: 10.1016/j.techfore.2023.122372

Belderbos, R., Cassiman, B., Faems, D., Leten, B., & Van Looy, B. (2014). Co-ownership of intellectual property: Explaining the value-appropriation and value-creation of the copatenting with different partners. *Research Policy*, *43*(5), 841–852.

Belderbos, R., Leten, B., & Suzuki, S. (2013). How global is R&D? Firm-level determinants of home-country bias in R&D. *Journal of International Business Studies*, *44*(8), 765–786.

Belitz, H. (2017). Companies with R&D abroad make Germany a strong research location. *DIW Economic Bulletin*, *7*, 477–487.

Belitz, H., & Lejpras, A. (2021). *R&D internationalization strategies of the world's top corporate R&D investors. DIW Discussion Papers, No. 1942, Deutsches Institut für Wirtschaftsforschung*. DIW.

Benford, R. D., & Snow, D. A. (2000). Framing processes and social movements: An overview and assessment. *Annual Review of Sociology*, *26*, 611–639.

Bennett, W. L., & Segerberg, A. (2015). Communication in movements. *The Oxford handbook of social movements*, 367-382.

Ben-Salha, O., Dachraoui, H., & Sebri, M. (2021). Natural resource rents and economic growth in the top resource-abundant countries: A PMG estimation. *Resources Policy*, *74*, 101229.

Bentz, J., & Ospina, S. M. (2021). Narrative meaning-making for systemic change: Addressing trauma and memory in post-conflict contexts. *Journal of Change Management*, *21*(2), 184–202.

Bergek, A., & Bruzelius, M. (2010). Are patents with multiple inventors from different countries a good indicator of international R&D collaboration? The case of ABB. *Research Policy*, *39*(10), 1321–1334.

Berger, J., & Milkman, K. L. (2012). What makes online content viral? *JMR, Journal of Marketing Research*, *49*(2), 192–205. DOI: 10.1509/jmr.10.0353

Berger, P., & Luckmann, T. (1966). The social construction of reality. In *Social theory re-wired* (pp. 110–122). Routledge.

Berg, F., Kölbel, J. F., & Rigobon, R. (2022). Aggregate Confusion: The Divergence of ESG Ratings. *Review of Finance*, *26*(6), 1315–1344. DOI: 10.1093/rof/rfac033

Berkes, F., & Ross, H. (2013). Community resilience: Toward an integrated approach. *Society & Natural Resources*, *26*(1), 5–20.

Bernanke, B. S. (2010). Monetary policy and the housing bubble. *American Economic Association Annual Meeting Speech.*

Bernanke, B. S. (2015). *The courage to act: A memoir of a crisis and its aftermath.* W. W. Norton & Company.

Berry, H. (2013). Global integration and innovation: Multi-country knowledge generation within MNCs. *Strategic Management Journal*, *35*(69), 869–890.

Bertay, A. C., Dordevic, L., & Sever, C. (2025). Gender inequality and economic growth: Evidence from industry-level data. *Empirical Economics*, •••, 1–36.

Beverland, M. B., Eckhardt, G. M., Sands, S., & Shankar, A. (2021). How brands craft national identity. *The Journal of Consumer Research*, *48*(4), 586–609.

Bhandari, M. P. (2023). The Fundamental Principles of Social Sciences. *Business Ethics and Leadership*, *7*(2), 73–86. DOI: 10.21272/bel.7(2).73-86.2023

Bhaumik, S. K., Estrin, S., & Narula, R. (2024). Integrating host-country political heterogeneity into MNE–state bargaining: Insights from international political economy. *Journal of International Business Studies*, *55*, 157–171.

Bhowmick, S. (2011). Social entrepreneurship: A review of literature. *International Journal of Management Research and Review*, *1*(3), 1–14.

Bhuiyan, M. S. (2024). The role of AI-Enhanced personalization in customer experiences. *Journal of Computer Science and Technology Studies*, *6*(1), 162–169.

Bitektine, A., & Haack, P. (2015). The "macro" and the "micro" of legitimacy: Toward a multilevel theory of the legitimacy process. *Academy of Management Review*, *40*(1), 49–75.

Blanchard, O. (2021). In defense of concerns about inflation. *Peterson Institute for International Economics*. https://www.piie.com/commentary/op-eds/defense-concerns-about-inflation

Blinder, A. S., Ehrmann, M., Fratzscher, M., De Haan, J., & Jansen, D.-J. (2008). Central bank communication and monetary policy: A survey of theory and evidence. *Journal of Economic Literature*, *46*(4), 910–945. DOI: 10.1257/jel.46.4.910

Block, F., & Somers, M. (2014). *The Power of Market Fundamentalism: Karl Polanyi's Critique*. Harvard University Press.

Bloomberg. (2021). "ESG Assets May Hit $53 Trillion by 2025."

Bloom, D. E., Canning, D., Kotschy, R., Prettner, K., & Schünemann, J. (2024). Health and economic growth: Reconciling the micro and macro evidence. *World Development*, *178*, 106575.

Blyth, M. (2013). *Austerity, the history of a dangerous idea*. Oxford University Press., https://global.oup.com/academic/product/austerity-9780199389445

BMBF. (2024). Federal Ministry of Education and Research. The Future Research and Innovation Strategy, available at: https://www.bmbf.de/bmbf/en/research/hightech-and-innovation/future-research-and-innovation-strategy/future-research-and-innovation-strategy_node.html (accessed 28 October 2024).

Bode, I., Huelss, H., Nadibaidze, A., & Watts, T. F. (2024). Cross-cultural narratives of weaponised artificial intelligence: Comparing France, India, Japan and the United States. *Big Data & Society*, *11*(4), 20539517241303151.

Boichenko, K., & Sierhieiev, M. (2024). ENTREPRENEURIAL ECOSYSTEMS IN THE CONTEXT OF DIGITALISATION. Economic scope. https://doi.org/.DOI: 10.32782/2224-6282/191-61

Boje, D. M. (1991). Consulting and change in the storytelling organisation. *Journal of Organizational Change Management, 4*(3), 7–17.

Boje, D. M. (2001). Narrative methods for organizational and communication research. *Sage (Atlanta, Ga.).*

Boje, D. M. (2008). Storytelling organizations. *Sage (Atlanta, Ga.).*

Boje, D. M. (2008). *Storytelling organizations*. SAGE Publications.

Boltanski, L., & Chiapello, È. (2005). *The New Spirit of Capitalism*. Verso.

Boltanski, L., & Thévenot, L. (2006). *On justification: Economies of worth*. Princeton University Press., DOI: 10.2307/j.ctv1zm2tzm

Borgatti, S. P., Mehra, A., Brass, D. J., & Labianca, G. (2009). Network analysis in the social sciences. *Science, 323*(5916), 892–895.

Bowker, G. C., & Star, S. L. (1999). *Sorting things out, classification and its consequences*. MIT Press. https://mitpress.mit.edu/9780262522958/sorting-things-out

Braga Campos, L. (2020). Knowledge cartograpies: supporting growth and sharing of different ontologies for the construction of the pluriverse.

Brand, S. (1999). *The clock of the long now: Time and responsibility*. Basic Books.

Branstetter, L., Li, G., & Veloso, F. (2013). The polyglot patent boom. *Scientific American, 309*(4), 62–63. PMID: 24137857

Braun, V., & Clarke, V. (2021). Thematic Analysis A Practical Guide. *Sage (Atlanta, Ga.).*

Brenner, C. (2022). Remarks at Urban Innovation Fund Forum.

Brown, A. D. (2006). A narrative approach to collective identities. *Journal of Management Studies, 43*(4), 731–753.

Brown, J., & Dillard, J. (2015). Dialogic Accountings for Stakeholders: On Opening Up and Closing Down Participatory Governance. *Journal of Management Studies, 52*(7), 961–985. DOI: 10.1111/joms.12153

Brown, T., & Wyatt, J. (2010). Design thinking for social innovation. *Stanford Social Innovation Review, 8*(1), 30–35.

Bruner, J. (1991). The narrative construction of reality. *Critical Inquiry, 18*(1), 1–21.

Brush, C. G., Ceru, D. J., & Blackburn, R. (2009). Pathways to entrepreneurial growth: The influence of management, marketing, and money. *Business Horizons*, *52*(5), 481–491.

Bruton, G. D., Ketchen, D. J.Jr, & Ireland, R. D. (2013). Entrepreneurship as a solution to poverty. *Journal of Business Venturing*, *28*(6), 683–689.

Buckley, P. J. (2022). Navigating three vectors of power: Global strategy in a world of intense competition, aggressive nation states, and antagonistic civil society. *Global Strategy Journal*, *12*(3), 543–554.

Buckley, P. J., & Casson, M. C. (1976). *The Future of the Multinational Enterprise*. Macmillan.

Bulturbayevich, M. B., & Jurayevich, M. B. (2020). The impact of the digital economy on economic growth. *International Journal of Business, Law, and Education*, *1*(1), 4–7.

Burchell, S., Clubb, C., Hopwood, A., Hughes, J., & Nahapiet, J. (1980). The roles of accounting in organizations and society. *Accounting, Organizations and Society*, *5*(1), 5–27. DOI: 10.1016/0361-3682(80)90017-3

Burgess, J., Foth, M., & Klaebe, H. (2006). Everyday creativity and digital storytelling. In *Proceedings of the ACM Multimedia Conference*.

Byrne, O., & Shepherd, D. A. (2015). Different strokes for different folks: Entrepreneurial narratives of emotion, cognition, and making sense of business failure. *Entrepreneurship Theory and Practice*, *39*(2), 375–405.

Cabral, C., & Dhar, R. L. (2019). Skill development research in India: A systematic literature review and future research agenda. *Benchmarking*, *26*(7), 2242–2266. DOI: 10.1108/BIJ-07-2018-0211

Cabral, L., Pandey, P., & Xu, X. (2022). Epic narratives of the Green Revolution in Brazil, China, and India. *Agriculture and Human Values*, *39*(1), 249–267. DOI: 10.1007/s10460-021-10241-x

Calás, M. B., Smircich, L., & Bourne, K. A. (2009). Extending the boundaries: Reframing "entrepreneurship as social change" through feminist perspectives. *Academy of Management Review*, *34*(3), 552–569.

Calderón-Fajardo, V., Molinillo, S., Anaya-Sánchez, R., & Ekinci, Y. (2023). Brand personality: Current insights and future research directions. *Journal of Business Research*, *166*, 114062.

Callon, M. (1998). Introduction: The embeddedness of economic markets in economics. *The Sociological Review*, *46*(1, suppl), 1–57.

Cantwell, J. A., & Mudambi, R. (2011). Physical attraction and the geography of knowledge sourcing in multinational enterprises. *Global Strategy Journal*, *1*(3–4), 206–232.

Cantwell, J., & Mudambi, R. (2000). The location of MNE R&D activity: The role of investment incentives. *Management International Review*, *40*(1), 127–148.

Cantwell, J., & Piscitello, L. (2014). Historical changes in the determinants of competence creation in MNC subunits: The increasing role of international knowledge. *Industrial and Corporate Change*, *23*(3), 633–660.

Capra, F., & Luisi, P. L. (2014). *The systems view of life: A unifying vision*. Cambridge University Press.

Cardon, M. S., Stevens, C. E., & Potter, D. R. (2011). Misfortunes or mistakes?: Cultural sensemaking of entrepreneurial failure. *Journal of Business Venturing*, *26*(1), 79–92.

Casella, B., & Bolwijn, R. (2024). Global economic fracturing and shifting investment patterns - A diagnostic of 10 FDI trends and their development implications (UNCTAD/DIAE/2024/1), 23 April, available at: https://unctad.org/publication/global-economic-fracturing-and-shifting-investment-patterns

Castellani, D., & Zanfei, A. (2002). Multinational experience and linkages with local firms. Evidence from the electronics industry. *Cambridge Journal of Economics*, *26*(1), 1–25.

Castells, M. (2010). Globalisation, networking, urbanisation: Reflections on the spatial dynamics of the information age. *Urban Studies (Edinburgh, Scotland)*, *47*(13), 2737–2745.

Castro-Sotomayor, J., & Minoia, P. (2023). Cultivating post-development: Pluriversal transitions and radical spaces of engagement. In *Challenging global development: Towards decoloniality and justice* (pp. 95–116). Springer Nature Switzerland.

Caviggioli, F., Colombelli, A., De Marco, A., Scellato, G., & Ughetto, E. (2023). Co-evolution patterns of university patenting and technological specialization in European regions. *The Journal of Technology Transfer*, *48*, 216–239.

Cayton, S. (2020). The "Green Patent Paradox" and fair use: The intellectual property solution to fight climate change. *Seattle Journal of Technology, Environmental & Innovation Law*, *11*(1), 8. https://digitalcommons.law.seattleu.edu/sjteil/vol11/iss1/8

Challoumis, C. (2024). Innovation and Economic Growth: A Comparative Study of Economocracy and Traditional Systems. SSRN *Electronic Journal.*

Chan, L., Ghali, M., & Prinsloo, P. (2023). Imagining higher education as infrastructures of care. *Higher Education for Good: Teaching and Learning Futures*, 111-136.

Chandler, D., & Reid, J. (2016). *The neoliberal subject: Resilience, adaptation and vulnerability.* Rowman & Littlefield.

Chatterjee, A. (2020). Financial inclusion, information and communication technology diffusion, and economic growth: A panel data analysis. *Information Technology for Development, 26*(3), 607–635.

Chaturvedi, R., & Verma, S. (2023). Opportunities and challenges of AI-driven customer service. Artificial Intelligence in customer service: The next frontier for personalized engagement, 33-71.

Chatzinikolaou, D. (2025). Integrating Sustainable Energy Development with Energy Ecosystems: Trends and Future Prospects in Greece. *Sustainability.* Advance online publication. DOI: 10.3390/su17041487

Chen, J. (2022). DAX Stock Index: Definition and member companies. *Investopedia,* April 5, available at https://www.investopedia.com/terms/d/dax.asp#:~:text=As%20of%20Jan.%2025%2C%202022%2C%20four%20months%20after,AG%20%28BMW%3AGR%29%208%20Brenntag%20SE%20%28BNR%3AGR%29%2A%20Weitere%20Elemente (accessed 1 March 2024).

Chen, J. (2022). DAX Stock Index: Definition and member companies. *Investopedia,* April 5, available at https://www.investopedia.com/terms/d/dax.asp#:~:text=As%20of%20Jan.%2025%2C%202022%2C%20four%20months%20after,AG%20%28BMW%3AGR%29%208%20Brenntag%20SE%20%28BNR%3AGR%29%2A%20Weitere%20Elemente (accessed 1 March 2025).

Chen, A., Lin, Y., Mariani, M., Shou, Y., & Zhang, Y. (2023). Entrepreneurial growth in digital business ecosystems: An integrated framework blending the knowledge-based view of the firm and business ecosystems. *The Journal of Technology Transfer, 48*, 1628–1653. DOI: 10.1007/s10961-023-10027-9

Chen, Y. C., & Vang, J. (2008). MNCs, global innovation networks and developing countries: Insights from Motorola in China. *International Journal of Business and Management Science, 1*(1), 11–30.

Chesbrough, H. (2003). The logic of open innovation: Managing intellectual property. *California Management Review, 45*(3), 33–58.

ChinaScope. (2024). Lianhe Zaobao: More German companies considering withdrawal from Chinese market, January 31, available at: https://chinascope.org/archives/34223#:~:text=Top%20concerns%20cited%20by%20German%20companies%20include%20increased,with%20a%20total%20of%20566%20member%20companies%20polled (accessed 2 April 2025).

Christensen, C. M. (1977). *The innovator's dilemma: when new technologies cause great firms to fail*. Harvard Business Review Press.

Christensen, H. B., Hail, L., & Leuz, C. (2021). Mandatory CSR and sustainability reporting: Economic analysis and literature review. *Review of Accounting Studies*, *26*(3), 1176–1248. DOI: 10.1007/s11142-021-09609-5

Chuan, C. H., Tsai, W. H. S., & Cho, S. Y. (2019, January). Framing artificial intelligence in American newspapers. In *Proceedings of the 2019 AAAI/ACM Conference on AI, Ethics, and Society* (pp. 339-344).

Clarida, R. (2022). Fed communications and market dynamics: 2021–2022 review. *American Economic Association Papers and Proceedings*, *112*, 15–21.

Clark, A., & Chalmers, D. (1998). The extended mind. *Analysis*, *58*(1), 7–19.

Clark, C. (1967). *The conditions of economic progress*. Alianza Editorial S.A.

Clarke, V., & Braun, V. (2014). Thematic analysis. In *Encyclopedia of critical psychology* (pp. 1947–1952). Springer.

Cohen, B., & Winn, M. I. (2007). Market imperfections, opportunity and sustainable entrepreneurship. *Journal of Business Venturing*, *22*(1), 29–49.

Cohen, W. M., & Levinthal, D. A. (1990). Absorptive capacity: A new perspective on learning and innovation. *Administrative Science Quarterly*, *35*(1), 128–152.

Conselho das Finanças Públicas. (2014). *Relatório sobre a execução orçamental*. https://www.cfp.pt/pt/publicacoes/evolucao-orcamental/analise-da-conta-das-administracoes-publicas-2014

Constantinou, S., Pamboris, A., Alexandrou, R., Kronis, C., Zeinalipour-Yazti, D., Papadopoulos, H., & Konstantinidis, A. (2022). EnterCY: A Virtual and Augmented Reality Tourism Platform for Cyprus. 2022 23rd IEEE International Conference on Mobile Data Management (MDM), 314-317. https://doi.org/DOI: 10.1109/MDM55031.2022.00069

Cooperrider, D. (1995). Introduction to appreciative inquiry. In French, W., & Bell, C. (Eds.), *Organization development* (5th ed., pp. 245–263). Prentice Hall.

Cope, J. (2011). Entrepreneurial learning from failure: An interpretative phenomenological analysis. *Journal of Business Venturing, 26*(6), 604–623.

Cornelissen, G., Pandelaere, M., Warlop, L., & Dewitte, S. (2008). Positive cueing: Promoting sustainable consumer behavior by cueing common environmental behaviors as environmental. *International Journal of Research in Marketing, 25*(1), 46–55.

Cornelissen, J. P., & Clarke, J. S. (2010). Imagining and rationalizing opportunities: Inductive reasoning and the creation and justification of new ventures. *Academy of Management Review, 35*(4), 539–557.

Cornelissen, J. P., Clarke, J. S., & Cienki, A. (2012). Sensegiving in entrepreneurial contexts: The use of metaphors in speech and gesture to gain and sustain support for novel business ventures. *International Small Business Journal, 30*(3), 213–241.

Cornelissen, J. P., Durand, R., Fiss, P. C., Lammers, J. C., & Vaara, E. (2015). Putting communication front and center in institutional theory and analysis. *Academy of Management Review, 40*(1), 10–27.

Cornelissen, J. P., Durand, R., Fiss, P. C., Lammers, J. C., & Vaara, E. (2015). Putting communication front and center in institutional theory and research. *Academy of Management Journal, 58*(4), 955–971.

Cornelissen, J., Bekkum, T. V., & Ruler, B. V. (2006). Corporate communications: A practice-based theoretical conceptualization. *Corporate Reputation Review, 9*(2), 114–133.

Couldry, N. (2010). *Why voice matters: Culture and politics after neoliberalism.* SAGE Publications.

Credit Suisse (2022). *Global Wealth Report.*

Creswell, J. W., & Clark, V. L. P. (2017). *Designing and conducting mixed methods research.* Sage publications.

Criscuolo, P., Narula, R., & Verspagen, B. (2005). Role of home and host country innovation systems in R&D internationalisation: A patent citation analysis. *Economics of Innovation and New Technology, 14*(5), 417–433.

Cuervo-Cazurra, A., & Genc, M. (2011). Obligating, pressuring, and supporting dimensions of the environment and the non-market advantages of developing-country multinational companies. *Journal of Management Studies, 48*(2), 441–455.

Czarniawska, B. (2004). Narratives in social science research. *Sage (Atlanta, Ga.).*

Czarniawska, B. (2004). *Narratives in social science research.* SAGE Publications.

Czarniawska, B. (2004). *Narratives in Social Science Research.* SAGE.

Czarniawska, B. (2004). On time, space, and action nets. *Organization, 11*(6), 773–791.

Czarniawska, B. (Ed.). (1997). *A narrative approach to organization studies.* Sage Publications.

Dachs, B. (2017). Internationalisation of R&D: A review of drivers, impacts and new lines of research. *Asian Research Policy, 8*(2), 17–39.

Dachs, B., Amoroso, S., Castellani, D., Papanastassiou, M., & von Zedtwitz, M. (2023). The internationalisation of R&D: Past, present and future. *International Business Review, 33*(1), 102191.

Dacin, P. A., Dacin, M. T., & Matear, M. (2010). Social entrepreneurship: Why we don't need a new theory and how we move forward from here. *The Academy of Management Perspectives, 24*(3), 37–57.

Dahiya, V. K., Dahiya, R., Dev, R., Sharma, A., Sharma, A., Kiran, P. B. N., . . . Yadav, S. (2024, October 24). Exploring the role of artificial intelligence in predictive analytics for financial markets. *2024 International Conference on Computing, Sciences and Communications (ICCSC)*, 1–5. Presented at the 2024 International Conference on Computing, Sciences and Communications (ICCSC), Ghaziabad, India. doi:DOI: 10.1109/iccsc62048.2024.10830442

Dahwa, C. (2023). Adapting and blending grounded theory with case study: A practical guide. *Quality & Quantity.* Advance online publication. DOI: 10.1007/s11135-023-01783-9

Daly, H. (1996). *Beyond Growth: The Economics of Sustainable Development.* Beacon Press.

Dana, L. P., & Anderson, R. B. (2007). A multidisciplinary theory of entrepreneurship as a function of cultural perceptions of opportunity. *International handbook of research on indigenous entrepreneurship, 1*, 595-603.

Daoust, G., & Selby, J. (2024). Climate change and migration: A review and new framework for analysis. *Wiley Interdisciplinary Reviews: Climate Change, 15*(4), e886. DOI: 10.1002/wcc.886

Daskalaki, M., Hjorth, D., & Mair, J. (2015). Are entrepreneurship, communities, and social transformation related? *Journal of Management Inquiry, 24*(4), 419–423.

Datt, G., & Mahajan, A. (2023). *Indian Economy* (72nd ed.). S Chand Publisher.

Davenport, T. H., & Kirby, J. (2016). Just how smart are smart machines? *MIT Sloan Management Review*, *57*(3), 21.

Davenport, T., & Harris, J. (2017). *Competing on analytics: Updated, with a new introduction: The new science of winning*. Harvard Business Press.

Davis, G. F. (2016). *The vanishing American corporation: Navigating the hazards of a new economy*. Berrett-Koehler Publishers.

De Beule, F., & Van Beveren, I. (2019). Sources of open innovation in foreign subsidiaries: An enriched typology. *International Business Review*, *28*(1), 135–147.

de Villiers, C., Rinaldi, L., & Unerman, J. (2014). Integrated Reporting: Insights, gaps and an agenda for future research. *Accounting, Auditing & Accountability Journal*, *27*(7), 1042–1067. DOI: 10.1108/AAAJ-06-2014-1736

Deaton, A. (2013). *The Great Escape : Health, Wealth, and the Origins of Inequality*. Princeton University Press.

Deaton, A., & Dreze, J. (2002). Poverty and Inequality in India: A Re-Examination. *Economic and Political Weekly*, *37*(36), 3729–3748.

Deephouse, D. L., & Carter, S. M. (2005). An examination of differences between organizational legitimacy and organizational reputation. *Journal of Management Studies*, *42*(2), 329–360.

Delbridge, R., & Edwards, T. (2008). Challenging conventions: Roles and processes during non-isomorphic institutional change. *Human Relations*, *61*(3), 299–325.

Demirbag, M., & Glaister, K. W. (2010). Factors determining offshore location choice for R&D projects: A comparative study of developed and emerging regions. *Journal of Management Studies*, *47*(8), 1534–1560.

Dempsey, J. (2020). *Enterprising nature: Economics, markets, and finance in global biodiversity politics (reprint edition)*. Wiley-Blackwell.

Denning, S. (2006). Effective storytelling: Strategic business narrative techniques. *Strategy and Leadership*, *34*(1), 42–48.

Denzin, N. K. (1978). *The research act, a theoretical introduction to sociological methods* (2nd ed.). McGraw Hill.

Desai, M. A. (2009). The decentering of the global firm. *World Economy*, *32*(9), 1271–1290.

Deshpande, A. (2021). How India's Caste Inequality Has Persisted—And Deepened in the Pandemic. *Current History (New York, N.Y.), 120*(825), 127–132. DOI: 10.1525/curh.2021.120.825.127

Dey, P., & Mason, C. (2018). Overcoming constraints of collective imagination: An inquiry into activist entrepreneuring, disruptive truth-telling and the creation of 'possible worlds'. *Journal of Business Venturing, 33*(1), 84–99.

Dey, P., & Steyaert, C. (2016). Rethinking the space of ethics in social entrepreneurship: Power, subjectivity, and practices of freedom. *Journal of Business Ethics, 133*(4), 627–641.

Diaz, J., Halkias, D., & Harkiolakis, T. (2025). *Digital Entrepreneurship and Disruptive Innovation: Intersections of Regional Development, Innovation Systems, and Sustainability*. Routledge.

DiMaggio, P. (1997). Culture and cognition. *Annual Review of Sociology, 23*(1), 263–287.

DiMaggio, P. J., & Powell, W. W. (1983). The iron cage revisited: Institutional isomorphism and collective rationality in organizational fields. *American Sociological Review, 48*(2), 147–160.

Dodge, M., & Kitchin, R. (2005). Codes of life: Identification codes and the machine-readable world. *Environment and Planning. D, Society & Space, 23*(6), 851–881.

Doğanalp, N., Ozsolak, B., & Aslan, A. (2021). The effects of energy poverty on economic growth: A panel data analysis for BRICS countries. *Environmental Science and Pollution Research International, 28*(36), 50167–50178.

Dossani, R., & Kenney, M. (2007). The next wave of globalization: Relocating service provision to India. *World Development, 35*(5), 772–791.

Drèze, J., & Sen, A. (2002). *India: Development and Participation* (2nd ed.). Oxford University Press. https://doi.org/DOI: 10.1093/acprof:oso/9780199257492.001.0001

Dryzek, J. S. (2013). *The politics of the earth: Environmental discourses* (3rd ed.). Oxford University Press.

Duan, C., Zhou, Y., Cai, Y., Gong, W., Zhao, C., & Ai, J. (2022). Investigate the impact of human capital, economic freedom and governance performance on the economic growth of the BRICS. *Journal of Enterprise Information Management, 35*(4/5), 1323–1347.

Dunning, J. H., & Lundan, S. M. (2008). *Multinational Enterprises and the Global Economy*. Edward Elgar.

Eccles, R. G., & Klimenko, S. (2019). The investor revolution. *Harvard Business Review*, *97*(3), 106–116.

Eccles, R. G., Lee, L. E., & Stroehle, J. C. (2020). The social origins of ESG: An analysis of Innovest and KLD. *Organization & Environment*, *33*(4), 575–596.

Eckhardt, G. M., & Bardhi, F. (2020). New dynamics of social status and distinction. *Marketing Theory*, *20*(1), 85–102.

Economic Survey, Govt. of India. (2010). Economic survey 2010-11 -statistical appendix. India Budget. Retrieved November 2, 2023, from https://www.indiabudget.gov.in/budget2011-2012/es2010-11/estat1.pdf

Entman, R. M. (1993). Framing: Toward clarification of a fractured paradigm. *Journal of Communication*, *43*(4), 51–58. DOI: 10.1111/j.1460-2466.1993.tb01304.x

Erlando, A., Riyanto, F. D., & Masakazu, S. (2020). Financial inclusion, economic growth, and poverty alleviation: Evidence from eastern Indonesia. *Heliyon*, *6*(10).

Ervits, I. (2018). Geography of corporate innovation: Internationalization of innovative activities by MNEs from developed and emerging markets. *Multinational Business Review*, *26*(1), 25–49.

Ervits, I. (2021). R&D by US multinationals in China: Geography and patterns of co-invention. *International Journal of Emerging Markets*, *16*(8), 1525–1555.

Ervits, I. (2023a). Co-patenting and technological value of patent applications: The case of Siemens. *International Journal of Innovation and Technology Management*, *21*(2), 2450009.

Ervits, I. (2023b). The effect of co-patenting as a form of knowledge meta-integration on technological differentiation at Siemens. *European Journal of Innovation Management*, *27*(8), 2575–2596.

Escobar, A. (2008). *Territories of difference: Place, movements, life, redes*. Duke University Press.

Escobar, A. (2012). *Encountering development: The making and unmaking of the Third World (reprint edition)*. Princeton University Press.

Espeland, W. N., & Stevens, M. L. (1998). Commensuration as a Social Process. *Annual Review of Sociology*, *24*(1), 313–343. DOI: 10.1146/annurev.soc.24.1.313

Espeland, W. N., & Stevens, M. L. (2008). A sociology of quantification. *Archives Européennes de Sociologie*, *49*(3), 401–436. DOI: 10.1017/S0003975609000150

Etim, E., & Daramola, O. (2020). The informal sector and economic growth of South Africa and Nigeria: A comparative systematic review. *Journal of Open Innovation*, *6*(4), 134.

Etzkowitz, H., & Klofsten, M. (2005). The innovating region: Toward a theory of knowledge-based regional development. *R & D Management*, *35*(3), 243–255.

Etzkowitz, H., & Leydesdorff, L. (2000). The dynamics of innovation: From national systems and "Mode 2" to a triple helix of university–industry–government relations. *Research Policy*, *29*(2), 109–123.

European Central Bank (ECB). (2023). *Narrative barometer pilot report*. https://www.ecb.europa.eu

European Union. (2023). *European Sustainability Reporting Standards, delegated act*. https://eur-lex.europa.eu/legal-content/EN/TXT/?uri=CELEX:32023R2772

Fahad, S., Nguyen-Thi-Lan, H., Nguyen-Manh, D., Tran-Duc, H., & To-The, N. (2023). Analyzing the status of multidimensional poverty of rural households by using sustainable livelihood framework: Policy implications for economic growth. *Environmental Science and Pollution Research International*, *30*(6), 16106–16119.

Fairclough, N. (2003). *Analysing discourse* (Vol. 270). Routledge.

Fairclough, N. (2003). *Analysing Discourse*. Routledge., DOI: 10.4324/9780203697078

Fancy, T. (2021). *The Secret Diary of a 'Sustainable Investor*. Medium.

Farooq, F., Yusop, Z., Chaudhry, I. S., & Iram, R. (2020). Assessing the impacts of globalization and gender parity on economic growth: Empirical evidence from OIC countries. *Environmental Science and Pollution Research International*, *27*(7), 6904–6917.

Fayolle, A. (2013). Personal views on the future of entrepreneurship education. *Entrepreneurship and Regional Development*, *25*(7–8), 692–701.

Federal Reserve Board. (2022). *Monetary policy report*. https://www.federalreserve.gov

Feldman, M. P., & Zoller, T. D. (2012). Dealmakers in place: Social capital connections in regional entrepreneurial economies. *Regional Studies*, *46*(10), 1497–1511.

Feldman, M., & Lowe, N. (2018). Policy and collective action in place. *Cambridge Journal of Regions, Economy and Society*, *11*(2), 335–351.

Felicetti, A., Corvello, V., & Ammirato, S. (2023). Digital innovation in entrepreneurial firms: A systematic literature review. *Review of Managerial Science*, •••, 1–48. DOI: 10.1007/s11846-023-00638-9

Fenton, C., & Langley, A. (2011). Strategy as practice and the narrative turn. *Organization Studies*, *32*(9), 1171–1196.

Ferguson, J. (1994). *The Anti-Politics Machine: Development, Depoliticization, and Bureaucratic Power in Lesotho*. University of Minnesota Press.

Fernandes, C., Ferreira, J., Veiga, P., Kraus, S., & Dabić, M. (2022). Digital entrepreneurship platforms: Mapping the field and looking towards a holistic approach. *Technology in Society*. Advance online publication. DOI: 10.1016/j.techsoc.2022.101979

Feukeu, K. E. (2024). A pluriversal definition of Futures Studies: Critical futures studies from margin to centre. In *Handbook of futures studies* (pp. 80–97). Edward Elgar Publishing.

Fioramonti, L. (2017). *Wellbeing economy: Success in a world without growth*. Pan Macmillan South Africa.

Fischer, F. (2003). *Reframing public policy: Discursive politics and deliberative practices*. Oxford University Press.

Fischer, F., & Forester, J. (Eds.). (1993). *The argumentative turn in policy analysis and planning*. Duke University Press.

Fisher, A. G. B. (1939). Production, Primary, Secondary and Tertiary. *The Economic Record*, *15*(1), 24–38. DOI: 10.1111/j.1475-4932.1939.tb01015.x

Fisher, G., Kuratko, D. F., Bloodgood, J. M., & Hornsby, J. S. (2017). Legitimate to whom? The challenge of audience diversity and new venture legitimacy. *Journal of Business Venturing*, *32*(1), 52–71.

Fisher, G., Stevenson, R., & Burnell, D. (2020). Permission to hustle: Igniting entrepreneurship in an organization. *Journal of Business Venturing Insights*, *14*, e00173.

Fisher, K., & Phelps, R. (2006). Recipe or performing art? Challenging conventions for writing action research theses. *Action Research*, *4*(2), 143–164.

Fisher, M. (2010). *Capitalist Realism: Is there no alternative?* Zero Books.

Fligstein, N., & McAdam, D. (2012). *A theory of fields*. Oxford University Press.

Florida, R. (2002). The economic geography of talent. *Annals of the Association of American Geographers*, *92*(4), 743–755.

Flower, J. (2015). The International Integrated Reporting Council: A story of failure. *Critical Perspectives on Accounting*, *27*, 1–17. DOI: 10.1016/j.cpa.2014.07.002

Flynn, Z., & Sastry, A. (2024). Macroeconomics with narratives. *Journal of Monetary Economics*. Advance online publication. DOI: 10.1016/j.jmoneco.2024.01.005

Folbre, N. (2001). *The Invisible Heart: Economics and Family Values*. New Press.

Foss, L., Henry, C., Ahl, H., & Mikalsen, G. H. (2019). Women's entrepreneurship policy research: A 30-year review of the evidence. *Small Business Economics*, *53*(2), 409–429.

Fourcade, M. (2011). Price and prejudice: On economics, and the enchantment/disenchantment of nature. In The worth of goods (pp. 41-62).

Foxworth, R. (2020, June 25). Reimagining Impact Investing. *The Nonprofit Quarterly*.

Franzosi, R. (2010). Quantitative narrative analysis. *Sage (Atlanta, Ga.)*.

Fraser, N. (2017). The end of progressive neoliberalism. *Dissent*, *64*(2), 130–137. DOI: 10.1353/dss.2017.0020

Fraser, N. (2017). The End of Progressive Neoliberalism. *Dissent*.

Friedman, M. (1970). "The Social Responsibility of Business is to Increase its Profits." *The New York Times Magazine*.

Friedman, M. (1970, September 13). The social responsibility of business is to increase its profits. *The New York Times Magazine*.

Fuenfschilling, L., & Truffer, B. (2016). Cross-border institutionalization of sustainability transitions? A case study of the Dutch and German energy transition. *Environmental Innovation and Societal Transitions*, *19*, 1–15.

Fukuyama, F. (1992). *The End of History and the Last Man*. The Free Press.

Fung, A. (2006). Varieties of participation in complex governance. *Public Administration Review*, *66*, 66–75.

Furnari, S., Crilly, D., Misangyi, V. F., Greckhamer, T., Fiss, P. C., & Aguilera, R. V. (2020). Capturing causal complexity: Heuristics for configurational theorizing. *Academy of Management Review*, *46*(4), 778–799.

Gabor, D. (1963). *Inventing the future*. Alfred A. Knopf.Gali, J. (2015). *Monetary policy, inflation, and the business cycle: An introduction to the new Keynesian framework* (2nd ed.). Princeton University Press.

Gabriel, Y. (2000). *Storytelling in organizations: Facts, fictions, and fantasies: Facts, fictions, and fantasies*. OUP Oxford.

Gabriel, Y. (2004). The voice of experience and the voice of the expert—Can they speak together? *Journal of Management Inquiry*, *13*(3), 238–244.

Gabriel, Y. (Ed.). (2004). *Myths, stories, and organizations: Premodern narratives for our times*. Oxford University Press.

Garcia, A., & Mirra, N. (Eds.). (2023). *Speculative pedagogies: Designing equitable educational futures*. Teachers College Press.

Garcia-Arias, J., Corbetta, S., & Baronnet, B. (2023). Decolonizing education in Latin America: Critical environmental and intercultural education as an indigenous pluriversal alternative. *British Journal of Sociology of Education*, *44*(8), 1394–1412.

Garud, R., & Gehman, J. (2012). Metatheoretical perspectives on sustainability journeys: Evolutionary, relational and durational. *Research Policy*, *41*(6), 980–995.

Garud, R., & Giuliani, A. P. (2013). A narrative perspective on entrepreneurial opportunities. *Academy of Management Review*, *38*(1), 157–160.

Garud, R., Jain, S., & Kumaraswamy, A. (2002). Institutional entrepreneurship in the sponsorship of common technological standards: The case of Sun Microsystems and Java. *Academy of Management Journal*, *45*(1), 196–214.

Garud, R., & Karnøe, P. (2003). Bricolage versus breakthrough: Distributed and embedded agency in technology entrepreneurship. *Research Policy*, *32*(2), 277–300.

Garud, R., Schildt, H. A., & Lant, T. K. (2014). Entrepreneurial storytelling, future expectations, and the paradox of legitimacy. *Organization Science*, *25*(5), 1479–1492.

Geels, F. W. (2002). Technological transitions as evolutionary reconfiguration processes: A multi-level perspective and a case-study. *Research Policy*, *31*(8-9), 1257–1274.

Gehman, J., Treviño, L. K., & Garud, R. (2013). Values work: A process study of the emergence and performance of organizational values practices. *Academy of Management Journal*, *56*(1), 84–112.

Gennaioli, N., & Shleifer, A. (2018). *A crisis of beliefs: Investor psychology and financial fragility*. Princeton University Press.

George, G., Howard-Grenville, J., Joshi, A., & Tihanyi, L. (2016). Understanding and tackling societal grand challenges through management research. *Academy of Management Journal*, *59*(6), 1880–1895.

Gephart, R. P. (2007). Crisis sensemaking and the public inquiry. *Journal of Organizational Change Management, 20*(4), 418–437.

Gershon, I. (2024). *Down and out in the new economy: How people find (or don't find) work today.* University of Chicago Press.

Ghani, E. (2011). *Reshaping Tomorrow: Is South Asia Ready for the Big Leap?* (1st ed.). Oxford University Press.

Giesler, M., & Thompson, C. J. (2016). A tutorial in consumer research: Process theorization in cultural consumer research. *The Journal of Consumer Research, 43*(4), 497–508.

GIIN. (2023). *Annual Impact Investor Survey.* Global Impact Investing Network.

Giotopoulos, K. (2024). *A Systematic Review in Digital Entrepreneurship Ecosystems.* Technium Social Sciences Journal., DOI: 10.47577/tssj.v61i1.11562

Giridharadas, A. (2018). *Winners Take All: The Elite Charade of Changing the World.* Knopf.

Global Impact Investing Network (GIIN). (2023). *Annual Impact Investor Survey.*

Glynn, M. A. (2008). Beyond constraint: How institutions enable identities. *The Sage handbook of organizational institutionalism, 41*, 3-430.

Godley, A., Buckley, T. & Joseph, M. (2025). Techno-nationalism and capability development in the global pharmaceuticals industry, 1918–1970. *Journal of International Business Policy,* doi: .DOI: 10.1057/s42214-025-00210-0

Goertler, T., Papert, M., Fischer, I., & Schmidt, M. (2025). *Building digital platform ecosystems: A synthetization of fundamental design topics from a literature review.* Digital Business., DOI: 10.1016/j.digbus.2025.100109

Goldstein, A. (2015). Structural change and industrial development in the BRICS. *International Affairs, 91*(6), 1427–1428. DOI: 10.1111/1468-2346.12470

Gollin, D., Hansen, C. W., & Wingender, A. M. (2021). Two Blades of Grass: The Impact of the Green Revolution. *Journal of Political Economy, 129*(8), 2344–2384. DOI: 10.1086/714444

Gond, J.-P., Cabantous, L., Harding, N., & Learmonth, M. (2016). What do we mean by performativity in organization and management studies? The uses and abuses of performativity. *International Journal of Management Reviews, 18*(4), 440–463.

Google Patents. (2024), available at: https://patents.google.com/ (accessed November 2020–October 2024).

Goti, A., Querejeta-Lomas, L., Almeida, A., de la Puerta, J. G., & López-de-Ipiña, D. (2023). Artificial intelligence in business-to-customer fashion retail: A literature review. *Mathematics, 11*(13), 2943.

Gottschall, J. (2012). *The storytelling animal: How stories make us human*. Houghton Mifflin Harcourt.

Goyal, N., & Ferrara, E. (2018). Graph-based approaches to identify misinformation and fake news on social media. arXiv preprint arXiv:1802.04291.

Graeber, D. (2001). *Toward an anthropological theory of value: The false coin of our own dreams*. Springer.

Greenhalgh, T., Jackson, C., Shaw, S., & Morrison, J. (2016). Achieving research impact through co-creation in community-based health services: Literature review and case study. *The Milbank Quarterly, 94*(2), 392–429.

Greenwood, R., Suddaby, R., & Hinings, C. R. (2002). Theorizing change: The role of professional associations in the transformation of institutionalized fields. *Academy of Management Journal, 45*(1), 58–80.

GRI. (2021). *The global standards for sustainability impacts*. https://www.globalreporting.org/standards/

Griliches, Z. (1990). Patent statistics as economic indicators: A survey. *NBER Working Papers,* Vol. 3301, National Bureau of Economic Research, Inc.

Guba, E. G., & Lincoln, Y. S. (1994). Competing paradigms in qualitative research. *Handbook of qualitative research, 2*(163-194), 105.

Gudi, A., Chinta, R., & Jin, A. (2024). Competencies and capabilities as determinants of digital entrepreneurship: An empirical validation. *Journal of International Technology and Information Management, 32*(1), 271–302.

Guha, A., Grewal, D., Kopalle, P. K., Haenlein, M., Schneider, M. J., Jung, H., & Hawkins, G. (2021). How artificial intelligence will affect the future of retailing. *Journal of Retailing, 97*(1), 28–41.

Habibi, F., & Zabardast, M. A. (2020). Digitalization, education and economic growth: A comparative analysis of Middle East and OECD countries. *Technology in Society, 63*, 101370.

Hadjimitsis, D., Themistocleous, K., Michaelides, S., Neocleous, K., Danezis, C., Kyriakides, N., . . . Philimis, P. (2024, July). Emerging Prospects in Earth Observation for Cyprus and the Eastern Mediterranean, Middle East and North Africa (EMMENA) Region Through the Eratosthenes Centre of Excellence and Excelsior H2020 Teaming Project. In *IGARSS 2024-2024 IEEE International Geoscience and Remote Sensing Symposium* (pp. 2863-2866). IEEE.

Hage, S., Hesse, M., Sauga, M., Müller-Arnold, B., & Giesen, C. (2022). German Companies Ignore major risks in China. *Spiegel International*, January 11, available at: https://www.spiegel.de/international/business/business-as-usual-german-companies-ignore-major-risks-in-china-a-27b66d20-f228-47d0-b2bf-95f3cc976e6c (accessed 2 April 2025).

Hage, S., Hesse, M., Sauga, M., Müller-Arnold, B., & Giesen, C. (2022). German companies ignore major risks in China. *Spiegel International*, November 1, available at: https://www.spiegel.de/international/business/business-as-usual-german-companies-ignore-major-risks-in-china-a-27b66d20-f228-47d0-b2bf-95f3cc976e6c

Hajer, M. A. (1995). *The politics of environmental discourse: Ecological modernization and the policy process*. Oxford University Press.

Hajer, M. A., & Wagenaar, H. (Eds.). (2003). *Deliberative policy analysis: Understanding governance in the network society*. Cambridge University Press.

Hale, J., Irish, A., Carolan, M., Clark, J. K., Inwood, S., Jablonski, B. B., & Johnson, T. (2023). A systematic review of cultural capital in US community development research. *Journal of Rural Studies*, *103*, 103113.

Halkias, D., & George-Ufot, D. (2021). Challenges of women micro-entrepreneurs in the Niger delta region of Nigeria for business survival: A narrative inquiry. The ISM Journal of International Business, 4(1), 11-22. https://www.proquest.com/scholarly journals/challenges-women-micro-entrepreneurs-niger-delta/docview/2609119503/se 2?accountid=14872

Halkias, D., Esposito, M., Diaz, J., & Harkiolakis, T. (2024). Young Digital Entrepreneurs Innovate the Greek Maritime Industry: A Case Study. In Ramadani, V., Kjosev, S., & Sergi, B. (Eds.), *Entrepreneurship Development in the Balkans Perspectives from Diverse Contexts*. Emerald.

Hall, Daneke, & Lenox. (2010). — Hall, J. K., Daneke, G. A., & Lenox, M. J. (2010). Sustainable development and entrepreneurship: Past contributions and future directions. *Journal of Business Venturing*, *25*(5), 439–448.

Hallett, T., & Ventresca, M. J. (2006). Inhabited institutions: Social interactions and organizational forms in Gouldner's Patterns of Industrial Bureaucracy. *Theory and Society*, *35*(2), 213–236.

Handbook of Statistics of Indian Economy. RBI. (2023, September 15). Macro-Economic Aggregates. Reserve Bank of India - Publications. Retrieved October 17, 2023, from https://rbi.org.in/Scripts/PublicationsView.aspx?id=21807

Hanushek, E. A., & Woessmann, L. (2020). Education, knowledge capital, and economic growth. *The economics of education*, 171-182.

Hanushek, E. A., & Woessmann, L. (2021). Education and economic growth. In *Oxford Research Encyclopedia of Economics and Finance*.

Hanushek, E. A., & Woessmann, L. (2023). *The knowledge capital of nations: Education and the economics of growth*. MIT press.

Hardi, I., Ringga, E. S., Fijay, A. H., Maulana, A. R. R., Hadiyani, R., & Idroes, G. M. (2023). Decomposed Impact of Democracy on Indonesia's Economic Growth. *Ekonomikalia Journal of Economics*, *1*(2), 51–60.

Hardy, C., Lawrence, T. B., & Grant, D. (2005). Discourse and collaboration: The role of conversations and collective identity. *Academy of Management Review*, *30*(1), 58–77.

Hardy, C., & Maguire, S. (2010). Discourse, field-configuring events, and change in organizations and institutional fields: Narratives of DDT and the Stockholm Convention. *Academy of Management Journal*, *53*(6), 1365–1392.

Hargadon, A. B., & Douglas, Y. (2001). When innovations meet institutions: Edison and the design of the electric light. *Administrative Science Quarterly*, *46*(3), 476–501.

Hargadon, A., & Sutton, R. I. (1997). Technology brokering and innovation in a product development firm. *Administrative Science Quarterly*, •••, 716–749.

Harkiolakis, T., & Komodromos, M. (2023). Supporting Knowledge Workers' Health and Well-being in the Post-Lockdown Era. Administrative Sciences - Special Issue "Innovative Leaders and Empowered Employees. *Insights into the Future of Work*, *13*(2), 49. Advance online publication. DOI: 10.3390/admsci13020049

Harris, T. (2021). Center for Humane Technology. https://www.humanetech.com

Hart, O., & Zingales, L. (2017). Companies should maximize shareholder welfare not market value. ECGI-Finance Working Paper, (521).

Hassan, N., Abdelraouf, M., & El-Shihy, D. (2025). *The moderating role of personalized recommendations in the trust–satisfaction–loyalty relationship: an empirical study of AI-driven e-commerce*. Future Business Journal.

Havel, V. (1990). *Disturbing the peace: A conversation with Karel Hvížďala*. Vintage.

Hawlitschek, F., & Hodapp, D. (2024). Challenges to Co-Creating Value in Nascent Platform Ecosystems: Gaps Between Theory and Practice. *International Journal of Electronic Commerce*, *28*, 156–185. DOI: 10.1080/10864415.2024.2332046

Hazir, C. S., LeSage, J., & Autant, B. C. (2018). The role of R&D collaboration networks on regional knowledge creation: Evidence from information and communication technologies. *Papers in Regional Science*, *97*(3), 549–567.

Heavey, C., & Simsek, Z. (2017). Distributed cognition in top management teams and organizational ambidexterity: The influence of transactive memory systems. *Journal of Management*, *43*(3), 919–945.

Heins, C. (2023). Artificial intelligence in retail–a systematic literature review. foresight.

Henriques, H. J. G., de Almeida, C. R., & Ramos, C. M. Q. (2024). The Application of artificial intelligence in the tourism industry: A systematic literature review based on Prisma methodology. Journal of Tourism. *Sustainability and Well-Being*, *12*(1), 65–86.

Heracleous, L., & Barrett, M. (2001). Organizational change as discourse: Communicative actions and deep structures in the context of information technology implementation. *Academy of Management Journal*, *44*(4), 755–778.

Herrendorf, B., Rogerson, R., & Valentinyi, Á. (2014). Growth and structural transformation. In Aghion, P., & Durlauf, S. N. (Eds.), *Handbook of Economic Growth* (Vol. 2, pp. 855–941). Elsevier., DOI: 10.1016/B978-0-444-53540-5.00006-9

Hill, A., & See, J., & Grassroots Collective. (2025). Grassroots modalities of learning: Creating and maintaining pluriversal knowing, being and doing for collective survival. *Local Environment*, ●●●, 1–10.

Hine, C. (2020). *Ethnography for the internet: Embedded, embodied and everyday*. Routledge.

Hines, R. D. (1988). Financial accounting: In communicating reality, we construct reality. *Accounting, Organizations and Society*, *13*(3), 251–261. DOI: 10.1016/0361-3682(88)90003-7

Hjorth, D., & Steyaert, C. (2004). *Narrative and discursive approaches in entrepreneurship*. Edward Elgar Publishing.

Hofstede, G. (2001). *Culture's consequences: Comparing values, behaviors, institutions, and organizations across nations* (2nd ed.). SAGE Publications.

Holt, D. B. (2004). *How brands become icons: The principles of cultural branding*. harvard business press.

Hong, Z., Zhang, Q., Xu, X., & Lyu, Z. (2024). Competition and cooperation in a platform-based business ecosystem within complementary partners. *International Journal of Production Economics*. Advance online publication. DOI: 10.1016/j.ijpe.2024.109337

Hopwood, A. G. (1987). The archeology of accounting systems. *Accounting, Organizations and Society*, *12*(3), 207–234. DOI: 10.1016/0361-3682(87)90038-9

Hospers, G.-J. (2003). Creative cities: Breeding places in the knowledge economy. *Inter Economics*, *38*(5), 260–269.

Huaman, E. S. (2022). How indigenous scholarship changes the field: Pluriversal appreciation, decolonial aspirations, and comparative indigenous education. *Comparative Education Review*, *66*(3), 391–416.

Huang, L., & Knight, A. P. (2017). Resources and relationships in entrepreneurship: An exchange theory of the development and effects of the entrepreneur-investor relationship. *Academy of Management Review*, *42*(1), 80–102.

Hulme, M. (2009). *Why we disagree about climate change: Understanding controversy, inaction and opportunity*. Cambridge University Press.

Hutchins, E. (1995). *Cognition in the Wild.* MIT press.

Ibarra, H., & Barbulescu, R. (2010). Identity as narrative: Prevalence, effectiveness, and consequences of narrative identity work in macro work role transitions. *Academy of Management Review*, *35*(1), 135–154.

IFRS Foundation. (2023a). *IFRS S1 General Requirements for Disclosure of Sustainability-related Financial Information*. https://www.ifrs.org/issued-standards/ifrs-sustainability-standards-navigator/ifrs-s1-general-requirements/

IFRS Foundation. (2023b). *IFRS S2 Climate-related Disclosures*. https://www.ifrs.org/issued-standards/ifrs-sustainability-standards-navigator/ifrs-s2-climate-related-disclosures/

Inavolu, S. M. (2024). Exploring AI-driven customer service: Evolution, architectures, opportunities, challenges and future directions. *International Journal of Engineering and Advanced Technology, 13*(3), 156–163.

Innes, J. E., & Booher, D. E. (2003). [Relevant collaborative work on collaborative planning and narrative sense-making; specific 2018-era extension or similar; e.g., *Journal of the American Planning Association, 62*(4), 460–472.]

IPCC (2023). *Sixth Assessment Report.*

Isenberg, D. J. (2010). How to start an entrepreneurial revolution. *Harvard Business Review, 88*(6), 40–50.

Istrefi, A., Zeqiri, J., Hasani, V., & Komodromos, M. (2025). The role of entrepreneurial marketing and digital transformation on women's entrepreneurial intentions in Kosovo, Int. *J. of Technology Enhanced Learning, 17*(2), 174–196. DOI: 10.1504/IJTEL.2024.10063470

Iwu, C. G., Opute, P. A., Nchu, R., Eresia-Eke, C., Tengeh, R. K., Jaiyeoba, O., & Aliyu, O. A. (2021). Entrepreneurship education, curriculum and lecturer-competency as antecedents of student entrepreneurial intention. *International Journal of Management Education, 19*(1), 100295.

Jackson, T. (2011). *Prosperity without growth: Economics for a finite planet.* Earthscan/Routledge.

Jacobides, M., Cennamo, C., & Gawer, A. (2024). Externalities and complementarities in platforms and ecosystems: From structural solutions to endogenous failures. *Research Policy*. Advance online publication. DOI: 10.1016/j.respol.2023.104906

Jallat, F., & Coronati, L. (2022). Why are leaders not taking political risks seriously enough despite a turbulent international business environment and geopolitical instability? *ESCP Impact Paper,* No. 2022-09-EN, available at: https://academ.escpeurope.eu/pub/IP%20N%C2%B02022-09-EN.pdf

Jamaliah, J., Rusmita, S., Elyta, E., Rosyadi, R., & Islahiyah, D. (2022). The role of crime in moderating the impact of economic growth and poverty: The case of West Kalimantan. [JEBIK]. *Jurnal Ekonomi Bisnis Dan Kewirausahaan, 11*(1), 84–99.

Jarzabkowski, P., Balogun, J., & Seidl, D. (2007). Strategizing: The challenges of a practice perspective. *Human Relations, 60*(1), 5–27.

Jenkins, A. S., Wiklund, J., & Brundin, E. (2014). Individual responses to firm failure: Appraisals, grief, and the influence of prior failure experience. *Journal of Business Venturing, 29*(1), 17–33.

Jennings, J. E., Edwards, T., Jennings, P. D., & Delbridge, R. (2015). Emotional arousal and entrepreneurial outcomes: Combining qualitative methods to elaborate theory. *Journal of Business Venturing*, *30*(1), 113–130.

Jennings, P. D., Greenwood, R., Lounsbury, M. D., & Suddaby, R. (2013). Institutions, entrepreneurs, and communities: A special issue on entrepreneurship. *Journal of Business Venturing*, *28*(1), 1–9.

Jessop, B. (2002). Liberalism, neoliberalism, and urban governance: A state–theoretical perspective. *Antipode*, *34*(3), 452–472.

Jonas, H. (1979). The Imperative of Responsibility. In *Search of an Ethics for the Technological Age*. University of Chicago Press.

Jones, C., Latham, J., & Betta, M. (2019). Narrative construction of the social entrepreneur: A case study. *International Journal of Entrepreneurial Behaviour & Research*, *25*(2), 288–307.

Jorgenson, D. W., & Fraumeni, B. M. (2020). Investment in education and US economic growth. In *The US savings challenge* (pp. 114–149). Routledge.

Jose, A. V. (2017). Agricultural wages in Indian states. *The Indian Journal of Labour Economics : the Quarterly Journal of the Indian Society of Labour Economics*, *60*(3), 333–345. DOI: 10.1007/s41027-018-0111-x

Joshi, V., & Little, I. M. D. (1994). *India : macroeconomics and political economy, 1964-1991*. World Bank Publications.

Kahneman, D. (2011). *Thinking, Fast and Slow*. Farrar, Straus and Giroux.

Kania, J., Hanleybrown, F., & Splansky Juster, J. (2018). The essentials of collective impact. *Stanford Social Innovation Review*, *16*(1), 39–41.

Kannan, K. P. (2022). India's Elusive Quest for Inclusive Development: An Employment Perspective. *The Indian Journal of Labour Economics : the Quarterly Journal of the Indian Society of Labour Economics*, *65*(3), 579–623. DOI: 10.1007/s41027-022-00393-7 PMID: 36246845

Kaperonis, S. (2025). *AI-Powered Personalization: Boosting User Engagement and Customer Experience*. AI Impacts on Branded Entertainment and Advertising. [REMOVED HYPERLINK FIELD]

Kaplan, B., & Duchon, D. (1988). Combining qualitative and quantitative methods in information systems research: A case study. *Management Information Systems Quarterly*, •••, 571–586.

Kaplan, S. (2008). Framing contests: Strategy making under uncertainty. *Organization Science, 19*(5), 729–752.

Kaplan, S., & Orlikowski, W. J. (2013). Temporal work in strategy making. *Organization Science, 24*(4), 965–995.

Kaplan, S., & Tripsas, M. (2008). Thinking about technology: Applying a cognitive lens to technical change. *Research Policy, 37*(5), 790–805.

Kaye, J. (1998). *Economy and Nature in the Fourteenth Century: Money, Market Exchange, and the Emergence of Scientific Thought*. Cambridge University Press.

Kelly, M. (2012). *Owning our future: The emerging ownership revolution*. Berrett-Koehler.

Kelly, M. (2012). *Owning Our Future: The Emerging Ownership Revolution*. Berrett-Koehler.

Kenney, M., & Mowery, D. C. (Eds.). (2014). *Public universities and regional growth: Insights from the University of California*. Stanford University Press.

Kenter, J. O., Raymond, C. M., van Riper, C. J., Azzopardi, E., Brear, M. R., Calcagni, F., Christie, I., Christie, M., Fordham, A., Gould, R. K., Ives, C. D., Hejnowicz, A. P., Gunton, R., Horcea-Milcu, A.-I., Kendal, D., Kronenberg, J., Massenberg, J. R., O'Connor, S., Ravenscroft, N., & Thankappan, S. (2019). Loving the mess: Navigating diversity and conflict in social values for sustainability. *Sustainability Science, 14*(5), 1439–1461. DOI: 10.1007/s11625-019-00726-4

Keynes, J. M. (1936). *The general theory of employment, interest and money*. Macmillan.

Keynes, J. M. (1936). *The general theory of employment, interest, and money*. Macmillan.

Keynes, J. M. (1936). *The General Theory of Employment*. Interest and Money.

Khan, M. A., Zubair, S. S., Rathore, K., Ijaz, M., Khalil, S., & Khalil, M. (2021). Impact of entrepreneurial orientation dimensions on performance of small enterprises: Do entrepreneurial competencies matter? *Cogent Business & Management, 8*(1), 1943241.

Kleissner, C. (2019). *Deep Impact Investing: A Case Study in Designing for Good*. Toniic Institute.

Koenigs, A. (2018). Storytelling for innovation and sustainability: Exploring narrative dynamics in entrepreneurial settings. *Sustainability, 10*(10), 3547.

Koleva, P. (2023). Critical analysis of grounded theory strategy in organisational and management research. *Qualitative Research Journal*. Advance online publication. DOI: 10.1108/qrj-08-2022-0118

Komodromos, M. (2025). A Literature Review on The Emergence of Entrepreneurship Education: Development, Trends, and Challenges in Masouras, A., Anastasiadou, S., Constantelou, A., & Apostolopoulos, S., pp. 21-46 (Eds.). (2025). Real-World Tools and Scenarios for Entrepreneurship Exploration. IGI Global. DOI: 10.4018/979-8-3693-9040-5

Komodromos, M. (2024). Gender Barriers Experienced by Women Entrepreneurs in Cyprus. In Dana, L. P., & Chhabra, M. (Eds.), *Women Entrepreneurship Policy*. Springer., DOI: 10.1007/978-981-97-3607-2_9

Komodromos, M., Anastasiadou, S., & Seremeti, L. (2026). Transforming Corporate Sustainability: Integrating Advanced Analytics and Business Intelligence Into ESG Implementation. In Komodromos, M., Seremeti, L., Anastasiadis, L., Liargovas, P., & Anastasiadou, S. (Eds.), *Data-Driven ESG Strategy Implementation Through Business Intelligence* (pp. 1–36). IGI Global Scientific Publishing., DOI: 10.4018/979-8-3373-5142-1.ch001

Komodromos, M., Harkiolakis, T., Khoshtaria, T., & Vassiliou, M. (2024). Adoption of digital technology and investigation of the impact in the marketing industry for marketing communication professionals: The case study of Cyprus and Greece. *International Journal of Technology Enhanced Learning*, *16*(2), 173–191.

Kopnina, H. (2020). Education for the future? Critical evaluation of education for sustainable development goals. *The Journal of Environmental Education*, *51*(4), 280–291.

Kornberger, M. (2017). The values of strategy: Valuation practices, rivalry and strategic agency. *Organization Studies*, *38*(12), 1753–1773.

Korsgaard, S. (2013). It's really out there: A review of the critique of the discovery view of opportunities. *International Journal of Entrepreneurial Behaviour & Research*, *19*(2), 130–148.

Kotabe, M. (1990). The relationship between offshore sourcing and innovativeness of US multinational firms: An empirical investigation. *Journal of International Business Studies*, *21*(4), 623–638.

Kozinets, R. V. (2022). Algorithmic branding through platform assemblages: Core conceptions and research directions for a new era of marketing and service management. *Journal of Service Management*, *33*(3), 437–452.

KPMG. (2024). Economic Key Facts Germany, available at: https://kpmg.com/de/en/home/insights/overview/economic-key-facts-germany.html (accessed 1 November 2024).

Kraatz, M. S., & Block, E. S. (2008). Organizational implications of institutional pluralism. In Greenwood, R., Oliver, C., Sahlin, K., & Suddaby, R. (Eds.), *The SAGE handbook of organizational institutionalism* (pp. 243–275). SAGE Publications.

Kraemer, C. (2023). German firms keep investing in China despite moves to 'de-risk´. *Reuters,* May 10, available at: https://www.reuters.com/business/german-firms-keep-investing-china-despite-moves-de-risk-2023-05-10/ (accessed April 3, 2024).

Kraemer, C. (2023). German firms keep investing in China despite moves to 'de-risk´. *Reuters,* May 10, available at: https://www.reuters.com/business/german-firms-keep-investing-china-despite-moves-de-risk-2023-05-10/ (accessed April 3, 2025).

Kramer, J. P., Revilla Diez, J., Marinelli, E., & Iammarino, S. (2010). Intangible assets and MNEs' locational strategies for innovation—or: Why the *regional* matters. [Review of Regional Research]. *Jahrbuch für Regionalwissenschaft*, *30*, 129–157.

Kreiterling, C. (2023). Digital innovation and entrepreneurship: A review of challenges in competitive markets. *Journal of Innovation and Entrepreneurship*, *12*, 1–13. DOI: 10.1186/s13731-023-00320-0

Krippendorff, K. (2019). *Content Analysis: An Introduction to Its Methodology*. SAGE Publications, Inc., DOI: 10.4135/9781071878781

Kshetri, N. (2011). The Indian environment for entrepreneurship and small business development. *Studia Universitatis Babes Bolyai-Negotia*, *56*(4), 35–52.

Kuemmerle, W. (1999). The drivers of foreign direct investment into research and development: And empirical investigation. *Journal of International Business Studies*, *30*(1), 1–24.

Kumar, V., Ashraf, A. R., & Nadeem, W. (2024). AI-powered marketing: What, where, and how? *International Journal of Information Management*, *77*, 102783.

Kusbandono, D., Zaakiyyah, H., & Junaidi, A. (2024). *THE ROLE OF INTERPERSONAL COMMUNICATION IN BUILDING LEADERSHIP THAT POSITIVELY IMPACTS BUSINESS PERFORMANCE IN THE ENTREPRENEURIAL CONTEXT. Technopreneurship and Educational Development Review*. TENDER., DOI: 10.61100/tender.v1i2.187

Lab, B. (2023). B Corporation Directory. https://www.bcorporation.net

Lab, B. (2024). *Certified B Corporation Directory*. https://www.bcorporation.net

Labov, W. (1972). *Language in the inner city*. University of Pennsylvania Press.

Lakoff, G., & Johnson, M. (1980). The metaphorical structure of the human conceptual system. *Cognitive Science*, *4*(2), 195–208.

Lamberton, C., & Stephen, A. T. (2016). A thematic exploration of digital, social media, and mobile marketing: Research evolution from 2000 to 2015 and an agenda for future inquiry. *Journal of Marketing*, *80*(6), 146–172.

Lamont, M. (2012). *How Professors Think: Inside the Curious World of Academic Judgment*. Harvard University Press.

Landis, J. R., & Koch, G. G. (1977). The measurement of observer agreement for categorical data. *Biometrics*, ●●●, 159–174.

Lapsley, I. (2009). New Public Management: The Cruellest Invention of the Human Spirit? [1]. *Abacus*, *45*(1), 1–21. DOI: 10.1111/j.1467-6281.2009.00275.x

Larsen, J., Boje, D. M., & Bruun, L. (2020). *True storytelling: Seven principles for an ethical and sustainable change-management strategy*. Routledge.

Lazarus, A. T., Zhang, L., Shitu, A. F., Wang, J., & Issifu, J. B. (2025). Decolonizing Global Education: A Sustainable Paradigm for Inclusive Knowledge Systems and Equitable Development in the 21st Century.

Le Bas, C., & Sierra, C. (2002). Location versus country advantages' in R&D activities: Some further results on multinationals' locational strategies. *Research Policy*, *31*(4), 589–609.

Lederach, J. P. (2005). *The Moral Imagination: The Art and Soul of Building Peace*. Oxford University Press.

Leigh, N. G. (2024). *Planning local economic development: Theory and practice*. SAGE publications.

Lê, J. K., & Pradies, C. (2021). How paradoxical tensions mobilize change in nonprofit organizations: Integrating institutional logics and institutional work. *Organization Studies*, *42*(5), 779–803.

Leonardi, P. M., & Vaast, E. (2017). Social media and their affordances for organizing: A review and agenda for research. *The Academy of Management Annals*, *11*(1), 150–188.

León-Gómez, A., Ruiz-Palomo, D., Fernández-Gámez, M. A., & García-Revilla, M. R. (2021). Sustainable tourism development and economic growth: Bibliometric review and analysis. *Sustainability*, *13*(4), 2270.

Letkemann, J. P. W. (2023, July). Sustainability in the Pluriverse: Learning from Global Futures. In *World Congress of Architects* (pp. 207-214). Cham: Springer International Publishing.

Levitt, B., & March, J. G. (1988). Organizational learning. *Annual Review of Sociology, 14*(1), 319–338.

LI, F. (2010). The Information Content of Forward-Looking Statements in Corporate Filings—A Naïve Bayesian Machine Learning Approach. *Journal of Accounting Research, 48*(5), 1049–1102. DOI: 10.1111/j.1475-679X.2010.00382.x

Liang, J., Wang, L., Luo, J., Yan, Y., & Fan, C. (2023). The relationship between student interaction with generative artificial intelligence and learning achievement: Serial mediating roles of self-efficacy and cognitive engagement. *Frontiers in Psychology, 14*, 1285392.

Lichterman, P. (2005). Civic culture at the grass roots. *The Blackwell companion to the sociology of culture*, 383-397.

Licoppe, C., & Smoreda, Z. (2005). Are social networks technologically embedded?: How networks are changing today with changes in communication technology. *Social Networks, 27*(4), 317–335.

Liedtka, J. (2008). Strategy making and the search for authenticity. *Journal of Business Ethics, 80*(2), 237–248.

Li, H., Wang, Y., & Qu, H. (2024, May). Where are we so far? understanding data storytelling tools from the perspective of human-ai collaboration. In *Proceedings of the 2024 CHI Conference on Human Factors in Computing Systems* (pp. 1-19).

Li, J., Shapiro, D. M., Ufimtseva, A., & Zhang, P. (2024). Techno-nationalism and cross-border acquisitions in an age of geopolitical rivalry. *Journal of International Business Studies, 55*, 1190–1203.

Lin, X., Zhao, Y., Ahmad, M., Ahmed, Z., Rjoub, H., & Adebayo, T. S. (2021). Linking innovative human capital, economic growth, and CO2 emissions: An empirical study based on Chinese provincial panel data. *International Journal of Environmental Research and Public Health, 18*(16), 8503.

Liu, Q. (2024). China on cusp of next-generation chip production despite US curbs. *The Financial Times*, February 6, available at: https://www.ft.com/content/b5e0dba3-689f-4d0e-88f6-673ff4452977 (accessed 16 March 2024).

Liu, Q. (2024). China on cusp of next-generation chip production despite US curbs. *The Financial Times*, February 6, available at: https://www.ft.com/content/b5e0dba3-689f-4d0e-88f6-673ff4452977 (accessed 16 March 2025).

Liu, D., Xu, C., Yu, Y., Rong, K., & Zhang, J. (2020). Economic growth target, distortion of public expenditure and business cycle in China. *China Economic Review*, *63*, 101373.

Lounsbury, M., & Glynn, M. A. (2001). Cultural entrepreneurship: Stories, legitimacy, and the acquisition of resources. *Strategic Management Journal*, *22*(6-7), 545–564.

Lounsbury, M., & Glynn, M. A. (2019). *Cultural entrepreneurship: A new agenda for the study of entrepreneurial processes and possibilities*. Cambridge University Press.

Lounsbury, M., Steele, C. W., Wang, M. S., & Toubiana, M. (2021). New directions in the study of institutional logics: From tools to phenomena. *Annual Review of Sociology*, *47*(1), 261–280.

Lounsbury, M., & Wang, M. S. (2020). Into the clearing: Back to the future of constitutive institutional analysis. *Organization Theory*, *1*(1), 2631787719891173.

Lowe, N. J., & Gertler, M. S. (2009). Building on diversity: Institutional foundations of hybrid strategies in Toronto's life sciences complex. *Regional Studies*, *43*(4), 589–603.

Lowrie, I. (2017). Algorithmic rationality: Epistemology and efficiency in the data sciences. *Big Data & Society*, *4*(1), 2053951717700925.

Lubinski, C., & Wadhwani, R. D. (2020). Geopolitical jockeying: Economic nationalism and multinational strategy in historical perspective. *Strategic Management Journal*, *41*, 400–421.

Luna-Cortés, G. (2017). The influence of symbolic consumption on experience value and the use of virtual social networks. *Spanish Journal of Marketing - ESIC*, *21*(1), 39–51.

Luo, Y. (2022). Illusions of techno-nationalism. *Journal of International Business Studies*, *53*, 550–567. PMID: 34511654

Luo, Y. (2024). Paradigm shift and theoretical implications for the era of global disorder. *Journal of International Business Studies*, *55*, 127–135.

Luo, Y., & Van Assche, A. (2023). The rise of techno-geopolitical uncertainty: Implications of the United States CHIPS and Science Act. *Journal of International Business Studies*, *54*, 1423–1440. PMID: 37359753

Lynn, L., & Salzman, H. (2023). Techno-nationalism or building a global science and technology commons? (but what about China?). *Global Policy*, *14*, 832–846.

MacKenzie, D. (2006). *An Engine, Not a Camera How Financial Models Shape Markets*. MIT Press.

MacKenzie, D. (2006). Is economics performative? Option theory and the construction of derivatives markets. *Journal of the History of Economic Thought*, *28*(1), 29–55.

Magala, S. (1997). The making and unmaking of sense. *Scandinavian Journal of Management*, *13*(4), 385–398.

Mahalik, M. K., Mallick, H., & Padhan, H. (2021). Do educational levels influence the environmental quality? The role of renewable and non-renewable energy demand in selected BRICS countries with a new policy perspective. *Renewable Energy*, *164*, 419–432.

Mair, J., & Martí, I. (2006). Social entrepreneurship research: A source of explanation, prediction, and delight. *Journal of World Business*, *41*(1), 36–44.

Mair, J., Wolf, M., & Seelos, C. (2016). Scaffolding: A process of transforming patterns of inequality in small-scale societies. *Academy of Management Journal*, *59*(6), 2021–2044.

Makri, C., & Neely, A. (2021). Grounded Theory: A Guide for Exploratory Studies in Management Research. *International Journal of Qualitative Methods*, *20*. Advance online publication. DOI: 10.1177/16094069211013654

Mallach, A. (2018). *The divided city: Poverty and prosperity in urban America*. Island Press.

Malthouse, E., & Copulsky, J. (2022). Artificial intelligence ecosystems for marketing communications. *International Journal of Advertising*, *42*, 128–140. DOI: 10.1080/02650487.2022.2122249

Manafe, M., Ohara, M., Gadzali, S., Harahap, M., & Ausat, A. (2023). Exploring the Relationship Between Entrepreneurial Mindsets and Business Success: Implications for Entrepreneurship Education. *Journal of Education*. Advance online publication. DOI: 10.31004/joe.v5i4.2238

Mansoori, Y., & Lackéus, M. (2020). Comparing effectuation to discovery-driven planning, prescriptive entrepreneurship, business planning, lean startup, and design thinking. *Small Business Economics*, *54*(3), 791–818.

Manzini, E. (2015). *Design, when everybody designs: An introduction to design for social innovation*. MIT Press.

Marlow, S., & McAdam, M. (2012). Analyzing the influence of gender upon high–technology venturing within the context of business incubation. *Entrepreneurship Theory and Practice*, *36*(4), 655–676.

Marlow, S., & Swail, J. (2014). Gender, risk and finance: Why can't a woman be more like a man? *Entrepreneurship and Regional Development*, *26*(1-2), 80–96.

Martens, M. L., Jennings, J. E., & Devereaux Jennings, P. (2007). Do the stories they tell get them the money they need? The role of entrepreneurial narratives in resource acquisition. *Academy of Management Journal*, *50*(5), 1107–1132.

Martí, I., & Fernández, P. (2013). The institutional work of oppression and resistance: Learning fromthe Holocaust. *Organization Studies*, *34*(8), 1195–1223.

Martin, N. (2024). German firms ignore calls to shrink China investments. *Deutsche Welle*. August 26, available at: https://www.dw.com/en/germanys-investments-in-china-increase-despite-warnings/a-69957810 (accessed 2 March 2025).

Mason, C., & Brown, R. (2014). Entrepreneurial ecosystems and growth oriented entrepreneurship. *Final report to OECD, Paris, 30*(1), 77-102.

Matthess, M., & Kunkel, S. (2020). Structural change and digitalization in developing countries: Conceptually linking the two transformations. *Technology in Society*, *63*, 1–13. DOI: 10.1016/J.TECHSOC.2020.101428

Mazutis, D. D., & Slawinski, N. (2015). Reconnecting business and society: Perceptions of authenticity in corporate social responsibility. *Journal of Business Ethics*, *131*(1), 137–150.

Mazzucato, M. (2018). *The value of everything: Making and taking in the global economy*. PublicAffairs.

McAdam, M., & Marlow, S. (2007). Building futures or stealing secrets? Entrepreneurial cooperation and conflict within business incubators. *International Small Business Journal*, *25*(4), 361–382.

McAdams, D. P. (1997). A conceptual history of personality psychology. In *Handbook of personality psychology* (pp. 3–39). Academic Press.

McCann, E., & Ward, K. (2010). Relationality/territoriality: Toward a conceptualization of cities in the world. *Geoforum*, *41*(2), 175–184.

McGinnv, N. F., Snodgrass, D. R., Kim, S. B., Kim, Q. Y., & Kim, Y. B. (2020). *Education and development in Korea* (Vol. 90). BRILL.

McGloin, R., Saxton, A., Rao, A., Hintz, E., Coletti, A., Hamlin, E., Turner, M., & Mathieu, J. (2023). Examining the Importance of Developing Entrepreneurial Communication Skills in Accelerator Programs: A Focus Group Based Approach. *Entrepreneurship Education and Pedagogy*, *7*, 468–493. DOI: 10.1177/25151274231183382

McGrath, R. G. (2010). Business models: A discovery driven approach. *Long Range Planning*, *43*(2-3), 247–261.

McMillan, M., Rodrik, D., & Verduzco-Gallo, Í. (2014). Globalization, structural change, and productivity growth, with an update on Africa. *World Development*, *63*, 11–32. DOI: 10.1016/j.worlddev.2013.10.012

Meadows, D. (1999). Leverage points. *Places to Intervene in a System*, *19*, 28.

Meadows, D. (2008). *Thinking in systems: International bestseller*. Chelsea Green Publishing.

Meadows, D. H. (2008). *Thinking in systems: A primer* (Wright, D., Ed.). Chelsea Green Publishing.

Meadows, D. H. (2008). *Thinking in Systems: A Primer*. Chelsea Green.

Mehrotra, S., Gandhi, A., Kumar, C., Saha, P., Sahoo, B. K., & Sharma, A. (2011). *India Human Development Report 2011 - Towards Social Inclusion* .

Merton, R. K. (1948). The self-fulfilling prophecy. *The Antioch Review*, *8*(2), 193–210.

Meyer, K. E., Mudambi, R., & Narula, R. (2011). Multinational enterprises and local contexts: The opportunities and challenges of multiple embeddedness. *Journal of Management Studies*, *48*(2), 235–252.

Michailova, S., & Zaidah, M. (2012). Subsidiary knowledge flows in multinational corporations: Research accomplishments, gaps, and opportunities. *Journal of World Business*, *47*(3), 383–396.

Milan, S., & Treré, E. (2019). Big data from the South (s): Beyond data universalism. *Television & New Media*, *20*(4), 319–335.

Miller, P., & Power, M. (2013). Accounting, Organizing, and Economizing: Connecting Accounting Research and Organization Theory. *The Academy of Management Annals*, *7*(1), 557–605. DOI: 10.5465/19416520.2013.783668

Milne, R. (2025). German companies are risking a 'lose-lose' decoupling with China. *Financial Times*, January 30, available at https://www.ft.com/content/f61d6b7b-0250-4edf-aac1-ba0ca18545cd (accessed 1 April 2025).

Minoia, P., & Castro-Sotomayor, J. (2024). Education and socio-environmental justice in the pluriverse: Decolonial perspectives. *Globalizations*, *21*(2), 303–312.

Mintzberg, H. (1973). Strategy-making in three modes. *California Management Review*, *16*(2), 44–53.

Miozzo, M., & Soete, L. (2001). Internationalization of Services. *Technological Forecasting and Social Change*, *67*(2–3), 159–185. DOI: 10.1016/S0040-1625(00)00091-3

Mishra, P. K., Mishra, S. K., & Sarangi, M. K. (2020). Do women's advancement and gender parity promote economic growth? Evidence from 30 Asian countries. *Millennial Asia*, *11*(1), 5–26.

Mishra, R. (1999). *Globalization and the Welfare State*. Edward Elgar Publishing Ltd.

Mohiuddin Babu, M., Akter, S., Rahman, M., Billah, M. M., & Hack-Polay, D. (2024). The role of artificial intelligence in shaping the future of Agile fashion industry. *Production Planning and Control*, *35*(15), 2084–2098.

Mohr, J. W., Wagner-Pacifici, R., Breiger, R. L., & Bogdanov, P. (2013). Graphing the grammar of motives in National Security Strategies: Cultural interpretation, automated text analysis and the drama of global politics. *Poetics*, *41*(6), 670–700.

Molyneux, L., & Holton, A. (2015). Branding (health) journalism: Perceptions, practices, and emerging norms. *Digital Journalism (Abingdon, England)*, *3*(2), 225–242.

Moore, M. L., & Westley, F. (2011). Surmountable chasms: Networks and social innovation for resilient systems. *Ecology and Society*, *16*(1).

Moulaert, F., MacCallum, D., & Hillier, J. (2013). Social innovation: Intuition, precept, concept, theory and practice. In The international handbook on social innovation: Collective action, social learning and transdisciplinary research (pp. 13-24).

Moulaert, F., & Sekia, F. (2003). Territorial innovation models: A critical survey. *Regional Studies*, *37*(3), 289–302.

Mulgan, G. (2014). *The locust and the bee: Predators and creators in capitalism's future*. Princeton University Press.

Munir, K. A. (2005). The social construction of events: A study of institutional change in the photographic field. *Organization Studies*, *26*(1), 93–112.

Myovella, G., Karacuka, M., & Haucap, J. (2020). Digitalization and economic growth: A comparative analysis of Sub-Saharan Africa and OECD economies. *Telecommunications Policy*, *44*(2), 101856.

Naeem, M., Ozuem, W., Howell, K., & Ranfagni, S. (2023). A Step-by-Step Process of Thematic Analysis to Develop a Conceptual Model in Qualitative Research. *International Journal of Qualitative Methods*, *22*. Advance online publication. DOI: 10.1177/16094069231205789

Nama, N. P. (2021). Enhancing user experience in mobile applications through AI-driven personalization and adaptive learning algorithms. World Journal of Advanced Engineering Technology and Sciences, 3(2), 083-094.

Nambisan, S., Lyytinen, K., Majchrzak, A., & Song, M. (2017). Digital innovation management. *Management Information Systems Quarterly*, *41*(1), 223–238.

Narula, R. (2014). Exploring the paradox of competence-creating subsidiaries: Balancing bandwidth and dispersion in MNEs. *International Journal of Strategic Management*, *47*(1–2), 4–15.

NASSCOM. (2020), The Technology Sector in India: Strategic Review 2020 (NASSCOM Strategic Review), NASSCOM, New Delhi. URL: https://nasscom.in/knowledge-center/publications/technology-sector-india-2020-techade-new-decade-strategic-review (Planning Commission of India. (2013). Twelfth Five Year Plan (2012–2017) Faster, More Inclusive and Sustainable Growth: Vol. Volume I. SAGE Publications India Pvt Ltd.

Navis, C., & Glynn, M. A. (2010). How new market categories emerge: Temporal dynamics of legitimacy, identity, and entrepreneurship in satellite radio, 1990–2005. *Administrative Science Quarterly*, *55*(3), 439–471.

Navis, C., & Glynn, M. A. (2011). Legitimate distinctiveness and the entrepreneurial identity: Influence on investor judgments of new venture plausibility. *Academy of Management Review*, *36*(3), 479–499.

Neil Alim, A., & Jones, S. (2024). German companies' dependence on China will last decades, warns Siemens. *Financial Times*, April 14, available at: https://www.ft.com/content/e8634f5c-5a28-4741-a13d-024665456f43?sharetype=blocked (accessed 2 April 2025).

Nicholls, A. (2010). The legitimacy of social entrepreneurship: Reflexive isomorphism in a pre-paradigmatic field. *Entrepreneurship Theory and Practice*, *34*(4), 611–633.

Nicholls, A., & Murdock, A. (Eds.). (2012). *Social innovation: Blurring boundaries to reconfigure markets*. Palgrave Macmillan.

Nicoli, N., & Komodromos, M. (2019). CSR Communication in the Digital Age. The Case of The Bank of Cyprus. In Alexandros, A., & Evi, D. (Eds.), *Cases on Corporate Social Responsibility and Contemporary Issues in Organizations* (pp. 71–89). IGI Global.

Nisbet, M. C. (2009). Communicating climate change: Why frames matter for public engagement. *Environment*, *51*(2), 14–18.

Nonaka, L., Takeuchi, H., & Umemoto, K. (1996). A theory of organizational knowledge creation. *International Journal of Technology Management*, *11*(7-8), 833–845.

North, D. C. (1990). *Institutions, Institutional Change and Economic Performance*. Cambridge University Press.

Novoa, M., & Vasudevan, R. (2025). Relational Engagements and Embodiments: Exploring Pluriversalities of Community-Based Scholarship in Planning. *Planning Theory & Practice*, *26*(1), 85–105.

O'Callaghan, C., Dwyer, J., & Schofield, P. (2024). Thematic analysis informed by grounded theory (TAG) in healthcare research: Foundations and applications. *Qualitative Research in Psychology*, *21*, 279–306. DOI: 10.1080/14780887.2024.2347580

O'Dwyer, B., & Owen, D. L. (2005). Assurance statement practice in environmental, social and sustainability reporting: A critical evaluation. *The British Accounting Review*, *37*(2), 205–229. DOI: 10.1016/j.bar.2005.01.005

Obama, B. (2009). *Remarks by the President on the American Recovery and Reinvestment Act*. The White House.

OECD. (2009). *Economic outlook* (No. 86). OECD Publishing.

OECD. (2021). *COVID-19 and the global economy: Initial insights and policy responses*. OECD Publishing.

Ofosu-Ampong, K. (2024). *Artificial intelligence research: A review on dominant themes, methods, frameworks and future research directions*. Telematics and Informatics Reports.

Ojala, A., Evers, N., & Rialp, A. (2018). Extending the international new venture phenomenon to digital platform providers: A longitudinal case study. *Journal of World Business*, *53*(5), 725–739.

Oliinyk, O., Bilan, Y., Mishchuk, H., Akimov, O., & Vasa, L. (2021). *The impact of migration of highly skilled workers on the country's competitiveness and economic growth*. Montenegrin Journal of Economics.

Oosthuizen, K., Botha, E., Robertson, J., & Montecchi, M. (2021). Artificial intelligence in retail: The AI-enabled value chain. *Australasian Marketing Journal, 29*(3), 264–273.

Orji, G. (2025). *Influence of Digital Ecosystems on Value Creation and Competitive Positioning in the E-Commerce Industry.* European Journal of Business and Strategic Management., DOI: 10.47604/ejbsm.3330

Orlikowski, W. J., & Scott, S. V. (2023). The digital undertow and institutional displacement: A sociomaterial approach. *Organization Theory, 4*(2).

Ospina, S. M., & Dodge, J. (2005). It's about time: Catching method up to meaning—The usefulness of narrative inquiry in public administration research. *Public Administration Review, 65*(2), 143–157.

Ospina, S., & Dodge, J. (2005). Narrative inquiry and the construction of meaning: A review essay. *Public Administration Review, 65*(5), 611–616.

Ostrom, E. (1990). *Governing the commons: The evolution of institutions for collective action.* Cambridge University Press.

Osuntuyi, B. V., & Lean, H. H. (2022). Economic growth, energy consumption and environmental degradation nexus in heterogeneous countries: Does education matter? *Environmental Sciences Europe, 34*(1), 48.

Ozawa, S., Laing, S. K., Higgins, C. R., Yemeke, T. T., Park, C. C., Carlson, R., & Omer, S. B. (2022). Educational and economic returns to cognitive ability in low-and middle-income countries: A systematic review. *World Development, 149*, 105668.

Ozturk, I., & Ullah, S. (2022). Does digital financial inclusion matter for economic growth and environmental sustainability in OBRI economies? An empirical analysis. *Resources, Conservation and Recycling, 185*, 106489.

Pache, A.-C., & Santos, F. (2010). When worlds collide: The internal dynamics of organizational responses to conflicting institutional demands. *Academy of Management Review, 35*(3), 455–476.

Palomeras, N., & Wehrheim, D. (2021). The strategic allocation of inventors to R&D collaborations. *Strategic Management Journal, 42*, 144–169.

Pandey, M., Yadav, P. S., & Pothireddy, R. R. (2024). Role of gamification in influencer marketing. In *Marketing and Gamification* (pp. 157–169). doi:DOI: 10.4324/9781032694238-13

Pandey, M., & Yadav, P. S. (2023). Understanding the role of individual concerns, attitude, and perceived value in green apparel purchase intention; the mediating effect of consumer involvement and moderating role of generation Z&Y. *Cleaner and Responsible Consumption, 9*(100120), 100120. DOI: 10.1016/j.clrc.2023.100120

Papanastassiou, M., Pearce, R., & Zanfei, A. (2020). Changing perspectives on the internationalization of R&D and innovation by multinational enterprises: A review of the literature. *Journal of International Business Studies, 51*, 623–664.

Parhankangas, A., & Ehrlich, M. (2014). How entrepreneurs seduce business angels: An impression management approach. *Journal of Business Venturing, 29*(4), 543–564.

Pascua, D. (2024). *Exploring the Application of Grounded Theory: A Systematic Review of Methodological Approaches*. International Journal of Multidisciplinary Research and Growth Evaluation., DOI: 10.54660/.ijmrge.2024.5.4.1362-1366

PatentScope. (2024). WIPO. Simple search, available at: https://patentscope.wipo.int/search/en/search.jsf (accessed November 2020-April 2024).

PatentScope. (2025). Browse by week, available at: https://patentscope.wipo.int/search/en/ resultWeeklyBrowse.jsf (accessed 29 March 2025).

Patton, M. Q. (2015). *Qualitative Research & EvaluationMethods* (4th ed.). SAGE.

Pentland, B. T., & Feldman, M. S. (2008). Issues in empirical field studies of organizational routines. In Handbook of organizational routines (p. 281).

Pentland, B. T. (1999). Building process theory with narrative: From description to explanation. *Academy of Management Review, 24*(4), 711–724.

Pentland, B. T., & Feldman, M. S. (2007). Narrative networks: Patterns of technology and organization. *Proceedings - Academy of Management, 2007*(1).

Peredo, A. M., & Chrisman, J. J. (2006). Toward a theory of community-based enterprise. *Academy of Management Review, 31*(2), 309–328.

Pereira, A. M., Moura, J. A. B., Costa, E. D. B., Vieira, T., Landim, A. R., Bazaki, E., & Wanick, V. (2022). Customer models for artificial intelligence-based decision support in fashion online retail supply chains. *Decision Support Systems, 158*, 113795.

Perry, M. (2021). Pluriversal literacies: Affect and relationality in vulnerable times. *Reading Research Quarterly, 56*(2), 293–309.

Perry, M. (2023). *Pluriversal literacies for sustainable futures: When words are not enough*. Routledge.

Petkova, A. P., Rindova, V. P., & Gupta, A. K. (2008). How can new ventures build reputation? An exploratory study. *Corporate Reputation Review, 11*(4), 320–341.

Phillips, N., Lawrence, T. B., & Hardy, C. (2004). Discourse and institutions. *Academy of Management Review, 29*(4), 635–652.

Pickering, A. (1995). Concepts and the mangle of practice: Constructing quaternions. *The South Atlantic Quarterly, 94*(2), 417–465.

Pillarisetty, R., & Mishra, P. (2022). A review of AI (artificial intelligence) tools and customer experience in online fashion retail. [IJEBR]. *International Journal of E-Business Research, 18*(2), 1–12.

Pinto Grunfeld, M., & Pinheiro Gomes, N. (2024). Semiotic protocols for cultural trend analysis in strategic communication: Brand cases from the telecommunications sector. *International Journal of Market Research, 66*(5), 567–588.

Polanyi, K. (1944). *The great transformation: The political and economic origins of our time*. Beacon Press.

Politis, D., & Gabrielsson, J. (2009). Entrepreneurs' attitudes towards failure: An experiential learning approach. *International Journal of Entrepreneurial Behaviour & Research, 15*(4), 364–383.

Polletta, F., Chen, P. C. B., Gardner, B. G., & Motes, A. (2011). The sociology of storytelling. *Annual Review of Sociology, 37*(1), 109–130.

Polletta, F., & Lee, J. (2006). Is telling stories good for democracy? Rhetoric in public deliberation after 9/11. *American Sociological Review, 71*(5), 699–721.

Pollock, T. G., & Rindova, V. P. (2003). Media legitimation effects in the market for initial public offerings. *Academy of Management Journal, 46*(5), 631–642.

Popp, A., & Holt, R. (2013). Entrepreneurship and being: The case of the Shaws. *Entrepreneurship and Regional Development, 25*(1–2), 52–68.

Powell, J. (2020). Remarks at the Jackson Hole Economic Symposium. *Federal Reserve Bank of Kansas City*.

Powell, M., Gillett, A., & Doherty, B. (2019). Sustainability in social enterprise: Hybrid organizing in public services. *Public Management Review, 21*(2), 159–186.

Powell, W. W., Oberg, A., Korff, V. P., & Brandsen, T. (2017). Innovating in the public interest: Managing complex coordination and institutional change in public sector ecosystems. *Public Management Review, 19*(5), 655–674.

Powell, W. W., & Rerup, C. (2017). Opening the black box: Micro-foundations of routines and capabilities. In Felin, F. A. (Ed.), *Microfoundations* (pp. 1–32). Emerald Publishing.

Power, M. (1997). *The Audit Society Rituals of Verification*. Oxford University Press.

Prabhu, J., & Jain, S. (2015). Innovation and entrepreneurship in India: Understanding jugaad. *Asia Pacific Journal of Management, 32*(4), 843–868.

Pradhan, B. (2002). *The role of education in wage inequality change in India: 1988–97*. National Council of Applied Economic Research.

Prasetyo, P. E., & Kistanti, N. R. (2020). Human capital, institutional economics and entrepreneurship as a driver for quality & sustainable economic growth. *Entrepreneurship and Sustainability Issues, 7*(4), 2575.

Purdy, J. (2015). *After nature: A politics for the Anthropocene*. Harvard University Press.

Quinn, G. (2011). PCT Basics: Understanding the international filing process, available at: https://www.ipwatchdog.com/2011/11/03/pct-basics-understanding-the-international-¯ling-process/id=19960

Quinn, G. (2011). PCT Basics: Understanding the international filing process, available at: https://www.ipwatchdog.com/2011/11/03/pct-basics-understanding-the-international-¯ling-process/id=19960/ (accessed 3 November 2021).

Qureshi, A. (2025). Glimpsing Speculative Utopias: Envisioning Futures. *HUB-Journal of Research in Art, Design and Society*, (4).

Ranis, G., Stewart, F., & Ramirez, A. (2000). Economic Growth and Human Development. *World Development, 28*(2), 197–219. DOI: 10.1016/S0305-750X(99)00131-X

Rao, D. T., Sethi, N., Dash, D. P., & Bhujabal, P. (2023). Foreign aid, FDI and economic growth in South-East Asia and South Asia. *Global Business Review, 24*(1), 31–47.

Ratnasari, A., Sofyan, A., & Yuniati, Y. (2023). *Business Communication in Improving the Entrepreneurial Skills of SME Members*. KnE Social Sciences., DOI: 10.18502/kss.v8i18.14252

Rawlins, B. (2008). Give the emperor a mirror: Toward developing a stakeholder measurement of organizational transparency. *Journal of Public Relations Research, 21*(1), 71–99.

Raworth, K. (2017). *Doughnut economics: Seven ways to think like a 21st-century economist.* Chelsea Green Publishing.

Reissner, S. C. (2005). Learning and innovation: A narrative analysis. *Journal of Organizational Change Management, 18*(5), 482–494.

Reserve Bank of India. (2023). *Handbook of Statistics on Indian Economy.* https://www.rbi.org.in/Scripts/PublicationsView.aspx?id=21807

Reuber, A. R., & Fischer, E. (2011). International entrepreneurship in internet-enabled markets. *Journal of Business Venturing, 26*(6), 660–679.

Reynolds-Cuéllar, P., Asoyarcocha, M., Matabonchoy, L., de la Cruz, Y. Y., Hidalgo, C., & Bonilla, O. (2024). Investigative Mingas: An Approach to Designing Sustainable, Pluriversal Futures. In *Routledge Handbook of Sustainable Design* (pp. 572-586). Routledge.

Reynolds, L., Koenig-Lewis, N., Doering, H., & Peattie, K. (2022). Competing for legitimacy in the place branding process:(re) negotiating the stakes. *Tourism Management, 91*, 104532.

Rhodes, C., & Brown, A. D. (2005). Narrative, organizations and research. *International Journal of Management Reviews*, *7*(3), 167–188.

Ribakova, E. (2024). Our experience with Russia holds lessons for future sanctions. *The Financial Times*, February 27, available at: https://www.ft.com/content/784eae1d-1fb8-455b-b3d3-2ddc4715499d (accessed 16 March 2024).

Ribakova, E. (2024). Our experience with Russia holds lessons for future sanctions. *The Financial Times*, February 27, available at: https://www.ft.com/content/784eae1d-1fb8-455b-b3d3-2ddc4715499d (accessed 16 March 2025).

Ricoeur, P. (1984). *Time and narrative* (Vol. 1; McLaughlin, K., & Pellauer, D., Trans.). University of Chicago Press.

Riessman, C. K. (2008). Narrative methods for the human sciences. *Sage (Atlanta, Ga.).*

Rincón, A. G., Barbosa, R. L. C., Mateus, M. A., & Saavedra, N. O. (2023). Symbolic consumption as a non-traditional predictor of brand loyalty in the sports industry, football club segment. *Heliyon, 9*(4). PMID: 37151644

Rindova, V. P., & Martins, L. L. (2021). Shaping possibilities: A design science approach to developing novel strategies. *Academy of Management Review, 46*(4), 800–822.

Rindova, V., Barry, D., & Ketchen, D. J.Jr. (2009). Entrepreneuring as emancipation. *Academy of Management Review*, *34*(3), 477–491.

Rindova, V., Dalpiaz, E., & Ravasi, D. (2011). A cultural quest: A study of organizational use of new cultural resources in strategy formation. *Organization Science*, *22*(2), 413–431.

Rinke, A., Waldersee, V., & Marsh, S. (2022). German business chiefs clash with Berlin over China policies. *Reuters*, October 14, available at: https://www.euronews.com/next/2022/10/14/germany-china-insight (accessed 15 April 2024).

Rinke, A., Waldersee, V., & Marsh, S. (2022). German business chiefs clash with Berlin over China policies. *Reuters*, October 14, available at: https://www.euronews.com/next/2022/10/14/germany-china-insight (accessed 15 November 2024).

Ritala, P. (2023). Grand challenges and platform ecosystems: Scaling solutions for wicked ecological and societal problems. *Journal of Product Innovation Management*. Advance online publication. DOI: 10.1111/jpim.12682

Rodrik, D. (2015). *Economics Rules: The Rights and Wrongs of the Dismal Science*. Norton.

Rodrik, D. (2016). Premature deindustrialization. *Journal of Economic Growth*, *21*(1), 1–33. DOI: 10.1007/s10887-015-9122-3

Roos, M. W., & Reccius, T. (2021). Narratives in macroeconomics: Definitions, typologies, and applications. *Journal of Economic Methodology*, *28*(3), 267–289. DOI: 10.1080/1350178X.2021.1901527

Ross, F. (2019). Kate Raworth - Doughnut Economics: Seven Ways to Think Like a 21st Century Economist (2017). *Regional and Business Studies*, *11*(2). Advance online publication. DOI: 10.33568/rbs.2409

Roundy, P. T. (2017). Hybrid organizations and the logics of entrepreneurial ecosystems. *The International Entrepreneurship and Management Journal*, *13*(4), 1221–1237.

Roundy, P. T., Bradshaw, M., & Brockman, B. K. (2018). The emergence of entrepreneurial ecosystems: A complex adaptive systems approach. *Journal of Business Research*, *86*, 1–10.

Ruebottom, T. (2013). The microstructures of rhetorical strategy in social entrepreneurship: Building legitimacy through heroes and villains. *Journal of Business Venturing*, *28*(1), 98–116.

Ruebottom, T., & Auster, E. R. (2018). Reflexive dis/embedding: Personal narratives, empowerment and the emotional dynamics of interstitial events. *Organization Studies, 39*(4), 467–490.

Rugman, A. M., & Verbeke, A. (2001). Subsidiary-specific advantages in multinational enterprises. *Strategic Management Journal, 22*(3), 237–250.

Saha, R. (2024). Empowering Absorptive Capacity in Healthcare Supply Chains Through Big Data Analytics and AI-Driven Collaborative Platforms: A Prisma-Based Systematic Review. Innovatech Engineering Journal.

Sahoo, G., Wani, A. M., Swamy, S. L., Roul, P. K., Dash, A. C., & Sharma, A. (2022). Livelihood Strategy and Sustainability Aspects in Industrialization as a Source of Employment in Rural Areas. In *Social Morphology* (pp. 643–670). Human Welfare, and Sustainability.

Saijo, T. (2020). Future Design: A New Way to Consider Future Generations. *Economics of Peace and Security Journal, 15*(2), 5–8.

Saijo, T. (2020). Future design: A new way to consider future generations. *Economics of Peace and Security Journal, 15*(2), 5–8. DOI: 10.15355/epsj.15.2.5

Saleh, H., Surya, B., Annisa Ahmad, D. N., & Manda, D. (2020). The role of natural and human resources on economic growth and regional development: With discussion of open innovation dynamics. *Journal of Open Innovation, 6*(4), 103.

Sandberg, J., Holmström, J., Lyytinen, K., & Hultin, L. (2019). Digital transformation of work: Will it change the skills employees need? In *Proceedings of the 25th Americas Conference on Information Systems*.

Sandoval-Llanos, J., Becerra, T., Ojeda-Ramirez, S., Hakim, N., & Turano, G. D. (2025). Pluriversal Frameworks for Equitable and Sustainable Learning Designs. Knowledge Co-construction Experiences along Lands, Territories, and AI in Colombia. In *Proceedings of the 19th International Conference of the Learning Sciences-ICLS 2025, pp. 664-672*. International Society of the Learning Sciences.

Santos, F. M. (2012). A positive theory of social entrepreneurship. *Journal of Business Ethics, 111*(3), 335–351.

Sarasvathy, S. D. (2005). Causation and effectuation: Toward a theoretical shift from economic inevitability to entrepreneurial contingency. *Academy of Management Review, 30*(2), 243–263.

Satar, M., Alshibani, S., & Alarifi, G. (2024). Effects of Firm-Level Entrepreneurship Orientation on Digital Transformation in SMEs: The Moderating Role of Strategic Agility. *Entrepreneurship Research Journal, 15*, 91–124. DOI: 10.1515/erj-2023-0267

Saxenian, A. (1996). *Regional advantage: Culture and competition in Silicon Valley and Route 128*. Harvard University Press.

Scharmer, O. (2018). *The essentials of Theory U: Core principles and applications*. Berrett-Koehler Publishers.

Scherngell, T., Schwegmann, K., and Zahradnik, G. (2023). The geographical dynamics of global R&D collaboration networks in robotics: Evidence from co-patenting activities across urban areas worldwide. *PLoS One*, Vol. 18 No. 4, doi: . pone.0281353.DOI: 10.1371/journal

Schick, A. (2014). The metamorphoses of performance budgeting. *OECD Journal on Budgeting, 13*(2), 49–79. DOI: 10.1787/budget-13-5jz2jw9szgs8

Schmitz, R. (2024). German government wants companies to ‘de-risk’ from China, but business is reluctant. National Public Radio, January 19, available at: https://www.kpbs.org/news/international/2024/01/19/german-government-wants-companies-to-de-risk-from-china-but-business-is-reluctant (accessed 1 November 2024).

Schoar, A. (2010). The divide between subsistence and transformational entrepreneurship. [). University of Chicago Press/NBER.]. *Innovation Policy and the Economy, 10*, 57–81.

Schoenmaker, D., & Schramade, W. (2019). *Principles of Sustainable Finance*.

Schoenmaker, D., & Schramade, W. (2019). *Principles of sustainable finance*. Oxford University Press.

Scholz, T. (2017). *Uberworked and underpaid: How workers are disrupting the digital economy*. Polity Press.

Schön, D. A. (1983). *The reflective practitioner: How professionals think in action*. New York: Scuotto, V., Magni, D., Garcia-Perez, A., & Pironti, M. (2024). The impact of innovation failure: Entrepreneurship adversity or opportunity? *Technovation, 131*, 102944.

Schön, D., & Rein, M. (1994). *Frame reflection: Toward the resolution of intractable policy controversies*. Basic Book.

Schumpeter, J. A. (1942). *Capitalism, Socialism and Democracy*. Harper & Row.

Schumpeter, J. A. (1942). Capitalism. *Socialism and Democracy*.

Scott, A. J. (Ed.). (2001). *Global city-regions: trends, theory, policy*. OUP Oxford.

Scott, S., Hughes, M., & Ribeiro-Soriano, D. (2021). Towards a network-based view of effective entrepreneurial ecosystems. *Review of Managerial Science*, *16*, 157–187. DOI: 10.1007/S11846-021-00440-5

Seelos, C., & Mair, J. (2007). How social entrepreneurs enable human, social, and economic development. *Business solutions for the global poor: Creating social and economic value*, 271-294.

Sen, A. (1960). Choice of techniques: An aspect of the theory of planned economic development. *Oxford Economic Papers*, *12*(I), 1–15.

Sen, A. (1987). *On Ethics and Economics*. Blackwell.

Sen, A. (1999). *Development as freedom*. Knopf.

Seo, M.-G., & Creed, W. E. D. (2002). Institutional contradictions, praxis, and institutional change: A dialectical perspective. *Academy of Management Review*, *27*(2), 222–247.

Seregina, A., & Weijo, H. A. (2017). Play at any cost: How cosplayers produce and sustain their ludic communal consumption experiences. *The Journal of Consumer Research*, *44*(1), 139–159.

Setia, N., Abhayawansa, S., Joshi, M., & Huynh, A. V. (2015). Integrated reporting in South Africa: Some initial evidence. *Sustainability Accounting. Management and Policy Journal*, *6*(3), 397–424. DOI: 10.1108/SAMPJ-03-2014-0018

Shanahan, E. A., McBeth, M. K., & Hathaway, M. R. (2011). Narrative policy framework: The influence of policy narratives on public opinion. *Policy Studies Journal: the Journal of the Policy Studies Organization*, *39*(3), 535–557.

Shane, S., & Venkataraman, S. (2000). The promise of entrepreneurship as a field of research. *Academy of Management Review*, *25*(1), 217–226.

Shankar, V., Kalyanam, K., Setia, P., Golmohammadi, A., Tirunillai, S., Douglass, T., & Waddoups, R. (2021). How technology is changing retail. *Journal of Retailing*, *97*(1), 13–27.

Shepherd, D. A. (2003). Learning from business failure: Propositions of grief recovery for the self-employed. *Academy of Management Review*, *28*(2), 318–328.

Shepherd, D. A., & Kuratko, D. F. (2009). The death of an innovative project: How grief recovery enhances learning. *Business Horizons*, *52*(5), 451–458.

Shepherd, D. A., & Patzelt, H. (2011). The new field of sustainable entrepreneurship: Studying entrepreneurial action linking "what is to be sustained" with "what is to be developed". *Entrepreneurship Theory and Practice*, *35*(1), 137–163.

Shiller, R. J. (2000). *Irrational exuberance*. Princeton University Press.

Shiller, R. J. (2017). Narrative economics. *The American Economic Review*, *107*(4), 967–1004. DOI: 10.1257/aer.107.4.967

Shiller, R. J. (2019). *Narrative economics: How stories go viral and drive major economic events*. Princeton University Press.

Shiller, R. J. (2019). *Narrative Economics: How Stories Go Viral and Drive Major Economic Events*. Princeton University Press.

Shimizutani, S., & Todo, Y. (2008). What determines overseas R&D activities? The case of Japanese multinational firms. *Research Policy*, *37*(3), 530–544.

Sica, G., Palazzo, M., Micozzi, A., & Ferri, M. A. (2025). Leveraging on cultural and creative industries to foster social innovation: A bibliometric analysis. *Journal of Innovation & Knowledge*, *10*(1), 100649.

Siemens (2019). Siemens catalyzes digital transformation of China with innovation ecosystem, May 15, available at: https://w1.siemens.com.cn/press/NewsDetail_en.aspx?ColumnId=9&ArticleId=8014 (accessed 1 November, 2024).

Siemens (2021). Press Releases. Siemens to establish electrical products China headquarters in Suzhou, China, June 21, available at: https://w1.siemens.com.cn/news_en/news_articles_en/17472.aspx (accessed 20 April 2025).

Siemens (2021). Press Releases. Siemens to establish electrical products China headquarters in Suzhou, China, June 21, available at: https://w1.siemens.com.cn/news_en/news_articles_en/17472.aspx (accessed 20 October 2024).

Siemens (2024a). Siemens R&D in China. Innovation for China's development, available at: https://www.siemens.com/cn/en/company/about/research.html (accessed 30 October 2024).

Siemens (2024b). Across borders. Siemens and China – Milestones in a successful partnership, available at: https://www.siemens.com/global/en/company/about/history/stories/siemens-in-china.html (accessed 1 November 2024).

Siemens (2025). Über Grenzen hinweg. Siemens und China – Meilensteine einer erfolgreichen Partnerschaft (Across borders. Siemens and China – Milestones of a Successful Partnership), available at: https://www.siemens.com/de/de/unternehmen/konzern/geschichte/stories/siemens-in-china.html (accessed 6 April 2025).

Simon, H. A. (1947). *Administrative Behavior* (1st ed.). Free Press.

Simon, J. A. (1984). Theoretical perspective on political risk. *Journal of International Business Studies, 15*, 123–143.

Singh, M., Michael, R. A., Saheb, S. S., Yadav, P. S., Kiran, P. B. N., & Malhi, A. (2025). A study on digital intelligence and influencer marketing for sustainable diversification of India's retail economy. In *Advances in Logistics, Operations, and Management Science* (pp. 1–22). doi:DOI: 10.4018/979-8-3693-8492-3.ch001

Singh, M., Syed, S. M., Saheb, S. S., Yadav, P. S., & Kiran, P. B. N. (2025). Role of influencer marketing in the Indian retail business sector. In *Advances in Marketing, Customer Relationship Management, and E-Services* (pp. 105–130). doi:DOI: 10.4018/979-8-3693-8542-5.ch005

Slovic, P., Finucane, M., Peters, E., & MacGregor, D. G. (2004). Risk as analysis and risk as feelings: Some thoughts about affect, reason, risk, and rationality. *Risk Analysis, 24*(2), 311–322. DOI: 10.1111/j.0272-4332.2004.00433.x

Smales, L. A. (2021). Investor attention and the GameStop short squeeze. *Finance Research Letters, 43*, 102140. DOI: 10.1016/j.frl.2021.102140

Snow, D. A., & Benford, R. D. (1988). Ideology, frame resonance, and participant mobilization. International social movement research, 1(1), 197-217.

Snow, D. A., & Benford, R. D. (1988). Ideology, frame resonance, and participant mobilization. *International Social Movement Research, 1*(1), 197–217.

Snow, D. A., Rochford, E. B.Jr, Worden, S. K., & Benford, R. D. (1986). Frame alignment processes, micromobilization, and movement participation. *American Sociological Review*, •••, 464–481.

Snowden, D. J. (2005). Multi-ontology sense making: A new simplicity in a complex world. In *Proceedings of the 2005 International Conference on Knowledge Management*.

Snowden, D. J. (2005). Multi-ontology sense making: A new simplicity in decision making. *Informatics in Primary Care, 13*(1), 45–53.

Socal, M. P., Sharfstein, J. M., & Greene, J. A. (2021). The pandemic and the supply chain: Gaps in pharmaceutical production and distribution. *American Journal of Public Health, 111*(4), 635–639. DOI: 10.2105/AJPH.2020.306138 PMID: 33507805

Somers, M. R. (1994). The narrative constitution of identity: A relational and network approach. *Theory and Society, 23*(5), 605–649.

Sonenshein, S. (2010). We're changing—Or are we? Untangling the role of progressive, regressive, and stability narratives during strategic change implementation. *Academy of Management Journal, 53*(3), 477–512.

Sonni, A. F., Irwanto, I., Hafied, H., & Putri, V. C. C. (2025). Cultural power dynamics and narrative transformation: A comparative analysis of Hollywood film remakes in contemporary Asian cinema. *Frontiers in Communication, 10*, 1568170.

Soros, G. (2013). *The alchemy of finance* (2nd ed.). Wiley.

Spanos, M. (2021). Brand storytelling in the age of artificial intelligence. *Journal of Brand Strategy, 10*(1), 6–13.

Spicer, A., Alvesson, M., & Kärreman, D. (2009). Critical performativity: The unfinished business of critical management studies. *Human Relations, 62*(4), 537–560.

Spigel, B. (2017). The relational organization of entrepreneurial ecosystems. *Entrepreneurship Theory and Practice, 41*(1), 49–72.

Spinuzzi, C., Altounian, D., Pogue, G., Zhu, L., & Cochran, R. (2016). Articulating problems and markets: A translation analysis of entrepreneurs' emergent value propositions. *Written Communication, 35*(1), 97–134.

Spivak, G. C. (1988). Can the subaltern speak? In C. Nelson & L. Grossberg (Eds.), *Marxism and the interpretation of culture* (pp. 271–313). University of Illinois Press.

Srinivasan, T. N. (2003). *Indian Economy: Current Problems and Future Prospects* (No. 173).

Stam, E., & Spigel, B. (2016). *Entrepreneurial ecosystems* (Vol. 16, No. 13, pp. 1-15). USE Discussion paper series.

Stam, E. (2015). Entrepreneurial ecosystems and regional policy: A sympathetic critique. *European Planning Studies, 23*(9), 1759–1769.

Stark, D. (2009). *The sense of dissonance: Accounts of worth in economic life.* Princeton University Press.

Star, S. L. (1999). The Ethnography of Infrastructure. *The American Behavioral Scientist, 43*(3), 377–391. DOI: 10.1177/00027649921955326

Statista - The Statistics Portal. (2023). Statista. Retrieved November 2, 2023, from https://www.statista.com/?kw=statista&crmtag=adwords&gclid=Cj0KCQiAyKurBhD5ARIsALamXaGbsE_fgFP9mYIbjcSxBBMW_tLHkjmjRuX6WkqEcttdVWVWsuso_QaAhptEALw_wcB

Steinfield, L., Holt, D., & Veresiu, E. (2021). The impact of AI on cultural production and consumption. *Journal of Cultural Economics, 14*(4), 456–473.

Stent, W., & Dowler, T. (2015). Early assessments of the gap between integrated reporting and current corporate reporting. *Meditari Accountancy Research, 23*(1), 92–117. DOI: 10.1108/MEDAR-02-2014-0026

Steyn, B. (2004). From strategy to corporate communication strategy: A conceptualisation. *Journal of Communication Management (London), 8*(2), 168–183.

Stiglitz, J. E., Sen, A., & Fitoussi, J.-P. (2009). *Report by the Commission on the Measurement of Economic Performance and Social Progress.*

Stiglitz, J. E. (2002). *Globalization and its discontents.* W. W. Norton & Company.

Stiglitz, J. E., Sen, A., & Fitoussi, J. P. (2009). *The measurement of economic performance and social progress revisited* (Vol. 33). Ofce.

Stoica, O., Roman, A., & Rusu, V. D. (2020). The nexus between entrepreneurship and economic growth: A comparative analysis on groups of countries. *Sustainability, 12*(3), 1186.

Stoneman, P. (1995). *Handbook of the Economics of Innovation and Technological Change.* Wiley-Blackwell.

Subramanian, A. (2009). *India's Turn: Understanding the Economic Transformation.* OUP Catalogue.

Suchman, M. C. (1995). Managing legitimacy: Strategic and institutional approaches. *Academy of Management Review, 20*(3), 571–610.

Suchman, M. C. (1995). Managing Legitimacy: Strategic and Institutional Approaches. *Academy of Management Review, 20*(3), 571–610. DOI: 10.5465/amr.1995.9508080331

Suddaby, R. (2010). Challenges for institutional theory. *Journal of Management Inquiry, 19*(1), 14–20.

Suddaby, R., Bruton, G. D., & Si, S. X. (2015). Entrepreneurship through a qualitative lens: Insights on the construction and/or discovery of entrepreneurial opportunity. *Journal of Business Venturing, 30*(1), 1–10.

Suddaby, R., Foster, W. M., & Trank, C. Q. (2010). Rhetorical history as a source of competitive advantage. In *The globalization of strategy research* (Vol. 27, pp. 147–173). Emerald Group Publishing Limited.

Suddaby, R., & Greenwood, R. (2005). Rhetorical strategies of legitimacy. *Administrative Science Quarterly*, *50*(1), 35–67.

Sukanya, R., & Tantia, V. (2023). *Urbanization and the Impact on Economic Development.*, DOI: 10.4018/978-1-6684-9261-1.ch019

Sundaram, K., & Tendulkar, S. D. (2003). Poverty among Social and Economic Groups in India in 1990s. *Economic and Political Weekly*, *38*(50), 5263–5276.

Sunstein, C. R. (2002). The law of group polarization. *Journal of Political Philosophy*, *10*(2), 175–195. DOI: 10.1111/1467-9760.00148

Surya, B., Menne, F., Sabhan, H., Suriani, S., Abubakar, H., & Idris, M. (2021). Economic growth, increasing productivity of SMEs, and open innovation. *Journal of Open Innovation*, *7*(1), 20.

Tambayong, L., Daly, A., & Singh, S. (2020). A narrative analytic approach for real-time social sensing using NLP. *Journal of Big Data*, *7*(1), 1–21.

Tandoc, E. C.Jr, Lim, D., & Ling, R. (2020). Diffusion of disinformation: How social media users respond to fake news and why. *Journalism*, *21*(3), 381–398.

Tao, Z., Zhang, Z., & Shangkun, L. (2022). Digital economy, entrepreneurship, and high-quality economic development: Empirical evidence from urban China. *Frontiers of Economics in China*, *17*(3), 393.

Tarkar, P. (2020). Impact of COVID-19 pandemic on education system. *International Journal of Advanced Science and Technology*, *29*(9), 3812–3814.

Tarra, V. K. (2024). Personalization in Salesforce CRM with AI: How AI/ML Can Enhance Customer Interactions through Personalized Recommendations and Automated Insights. *International Journal of Emerging Research in Engineering and Technology*, *5*(4), 52–61.

Tavory, I., & Timmermans, S. (2014). *Abductive Analysis*. University of Chicago Press., DOI: 10.7208/chicago/9780226180458.001.0001

Taylor, C. (1989). *Sources of the self: The making of the modern identity*. Harvard University Press.

Taylor, C. (1989). *Sources of the Self: The Making of the Modern Identity*. Harvard University Press.

Taylor, F. W. (1911). Principles and methods of scientific management. *Journal of Accountancy*, *12*(3), 3.

Tedmanson, D., Verduijn, K., Essers, C., & Gartner, W. B. (2012). Critical perspectives in entrepreneurship research. *Organization, 19*(5), 531–541.

Thathsarani, U. S., Wei, J., & Samaraweera, G. R. S. R. C. (2021). Financial inclusion's role in economic growth and human capital in South Asia: An econometric approach. *Sustainability, 13*(8), 4303.

The Conference Board. (2009). *U.S. consumer confidence survey.* https://conference-board.org

The Economic Times. (2025, April 16). India's trade deficit with China widens to record $99.2 bln amid dumping concerns. *The Economic Times.*

Thomke, S. H. (2003). *Experimentation matters: unlocking the potential of new technologies for innovation.* Harvard Business Press.

Thornton, P. H., Ocasio, W., & Lounsbury, M. (2012). *The institutional logics perspective: A new approach to culture, structure, and process.* Oxford University Press.

Timmermans, S., & Epstein, S. (2010). A World of Standards but not a Standard World: Toward a Sociology of Standards and Standardization. *Annual Review of Sociology, 36*(1), 69–89. DOI: 10.1146/annurev.soc.012809.102629

Tooze, A. (2021). *Shutdown: How Covid shook the world's economy.* Viking.

Tracey, P., & Phillips, N. (2016). Managing the consequences of organizational stigmatization: Identity work in a social enterprise. *Academy of Management Journal, 59*(3), 740–765.

Tracey, P., Phillips, N., & Jarvis, O. (2011). Bridging institutional entrepreneurship and the creation of new organizational forms: A multilevel model. *Organization Science, 22*(1), 60–80.

Tracey, P., & Stott, N. (2017). Social innovation: A window on alternative ways of organizing and innovating. Innovation. *Organization & Management, 19*(1), 51–60.

Tran, M. T. (2024). Unlocking the AI-powered customer experience: Personalized service, enhanced engagement, and data-driven strategies for e-commerce applications. In *Enhancing and Predicting Digital Consumer Behavior with AI* (pp. 375–382). IGI Global. [REMOVED HYPERLINK FIELD]

Tribunal de Contas. (2019). *Relatório de auditoria ao cumprimento das regras orçamentais.* https://www.tcontas.pt/pt-pt/ProdutosTC/Relatorios/RelatoriosAuditoria/Pages/Anos.aspx

Tsoukas, H. (2005). Afterword: Why language matters in the analysis of organizational change. *Journal of Organizational Change Management*, *18*(1), 96–104.

Tsukada, N., & Nagaoka, S. (2015). International Research Collaboration. *Asian Economic Policy Review*, *10*, 96–119.

Tuckett, D., & Nikolic, M. (2017). The role of conviction and narrative in decision-making under radical uncertainty. *Theory & Psychology*, *27*(4), 501–523.

Tung, R. L., Zander, I., & Fang, T. (2023). The Tech Cold War, the multipolarization of the world economy, and IB research. *International Business Review*, *32*(6), 102195.

Turchin, P. (2003). *Historical dynamics: Why states rise and fall*. Princeton University Press.

Turkina, E., Van Assche, A., & Mudambi, R. (2025). Cross-border innovation for global value chain orchestration. *International Business Review*, *102450*. Advance online publication. DOI: 10.1016/j.ibusrev.2025.102450

Tversky, A., & Kahneman, D. (1974). Judgment under uncertainty: Heuristics and biases. *Science*, *185*(4157), 1124–1131. DOI: 10.1126/science.185.4157.1124

Ucbasaran, D., Shepherd, D. A., Lockett, A., & Lyon, S. J. (2013). Life after business failure: The process and consequences of business failure for entrepreneurs. *Journal of Management*, *39*(1), 163–202.

Ugochukwu, S. D., & Oruta, L. I. (2021). Government expenditure and economic growth in Nigeria: A disaggregated analysis. *Traektoria Nauki= Path of Science*, *7*(11), 4022-4035.

University of Michigan Surveys of Consumers. (2009). *Consumer sentiment index*. https://data.sca.isr.umich.edu

Vaara, E., Sonenshein, S., & Boje, D. (2016). Narratives as sources of stability and change in organizations: Approaches and directions for future research. *The Academy of Management Annals*, *10*(1), 495–560.

Vaara, E., & Tienari, J. (2008). A discursive perspective on legitimation strategies in multinational corporations. *Academy of Management Review*, *33*(4), 985–993.

Vaara, E., & Whittle, A. (2022). Common sense, new sense or non-sense? A critical discursive perspective on power in collective sensemaking. *Journal of Management Studies*, *59*(3), 755–781.

Vallabhaneni, A. S., Perla, A., Regalla, R. R., & Kumari, N. (2024). The Power of Personalization: AI-Driven Recommendations. In Minds Unveiled (pp. 111-127). Productivity Press.

Van Hulst, M. (2012). Storytelling, a model of and a model for planning. *Planning Theory, 11*(3), 299–318.

van Lente, H., & Rip, A. (1998). Expectations in technological developments: An example of prospective structures to be filled in by agency. In Disco, C., & van der Meulen, B. J. R. (Eds.), *Getting new technologies together: Studies in making sociotechnical order* (pp. 203–229). Walter de Gruyter.

Vargo, S. L., & Lusch, R. F. (2004). Evolving to a New Dominant Logic for Marketing. *Journal of Marketing, 68*(1), 1–17. DOI: 10.1509/jmkg.68.1.1.24036

Vargo, S. L., & Lusch, R. F. (2016). Institutions and axioms: An extension and update of service-dominant logic. *Journal of the Academy of Marketing Science, 44*(1), 5–23. DOI: 10.1007/s11747-015-0456-3

Venizelou, V., & Poullikkas, A. (2025). Navigating the Evolution of Cyprus' Electricity Landscape: Drivers, Challenges and Future Prospects. *Energies.* Advance online publication. DOI: 10.3390/en18051199

Verbeke, A. (2009). *International Business Strategy.* Cambridge University Press.

Verduijn, K., Dey, P., Tedmanson, D., & Essers, C. (2014). Emancipation and/or oppression? Conceptualizing dimensions of criticality in entrepreneurship studies. *International Journal of Entrepreneurial Behaviour & Research, 20*(2), 98–107.

Verduijn, K., & Essers, C. (2013). Questioning dominant entrepreneurship assumptions: The case of female ethnic minority entrepreneurs. *Entrepreneurship and Regional Development, 25*(7-8), 612–630.

Villiers, C., & Farooq, M. (2019). Farooq, M.B. & De Villiers, C. 2019. Understanding how managers institutionalise sustainability reporting: Evidence from Australia and New Zealand, Accounting, Auditing & Accountability Journal, 32(5), 1240-1269. *Accounting, Auditing & Accountability Journal, 32,* 1240–1269.

Vivek, R., & Nanthagopan, Y. (2023). Storytelling as a qualitative approach for organizational management. *European Journal of Management Issues, 31*(2), 113–122.

Volkswagen Group China. (2025). Chronology", available at: https://volkswagengroupchina.com.cn/en/volkswagengroupchina/chronology (accessed 6 April 2025).

Vu, K., Hanafizadeh, P., & Bohlin, E. (2020). ICT as a driver of economic growth: A survey of the literature and directions for future research. *Telecommunications Policy*, *44*(2), 101922. DOI: 10.1016/j.telpol.2020.101922

Vygotsky, L. S. (1978). *Mind in society: The development of higher psychological processes* (Cole, M., John-Steiner, V., Scribner, S., & Souberman, E., Eds.). Harvard University Press.

Waddell, S. (2016). *Change for the audacious: A doer's guide to large systems change for flourishing futures*. NetworkingAction.

Watson, T. J. (2009). Narrative, life story and manager identity: A case study in autobiographical identity work. *Human Relations*, *62*(3), 425–452.

Weick, K. E. (1995). Sensemaking in organizations: Vol. 3. *No. 10.1002*. Sage publications.

Weick, K. E., & Roberts, K. H. (1993). Collective mind in organizations: Heedful interrelating on flight decks. *Administrative Science Quarterly*, •••, 357–381.

Welter, F., Baker, T., Audretsch, D. B., & Gartner, W. B. (2017). Everyday entrepreneurship: A call for entrepreneurship research to embrace entrepreneurial diversity. *Entrepreneurship Theory and Practice*, *41*(3), 311–321.

Wen, Y., & Laporte, S. (2025). Experiential narratives in marketing: A comparison of generative AI and human content. *Journal of Public Policy & Marketing*, *44*(3), 392–410.

Westley, F. R., Tjornbo, O., Schultz, L., Olsson, P., Folke, C., Crona, B., & Bodin, Ö. (2013). A theory of transformative agency in linked social-ecological systems. *Ecology and Society*, *18*(3).

Westley, F., Zimmerman, B., & Quinn Patton, M. (2011). *Getting to maybe: How the world is changed*. Random House Canada.

Wilson, E. O. (1998). *Consilience: The unity of knowledge*. Alfred A. Knopf.

Wilson, S. (2020). New Zealand's elimination strategy for the COVID-19 pandemic and what other countries can learn. *BMJ (Clinical Research Ed.)*, *370*, m2989. DOI: 10.1136/bmj.m2989

Wodak, R. (2015). *Methods of Critical Discourse Studies* (3rd ed.).

Wood, A. J., Graham, M., Lehdonvirta, V., & Hjorth, I. (2019). Good gig, bad gig: Autonomy and algorithmic control in the global gig economy. *Work, Employment and Society*, *33*(1), 56–75.

Woodford, M. (2005). Central bank communication and policy effectiveness. In *The Greenspan era: Lessons for the future* (pp. 399–474). Federal Reserve Bank of Kansas City. * Corresponding author. Tel.: + 357 26843508; Fax: +357 931944. E-mail address: c.volos@nup.ac.cy.

World Development Indicators. Washington, D.C.: The World Bank

World Resources Institute. (2011). *World Resources Institute, & World Business Council for Sustainable Development.* https://www.wri.org/research/greenhouse-gas-protocol-corporate-value-chain-scope-3-accounting-and-reporting-standard

Wry, T., Lounsbury, M., & Glynn, M. A. (2011). Legitimating nascent collective identities: Coordinating cultural entrepreneurship. *Organization Science*, *22*(2), 449–463.

Xinhua (2023). Siemens expands China business. June 16. 2023, available at: https://global.chinadaily.com.cn/a/202306/16/WS648bee18a31033ad3f7bcac3.html (accessed 29 October 2024).

Xinhua (2025). German enterprises double down on China investment amid growing global uncertainties. *China Daily*, March 28, available at: https://www.chinadaily.com.cn/a/202503/28/WS67e6102ea3101d4e4dc2b69e.html (accessed 2 April 2025).

Yadav, P. S., Kiran, P. B. N., Vemula, R., Kumari, P., Wazir, R., & Sharma, A. (2025). A deep understanding of virtual reality's role in human resource management, with a specific focus on recruitment. In *The Future of HRM in a World of Persistent Virtual Reality* (pp. 87–107). doi:DOI: 10.1108/978-1-83662-110-220251004

Yadav, P. S., & Pandey, M. (2021). Influencer Marketing: Social-media-Influencer On Shopper stress Deviation and Purchase Intention. *Turkish Online Journal of Qualitative Inquiry*, *12*(8), 6926–6947.

Yakhshiboeva, L., & Eshbayev, O. (2023). *The Synergistic Impact of Artificial Intelligence and Big Data Analytics on Marketing Communication Strategies in Entrepreneurial Ecosystems: A Data-Driven Approach.* Financial Technology and Innovation., DOI: 10.54216/fintech-i.030202

Yamakawa, Y., Peng, M. W., & Deeds, D. L. (2015). Rising from the ashes: Cognitive determinants of venture growth after entrepreneurial failure. *Entrepreneurship Theory and Practice*, *39*(2), 209–236.

Yamin, M., & Otto, J. (2004). Patterns of knowledge flows and MNE innovative performance. *Journal of International Management*, *10*(2), 239–258.

Yan, J., Wang, S., Xiong, J., Scaringella, L., & Chen, X. (2024). Value Creation Reflecting CVC Strategic Orientations in Internet Platform Business Ecosystems: The Case of Tencent. *IEEE Transactions on Engineering Management, 71*, 1531–1541. DOI: 10.1109/TEM.2023.3336407

Yanow, D. (2000). *Conducting interpretive policy analysis.* SAGE Publications.

Yashashwini, A. (2024). *Unveiling the Role of Artificial Intelligence and its Impact on Consumer Buying Behaviour in Online Fashion Retail: A Comprehensive Study.* RVIM Journal of Management Research.

Ybema, S., Yanow, D., Velsink, H., & Buytelaar, M. (2009). *Organizational ethnography: Studying the complexities of everyday life.* SAGE Publications.

Yeo, S. F., Tan, C. L., Kumar, A., Tan, K. H., & Wong, J. K. (2022). Investigating the impact of AI-powered technologies on Instagrammers' purchase decisions in digitalization era–A study of the fashion and apparel industry. *Technological Forecasting and Social Change, 177*, 121551.

Yoo, Y., Henfridsson, O., & Lyytinen, K. (2010). The new organizing logic of digital innovation: An agenda for information systems research. *Information Systems Research, 21*(5), 724–735.

Yunus, M. (2007). *Creating a world without poverty: Social business and the future of capitalism.* PublicAffairs.

Zafar, M. W., Saleem, M. M., Destek, M. A., & Caglar, A. E. (2022). The dynamic linkage between remittances, export diversification, education, renewable energy consumption, economic growth, and CO2 emissions in top remittance-receiving countries. *Sustainable Development, 30*(1), 165–175.

Zafar, M. W., Shahbaz, M., Sinha, A., Sengupta, T., & Qin, Q. (2020). How renewable energy consumption contribute to environmental quality? The role of education in OECD countries. *Journal of Cleaner Production, 268*, 122149.

Zafar, M. W., Sinha, A., Ahmed, Z., Qin, Q., & Zaidi, S. A. H. (2021). Effects of biomass energy consumption on environmental quality: The role of education and technology in Asia-Pacific Economic Cooperation countries. *Renewable & Sustainable Energy Reviews, 142*, 110868.

Zahra, S. A., Gedajlovic, E., Neubaum, D. O., & Shulman, J. M. (2009). A typology of social entrepreneurs: Motives, search processes and ethical challenges. *Journal of Business Venturing, 24*(5), 519–532.

Zakzouk, F.A., Sumadi, A. M., & Khadra, T, & Komodromos.m. (2023). Factors affecting women entrepreneurs in Jordan. International Journal of Environment. *Workplace and Employment*, *7*(2), 164–185. DOI: 10.1504/IJEWE.2023.134532

Zanfei, A. (2000). Transnational firms and the changing organisation of innovative activities. *Cambridge Journal of Economics*, *24*(5), 515–542.

Zelizer, V. (1994). *The Social Meaning of Money*. Basic Books.

Zhang, D., Mohsin, M., Rasheed, A. K., Chang, Y., & Taghizadeh-Hesary, F. (2021). Public spending and green economic growth in BRI region: Mediating role of green finance. *Energy Policy*, *153*, 112256.

Zhang, L., Godil, D. I., Bibi, M., Khan, M. K., Sarwat, S., & Anser, M. K. (2021). Caring for the environment: How human capital, natural resources, and economic growth interact with environmental degradation in Pakistan? A dynamic ARDL approach. *The Science of the Total Environment*, *774*, 145553.

Zhang, S., Zhao, X., Zhou, T., & Kim, J. H. (2024). Do you have AI dependency? The roles of academic self-efficacy, academic stress, and performance expectations on problematic AI usage behavior. *International Journal of Educational Technology in Higher Education*, *21*(1), 34.

Zhang, Z., Khan, R., & Nandhakumar, J. (2024). Exploring liminal experience of entrepreneurs in information and communication technology innovation in underdeveloped regions. *Information & Management*, *62*, 104067. DOI: 10.1016/j.im.2024.104067

Zhao, E. Y., & Wry, T. (2016). Not all inequality is equal: Deconstructing the societal logic of patriarchy to understand microfinance lending to women. *Academy of Management Journal*, *59*(6), 1994–2020.

Zhao, J., & Zhou, N. (2021). Impact of human health on economic growth under the constraint of environment pollution. *Technological Forecasting and Social Change*, *169*, 120828.

Zhao, L. (2022). On the grant rate of Patent Cooperation Treaty applications: Theory and evidence. *Economic Modelling*, *117*, 106051.

Zhao, S., Tan, H., & Papanastassiou, M.. (2020). The internationalization of innovation towards the South: A historical case study of a global pharmaceutical corporation in China (1993–2017). *Asia Pacific Journal of Management*, *37*, 553–585.

Zhao, S., Zhang, Y., Iftikhar, H., Ullah, A., Mao, J., & Wang, T. (2022). Dynamic influence of digital and technological advancement on sustainable economic growth in Belt and road initiative (BRI) countries. *Sustainability, 14*(23), 15782.

Zhou, F. (2024). Utilizing Social Media Platforms to Foster Entrepreneurial Communication among Start-up Communities. *El Profesional de la Información*. Advance online publication. DOI: 10.3145/epi.2024.ene.0525

Zhou, W., Zhang, C., Wu, L., & Shashidhar, M. (2023). ChatGPT and marketing: Analyzing public discourse in early Twitter posts. *Journal of Marketing Analytics, 11*(4), 693–706.

Zilber, T. B. (2009). Institutional maintenance as narrative acts. In Institutional work: Actors and agency in institutional studies of organizations (pp. 205-235).

Zilber, T. B. (2006). The work of the symbolic in institutional processes: Translations of rational myths in Israeli high tech. *Academy of Management Journal, 49*(2), 281–303.

Zilber, T. B. (2007). Stories and the discursive dynamics of institutional entrepreneurship: The case of Israeli high-tech after the bubble. *Organization Studies, 28*(7), 1035–1054.

Zimand-Sheiner, D. (2024). Four Dimensions of Brand Storytelling: Framework for Managing and Analyzing Online Brand Stories. *International Journal of Business Communication*, •••, 23294884241261362.

Zimmerman, M. A., & Zeitz, G. J. (2002). Beyond survival: Achieving new venture growth by building legitimacy. *Academy of Management Review, 27*(3), 414–431.

Žižek, S. (2009). *First as Tragedy, Then as Farce*. Verso.

Zott, C., Amit, R., & Massa, L. (2011). The business model: Recent developments and future research. *Journal of Management, 37*(4), 1019–1042.

Zott, C., & Huy, Q. N. (2007). How entrepreneurs use symbolic management to acquire resources. *Administrative Science Quarterly, 52*(1), 70–105.

About the Contributors

Christos N. Christodoulou-Volos is an Associate Professor of Economics and Finance and the Head of the Department of Economics and Business . He received an MA in economics from the City College of the City University of New York (CUNY), an MPhil and a Ph.D. in economics in 1997 from the Graduate School of the CUNY. His areas of specialization are Macroeconometrics and Monetary and Financial Economics. His teaching interests are in the areas of Macroeconomics, Financial Economics, Finance, Statistics, and Econometrics and he has taught in many universities worldwide. He has published many scientific articles in leading academic journals and his current research interests are in the broad field of empirical macroeconomics and financial economics, including the problem of pricing and hedging of financial assets, decisions making under uncertainty, and the application of econometric models, such as, univariate and multivariate GARCH models, FIGARCH models, and correlated unobserved components model, and the effects of economic uncertainty on real estate. For a number of years, he held managerial positions in well-known companies in the United States, as a senior economist for a trade association and as Economist for a financial company dealing with litigation economics, in USA.

Andreas Masouras is an Associate Professor of Communication and Marketing in the Department of Economics & Business at Neapolis University. He holds a PhD with distinction from the Department of Political Science and International Relations at the University of the Peloponnese. The title of his doctoral dissertation is "Entrepreneurship and Competitiveness through the Lens of Institutional Analysis: The Case of Cyprus," which is available in the National Archive of Doctoral Dissertations. He coordinates the following programs: the conventional MBA, the distance MBA (English-speaking), and the MBA in Tourism (distance learning). He is actively writing scientific articles and participates in international conferences.

His book, “Entrepreneurship in Small and Medium-Sized Enterprises”, published by the American publishing house Nova, has been indexed in the Scopus book list.

***

Ashraf Alam currently holds the position of Research Faculty at the Centre of Excellence in Public Policy, Sustainability and ESG Research, housed within the Faculty of Law and Policy Studies at Alliance University, Bangalore. Renowned for his transdisciplinary acumen spanning education, technology, and policy studies, Dr. Alam has been distinguished as one of the World’s Top 2% Scientists by Stanford University and Elsevier, an accolade that highlights the breadth and depth of his scholarly impact. His prolific academic output includes over fifty SCOPUS-indexed publications and more than one hundred conference presentations at esteemed national and international academic forums. His educational trajectory reflects both depth and versatility: a Ph.D. in Education from the Indian Institute of Technology (IIT) Kharagpur, an M.Ed. with a concentration in Mathematics Education, and a B.S. in Computer Science from the University of Delhi. He also holds postgraduate degrees in Public Administration (M.A.) and Mathematics (M.Sc.), supplemented by a suite of postgraduate diplomas in Higher Education, Guidance and Counselling, Educational Technology, Criminology and Police Administration, and Value Education. Dr. Alam is an active participant in global scholarly discourse. As Associate Editor of Frontiers in Education, he plays a pivotal role in curating and guiding intellectual contributions across the Educational, Pedagogical, and Learning Sciences. Concurrently, he serves as Founder and Director of Academic Chronicles Private Limited, where he orchestrates editorial board formation and develops strategic frameworks for peer-reviewed publishing in the Social, Behavioural, and Humanistic Sciences. His research has been disseminated through elite academic platforms, including Elsevier, Springer, Taylor & Francis, Wiley, Routledge, Sage, Emerald, and IEEE, and addresses a remarkably diverse array of topics such as pedagogical theories, educational administration, educational psychology, educational sociology, educational policy, technology-integrated learning, mathematics education, educational philosophy, educational technology, teacher education, and AI/ML in learning environments. Dr. Alam’s scholarly rigor is further demonstrated by his success in qualifying multiple prestigious examinations including UGC-NET-JRF in both Education and Adult and Continuing Education, UGC- NET in Public Administration, and GATE in Humanities and Social Sciences. At the heart of Dr. Alam’s intellectual ethos lies a steadfast belief in education as a catalytic and emancipatory force capable of engendering social justice, human flourishing, societal progress, and sustainable transformation. His recent explorations into 'Happiness Engineering' and holistic well- being in school ecosystems articulate a vision of education that transcends instrumentalist paradigms, prioritizing emotional intelligence, global consciousness, empathy, and psychosocial resilience among learners. His research agenda seeks to conceive educational praxis through the affordances of inclusive digital learning, gamified instructional design, and artificial intelligence (AIEd), thereby constructing robust, ethically grounded frameworks for 21st century education. At Alliance University, Dr. Alam remains committed to architecting pedagogical models that not only cultivate intellectual dexterity and cognitive excellence but also prepare learners for morally responsive leadership, intellectual nimbleness, and responsible global citizenship in an increasingly volatile, uncertain, complex, and ambiguous (VUCA) world.

Sofia Anastasiadou is a Professor of Statistics and Research Methodology, Biostatistics Biomedical Research Methodology and at the School Health Sciences of the University of Western Macedonia (UOWM). She is the Deam of the Health Sciences of UOWM, Senate Member, and Member of post graduate studies committee. She is also a Member of the Senate and a MODIP Member of the same University. Her teaching focuses in the field of cognitive Research Methodology (Quantitative and Qualitative), Educational Research and Statistical Data Analysis, Big Data. She has written three books on Statistics and Research Methodology and has published over 300 articles in International and Greek Journals as well as in International and Greek Conferences. She has participated in scientific dialogues/ round tables in more than 200 lectures at International and Greek Conferences. She has organized and co-organized Greek and International Scientific Conferences. She is also a Member of a number of Scientific Committees of International Conferences and is an Associate Editor in a number of International Journals.

Deepak Bisht is an academician and researcher with over eight years of teaching experience at undergraduate and postgraduate levels, currently serving as an Assistant Professor at Lovely Professional University, Punjab. His research interests span marketing, tourism studies, and entrepreneurship, and he has published research papers in reputed peer-reviewed journals and edited volumes.

Irina Ervits is a Professor of Economics and Empirical Research at the CBS International Business School in Cologne, Germany. Her areas of research include International Economics and Innovation Management. She is interested in the topics of co-invention, innovation strategies by multinational enterprises and international patenting. She is also Dean of Economics and Finance department.

Oindrilla Ghosh is a Research Scholar hailing from the School of Law at Christ (Deemed to be University) Bangalore, India, focuses her research on health of tribal communities. Her work includes total 36 publications on Neuroscience on Organizational Behavior at IGI Global and also Environmental Disasters that are indexed in SCOPUS and Web of Sciences conference proceedings. She completed her early education at Loreto Day School, Bowbazar and went on to achieve distinction in her undergraduate (B.A/LL.B) and postgraduate (LL.M) studies at Calcutta University. Having served as an Assistant Professor of Law at Kingston Law College,Barasat, Kolkata, Ghosh has actively participated in numerous National Conferences, International Conferences, UN Global Summits, webinars, and seminars organized by prestigious academic institutions. Her expertise lies in health perspectives, environmental studies, human behavior, and related areas. Additionally, she has conducted various workshops, lectures in educational institutions, and is engaged with NGO's to help the needy.

Tatiana Harkiolakis, M.Sc., is a DBA candidate at ISC Paris Business School and a Research Fellow at the Center for Policy and Competitiveness at Ecole des Ponts ParisTech Business School. She is also the Director of Communications at Executive Coaching Consultants and holds an M.Sc. in Communications and Media from the London School of Economics and Political Science. Ms. Harkiolakis is a published journalist and academic researcher with several publications in the areas of emerging digital technologies, entreprenuerial ecosystems, and media and communications.

M. Robinson Joel received the Doctorate Ph.D degree in Computer and Information Technology from the Manonmaniam Sundaranar University, Tirunelveli in March 2019. He completed his M.Tech and MCA from the same university year 2008 and 2002 respectively. He is currently working as an Associate Professor in the Department of Computer Science and Engineering at KCG College of Technology, Chennai. He has 16 years of teaching experience.

Marcos Komodromos is a Professor of Marketing Communications and Management at the University of Nicosia and a Chartered PR Practitioner in the United Kingdom. He holds two Master's degrees (in Communications and in Management) and obtained his PhD with research focused on organizational justice, trust, and the management of change. Dr. Komodromos is an active consultant in strategic communication, digital media, and marketing communication for organizations in Cyprus and internationally. In addition, he is a certified trainer and coach specializing in communication, public relations, marketing, social media, leadership, media training, and management. With over 15 years of professional experience in the media industry, Dr. Komodromos has delivered more than 10,000 hours of training to working professionals. He is a Member of the Chartered

Institute of Public Relations (CIPR) in Great Britain, a Chartered PR Practitioner, and an active participant in numerous professional associations and organizations in both Cyprus and the UK

Binod Kumar is a graduate student known for his academic excellence, with his total 25 scholarly works indexed in Scopus database on Neuroscience in Organizational Behavior, Neuromarketing and AI Applications on Wildlife Conservation. Hailing from Ranchi, Jharkhand in India, Binod is currently pursuing a Master of Laws (LL.M) at the prestigious Radha Govind University. His passion for technology and innovation along with climate changes is evident through his high academic performance and numerous projects. Binod has a keen interest in artificial intelligence, machine learning, neuropsychology, climate changes and disaster management, fields in which he has undertaken several research projects. He is particularly fascinated by the potential of AI to solve real-world problems and is dedicated to learning and applying new techniques in these areas. He volunteers at local NGOs, teaching underprivileged children basic computer skills. With a strong foundation in law and a commitment to making a positive impact, Binod Kumar is poised to become a future leader in the legal industry.

Pravin Kumar is an experienced academic with over a decade of teaching Marketing Management and related subjects at reputed institutions across India. He currently serves as Assistant Professor at Brainware University, Kolkata, and has previously taught at different Management Institutions. He holds a Ph.D. in Management from T.M. Bhagalpur University, focusing on visual merchandising on consumer behavior in branded apparel. He also holds an MBA in Marketing and an M.A. in English, reflecting his interdisciplinary background. Dr. Kumar has published widely in Scopus, UGC-CARE, and peer-reviewed journals on topics including consumer behavior, brand personality, and AI in sales. He has presented at numerous national and international seminars and FDPs on subjects like cross-cultural marketing and leadership. His industry experience in the automobile and pharmaceutical sectors adds practical value to his teaching. Known for his engaging teaching style and commitment to Universal Design for Learning.

Preeti Kumari is an accomplished academic and corporate professional with over four and a half years of experience as an Assistant Professor at Lovely Professional University. With dual expertise in Commerce and Management, she brings a dynamic blend of theoretical knowledge and practical insight to her teaching. In addition to her academic role, she is a qualified Company Secretary, further enriching her understanding of corporate governance and business law. Her commitment to education and her interdisciplinary proficiency make her a valuable contributor to academic excellence and professional development in field of commerce and management.

Rajakumar Muthusamy Palani is currently working Department of Artificial Intelligence and Data Science, St. Joseph's College of Engineering, Old Mamallapuram Road, Chennai- 600 119, Tamil Nadu, India. His research interests include soft computing, machine learning techniques.

Domitilla Magni Ph.D., is Associate Professor at Catholic University of the Sacred Heart, Department of Economics and Business Management Sciences, where she teaches International Business Management and Innovation. She has been a Visiting Professor at prestigious universities worldwide, including New York University (USA), Universidad Panamericana (Mexico), Montpellier Business School (France), and SouthWestern University of Finance and Economics (China). She obtained her Ph.D. in "Management, Banking and Commodity Sciences" at Sapienza University of Rome. Specialized in innovative business models and sustainability, she has published numerous articles in leading academic journals, such as Business Strategy and the Environment and Technovation. Her research interests focus on International Business, Business Models, Knowledge Management, Innovation, Technology Management, and Open Innovation. She is a member of the Editorial Board for European Management Journal, Electronic Commerce Research, Energy Research & Social Science, and the Editorial Review Board of Business Ethics, the Environment & Responsibility. Additionally, she serves as Associate Editor for the Journal of the Knowledge Economy.

Stylianos Papalexandris is an orthopaedic surgeon with specialty and fellowship training in Greece and the UK, and extensive experience in the fields of lower limb arthroplasty, sports injuries and orthopaedic trauma. He is a PhD candidate at the Democritus University of Thrace. His thesis research project is in Marketing and Strategic Management of Healthcare Organisations. He possesses Master degrees in "Healthcare Management", "Adult Education", "Research Methodology in Biomedicine, Biostatistics and Clinical Bioinformatics" and "Contemporary Medical Acts: Legal Regulation and Bioethical Dimension". He has authored a first aid manual for lay persons and chapters in a book of orthopaedic trauma. His scientific papers have been published in Greek and international peer-reviewed journals and presented in national and international meetings and congresses.

Vinston Raja R. is an Assistant Professor with 12 years of experience successfully contributing to Information Technology and Computer Science curriculum development and delivery. Driven to contribute to program outcomes by facilitating engagement and supporting learning objectives. Enthusiastic professional with background in academic advisement. CAMPUS: Department of Computational Intelligence, Faculty of Engineering & Technology, Kattankulathur - Chennai EXPERIENCE: 15 years of experience RESEARCH INTEREST: Deep learning IOT

Premasish Roy is currently associated as Dean, Faculty of Management Studies, Bharati Vidyapeeth (Deemed to be University), Pune. He is also the Director - Department of Management Studies, Bharati Vidyapeeth (Deemed to be University), Navi Mumbai Campus. He is an alumnus of Jadavpur University, Kolkata. He has Ph.D (Management), PGDM (HR/OB specialization), PGDMC (Mass Communication), MA (English) with twenty two years of full-time Teaching, Academic Administration, General Administration, Consultancy and Research experience in Business Schools & Universities in pan India location. He has guided two Ph.D scholars and 98 master thesis dissertations. He is currently supervising two doctoral research scholarsHe has presented over 30 research papers in various International and National conferences. He has published around 22 articles in various journals of repute. He has published over 7 chapters in different edited books of repute. He is a Mentor to a startup working at one of the Atal Incubation Centres (Centre supported by NITI Aayog, Government of India). He is an Innovation Evangelist – organized 20 entrepreneurship boot camps & 01 training programme supported by DST, GoI.

Pinnika Syam Yadav is a distinguished researcher and academician, Whose Expertise and dedication have significantly advanced the field of business and Marketing analytics. Withholding a PhD and over two years of teaching experience at Mittal School of Business, Lovely Professional University, He has made notable contributions to Academia with his work. Currently working as an Assistant Professor at School of Business and Management, Christ University, Dr. Yadav, PS. He continues to excel in research and education, demonstrating a profound commitment to his field area by guiding and awarding 2 PhD Scholars under his guidance. His Innovative approach and extensive knowledge have earned him recognition as a leading figure in his field, inspiring young research scholars to follow in his footsteps

Index

A

B

C

D

E

N

O

P

R

S

T

V

www.ingramcontent.com/pod-product-compliance
Lightning Source LLC
LaVergne TN
LVHW061526290926
849162LV00004B/685